RENEWALS 691-4574

DATE DUE

AUG 0 8			
DEC 15 2008			

INTRODUCTION TO
WAVE PHENOMENA

INTRODUCTION TO WAVE PHENOMENA

AKIRA HIROSE
University of Saskatchewan

KARL E. LONNGREN
University of Iowa

A WILEY-INTERSCIENCE PUBLICATION

JOHN WILEY & SONS
New York • Chichester • Brisbane • Toronto • Singapore

Library of Congress Cataloging in Publication Data:

Hirose, Akira, 1941–
   Introduction to wave phenomena.

   "A Wiley-Interscience publication."
   Bibliography: p.
   Includes index.
   1. Waves. 2. Wave-motion, Theory of.   I. Lonngren,
Karl E. (Karl Erik), 1938–   . II. Title.

QC157.H56 1985   531′.1133   84-15365
ISBN 0-471-81440-7

Printed in the United States of America

10 9 8 7 6 5 4 3 2 1

In honour of our parents

Genji
Katsuyo
Bruno
Edith

PREFACE

Traditionally, the subjects on wave phenomena have been taught in different disciplines. Indeed, sound waves and electromagnetic waves are entirely different physical phenomena. However, once one realizes that all waves, mechanical or electromagnetic, carry energy (and momentum) and that the observable quantities are necessarily associated with energy propagation, then one can formulate a unified understanding of wave phenomena in its totality. To this end we have formulated a textbook that elucidates the general properties of wave phenomena, both linear and nonlinear, and illustrates the physical contexts in which wave phenomena arise.

This book consists of sixteen chapters. We begin in Chapter 1 with a review of oscillations, both mechanical and electromagnetic. Oscillation systems can be the sources of harmonic waves since the mass in a mechanical oscillation system and charges in an electromagnetic oscillation system are all under acceleration, which is the fundamental requirement for creating waves.

In Chapter 2 we study general properties of wave motion without specifying mechanical and electromagnetic waves. A mathematical expression for a sinusoidal wave and the wave equation is introduced. Phase and group velocities are described.

Chapter 3 provides mathematical preparations that are needed in deriving the wave equation from first principles

$$\frac{\partial^2 \xi}{\partial t^2} = c_w^2 \frac{\partial^2 \xi}{\partial x^2}$$

for various kinds of waves. It can be skipped if the reader is already familiar with Taylor expansion techniques.

In Chapters 4 and 5 we study mechanical waves, including waves on springs (longitudinal), stretched string (transverse), and sound waves (longitudinal) in solids, liquids, and gases. We show how Newton's equation of motion can be converted into the wave equation for mechanical waves and formulate the amount of energy and momentum that can be carried by these waves.

In Chapter 6 we see how mechanical waves are reflected at boundaries. The wave equation allows two solutions, which propagate in opposite directions. When these two waves coexist, standing waves are created. Standing waves play important roles in musical instruments, which become sources of sound waves in air. The concept of mechanical impedance is introduced as an analogy to the impedance in electromagnetic phenomena. Wave reflection is explained in terms of energy (and momentum) reflection.

When a loudspeaker creates sound waves in air, the wave amplitude becomes smaller with the distance from the speaker. This effect is caused by geometrical dispersion of wave energy as shown in Chapter 7. The variation of wave amplitude can also occur in one-dimensional waves if the medium is not uniform or the wave velocity varies from point to point. This interesting (but difficult to analyze) phenomenon is briefly discussed.

In Chapter 8, the Doppler effect of sound waves is studied. Whenever a sound source and/or an observer are moving relative to air, the observer hears a frequency different from the true frequency. We also see what happens if an object travels faster than sound (shock waves).

In Chapters 9 and 10 we study the propagation and radiation of electromagnetic waves. We start with an LC transmission line that is an analogue to the mass–spring mechanical transmission line used in Chapter 3. The reflection of electromagnetic waves is discussed in terms of the impedance mismatching, as we did in Chapter 6 for mechanical waves. Electromagnetic waves in conducting materials such as metals and plasma require special treatment. In such media the wave equation is drastically modified and waves become strongly dispersive. The radiation of electromagnetic waves is described, in terms of charge acceleration or deceleration.

In Chapter 11 interference and diffraction, which are caused by more than one wave propagating in the same direction, are described. Depending on the phase of each wave, the net wave amplitude is either strengthened or weakened. Light does not always travel along a straight line because of its diffractive nature. This diffraction phenomena can be discussed in terms of interference among many waves.

Geometrical optics (Chapter 12) is one branch of optics in which we largely (not entirely) neglect the wave nature of light. We assume that light travels along a straight line in a given uniform medium. Of course, when light hits a boundary interfacing two media, light changes its propagation direction (refraction). Optical devices such as mirrors and lenses are discussed, followed by a discussion of optical instruments (microscope and telescope).

Chapter 13 is an introduction to Fourier series and Laplace transformation. The concept of frequency spectrum is explained. Also, it is shown how Laplace transformation can convert a differential equation into an algebraic equation.

Chapter 14 is an introduction to modern (quantum) physics. It is shown that under certain circumstances, light (electromagnetic waves) behaves as a collection of particles (photons). Briefly discussed is the fact that energetic particles, such as accelerated electrons, have a wave nature (wave–particle duality). Hence electron microscopes can have a higher resolving power than optical microscopes.

The material covered in the previous chapters made the tacit assumption, where appropriate, that the equations describing wave phenomena could be linearized. For example, the wave equation for sound waves resulted from a linearized (small-amplitude) approximation. Often, the wave amplitudes may become large and a linearization procedure is no longer valid. In Chapter 15 we introduce the reader to some nonlinear mathematical techniques to obtain some analytical solutions to nonlinear wave equations.

Finally, in Chapter 16 the reader is introduced to two topics of nonlinear wave phenomena that are receiving much current attention in various disciplines, namely, solitons and shock waves, in which nonlinearity is balanced with dispersion or dissipation.

The MKS unit system is used throughout the book. Some traditional and some conventional units are used, however. Examples are the angstrom $(1 \text{ Å} = 10^{-10} \text{ m})$ in optics and the electron volt $(1 \text{ eV} = 1.6 \times 10^{-19} \text{ J})$ for the energy of elementary particles. In physics, a traveling wave is conventionally written as $A \sin (kx - \omega t)$ (or $A e^{i(kx - \omega t)}$) and in engineering it is written as $A \sin (\omega t - kx)$ or $(A e^{j(\omega t - kx)})$. We use the former representation but there are no fundamental differences between the two.

We have greatly benefited from frequent discussions with our colleagues and students. We are particularly grateful to Ray Montalbetti, Harvey Skarsgard, Ludwig Schott, and Adrian Korpel, who went over the entire manuscript and gave us invaluable comments. Portions of this work were supported by the Natural Sciences and Engineering Research Council of Canada and the National Science Foundation of the United States.

Special thanks go to Mrs. Vera Cyr, who typed the original manuscript with remarkable accuracy and efficiency. Finally, we thank our wives, Kimiko and Vicki, for tireless support.

AKIRA HIROSE
KARL E. LONNGREN

Saskatoon, Saskatchewan, Canada
Iowa City, Iowa
October 1984

CONTENTS

INTRODUCTION TO WAVE PHENOMENA

CHAPTER 1

Review of Oscillations

1.1. Introduction

Most waves, either mechanical or electromagnetic, we encounter are created by something vibrating or oscillating. In the classroom the instructor's voice reaches your ears as sound waves in air. To create the waves the instructor uses his vocal chords, which are forced to vibrate by the airflow through his throat. Similarly, radio waves emitted from a radio station also originate from something oscillating. In this case free electrons in a vertically erected antenna execute up and down oscillatory motion with a certain frequency, which is determined by an electrical oscillator connected to the antenna. Whenever physical objects oscillate or vibrate, there is a possibility that waves are created in the medium surrounding those objects.

In this chapter we review oscillation phenomena, both mechanical and electromagnetic, since oscillations and waves have many common properties, hence understanding oscillations can greatly help us understand wave phenomena. More important, harmonic (or sinusoidal) waves we frequently encounter in daily life are created by physical objects undergoing oscillatory motions. It is recommended that you refresh your knowledge (and skills) of properties of trigonometric functions, such as

$$\frac{d}{dx} \sin ax = a \cos x, \qquad \frac{d}{dx} \cos ax = -a \sin x,$$

and so on.

1.2. Mass–Spring System

Consider a mass M (kg) on a frictionless plane connected to a spring with a spring constant k (N/kg), and a natural length l (Fig. 1.1). Without any external disturbance the mass would stay at the equilibrium position, $x = 0$. Suppose one pulls the mass a certain distance and then releases the mass. The mass would start oscillating with a certain frequency. If one pushes the mass and then releases it, the mass would start oscillating with the *same* frequency. Otherwise, one could hit the mass with a hammer to give a sudden impulse to let the mass

1

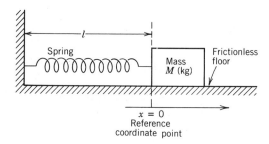

Fig. 1.1. Mass–spring system in equilibrium. The spring has a spring constant k, and l is the natural length of the spring.

start oscillating. No matter how the oscillation is started, the frequency is the same.

One of the major objectives in studying oscillation is to find oscillation frequencies that are determined by physical quantities. As we will see later, the mass M and the spring constant k determine the oscillation frequency in the preceding example.

What makes the spring–mass system oscillate? When one pulls the mass the spring must be elongated, and it tends to pull the mass back to its equilibrium position, $x=0$. Therefore on being released, the mass starts moving to the left, being pulled by the spring. This pulling force is given by *Hooke's law*,

$$F = kx \quad \text{(directed to the } left) \tag{1.1}$$

since x is the deviation of the spring length from the natural length, l. The pulling force provided by the spring disappears at the instant when the mass reaches the equilibrium position, $x=0$. By this time, however, the mass has acquired a kinetic energy (which will be shown to be equal to the potential energy initially stored in the spring). Because of its inertia, it cannot stop at the equilibrium position, but keeps moving, overshooting or passing the equilibrium position. The spring is then squeezed and tends to push the mass back to the equilibrium position. This time the force is given by

$$F = -kx \quad \text{(directed to the } right) \tag{1.2}$$

since x is now a negative quantity. The mass keeps moving to the left until the kinetic energy, which the mass had when it passed through the equilibrium position, $x=0$, is all converted into the potential energy stored in the spring, and after this instant, the mass again starts moving to the right toward the equilibrium position. This process continues and appears as an oscillation.

The key agent in the oscillatory motion in the mass--spring system is the force provided by the spring. This force always acts on the mass so as to make it seek its equilibrium position, $x=0$. Such a force is called a *restoring force*. In any mechanical oscillating system, there is always a restoring force (or torque). In the case of a grandfather clock, gravity provides a restoring force, and for a wheel balance in a watch, a spiral hair-spring does the job by providing a restoring torque. If the spring constant k is larger (stronger spring), the spring can pull or push the mass more quickly, and we expect that the oscillation frequency is larger. On the other hand, if the mass is larger, the mass should move more slowly, and we expect that the oscillation frequency is smaller. Indeed, as we will see, the oscillation frequency for the mass–spring is given by

$$v = \frac{1}{2\pi}\sqrt{\frac{k}{M}} \quad \text{(cycles/sec)}. \tag{1.3}$$

Let us now find out what kind of mathematical expression can describe the oscillatory motion of the mass–spring system. We assume that the mass is gradually pulled a distance x_0 to the right from the equilibrium position, $x=0$, and then released at time $t=0$. Suppose that at a certain time, the mass is a distance x away from the equilibrium position, $x=0$ (Figs. 1.2 and 1.3). The instantaneous velocity of the mass is given by

$$v = \frac{dx}{dt} \quad \text{(m/sec)} \tag{1.4}$$

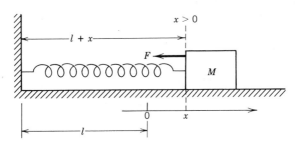

Fig. 1.2.　Displacement $x>0$. The spring is stretched and pulls the mass toward the equilibrium position, $x=0$.

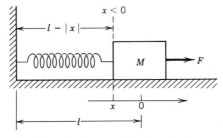

Fig. 1.3.　Displacement $x<0$. The spring is squeezed and pushes the mass toward the equilibrium position.

and the acceleration is

$$a = \frac{dv}{dt} = \frac{d^2x}{dt^2} \quad (\text{m/sec}^2).$$ (1.5)

But the force acting on the mass is [from Eqs. (1.1) and (1.2)]

$$F = -kx \quad (\text{N}).$$ (1.6)

[Note that in Eq. (1.1), the force was given by kx directed to the *left*, which is equivalent to a force $-kx$ directed to the *right*. Therefore the restoring force can be generalized as $-kx$ regardless of the sign of x.] Applying Newton's second law (mass × acceleration = force), we find

$$M \frac{d^2x}{dt^2} = -kx.$$ (1.7)

This is the equation of motion for the mass. The position x of the mass is to be found this *differential equation* as a function of time t. Remember that the mass is located at $x = x_0$ at $t = 0$, when the oscillation starts, or

$$x(0) = x_0,$$ (1.8)

where $x(0)$ means the value of x at $t = 0$. We know that the second derivatives of sinusoidal functions, $\sin a\theta$ and $\cos a\theta$, are

$$\frac{d^2}{d\theta^2} \sin a\theta = -a^2 \sin a\theta,$$

$$\frac{d^2}{d\theta^2} \cos a\theta = -a^2 \cos a\theta.$$

Therefore it is very likely that Eq. (1.7) has a sinusoidal solution. Let the solution for $x(t)$ be

$$x(t) = A \cos \omega t,$$ (1.9)

where A and ω are constants to be determined. Since $\cos 0 = 1$, we find

$$x(0) = A.$$ (1.10)

Comparing Eq. (1.8) with Eq. (1.10), we must have $A = x_0$. This quantity is called the amplitude of oscillation.

To find ω we calculate the second derivative of $x(t) = x_0 \cos \omega t$:

$$\frac{dx}{dt} = x_0 \frac{d}{dt} \cos \omega t = -x_0 \omega \sin \omega t$$

$$\frac{d^2x}{dt^2} = x_0 \frac{d^2}{dt^2} \cos \omega t = -x_0 \omega \frac{d}{dt} \sin \omega t = -x_0 \omega^2 \cos \omega t$$ (1.11)

Substituting Eqs. (1.11) into Eq. (1.7), we find

$$-M\omega^2 x_0 \cos \omega t = -kx_0 \cos \omega t,$$

which yields

$$\omega = \sqrt{\frac{k}{M}}. \tag{1.12}$$

This quantity ω is called the angular frequency and has the dimensions of radians/sec. Since the function $\cos \omega t$ has a period of 2π rad, the temporal period T is given by

$$T = \frac{2\pi}{\omega} \quad (\text{sec}). \tag{1.13}$$

In 1 sec, the oscillation repeats $1/T$ times (Fig. 1.4). This number is defined as the frequency, ν (cycles/sec).* It is obvious that

$$\nu = \frac{1}{T} = \frac{\omega}{2\pi} \tag{1.14}$$

Example 1. Show that the quantity $\sqrt{k/M}$ indeed has the dimensions of 1/sec.

Since the spring constant k has the dimensions of $N/m = (kg \cdot m/sec^2)/m = kg/sec^2$, and the mass kg, we find $\sqrt{k/M}$ has the dimensions of

$$\sqrt{\frac{kg/sec^2}{kg}} = \frac{1}{sec}.$$

Note that the angle (radian) is a dimensionless quantity.

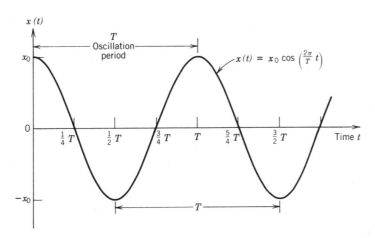

Fig. 1.4. The displacement $x(t)$ is shown as a function of time t

$$T = 2\pi/\omega = 1/\nu$$

is the oscillation period.

*Cycles/second = Hz (hertz).

Example 2. A spring 2 m long hangs from the ceiling as shown in Fig. 1.5. When a mass of 1.5 kg is suspended from the spring, the spring is elongated by 30 cm in equilibrium. The mass is then pulled down an additional 5 cm and released. Neglecting the mass of the spring, find an equation to describe the oscillatory motion of the mass.

The mass is expected to oscillate about the equilibrium position, $x=0$ ($x>0$ upward). The spring constant k is found from

$$Mg = k\,\Delta l,$$

where $M=1.5$ kg, $\Delta l = 0.3$ m. Then

$$k = \frac{1.5\ kg \times 9.8\ \text{N/kg}}{0.3\ \text{m}} = 49.0\ \text{N/m}$$

and

$$\omega = \sqrt{\frac{k}{M}} = \sqrt{\frac{49\ \text{N/m}}{1.5\ \text{kg}}} = 5.7\ \text{rad/sec}.$$

Since the initial position is $x_0 = -0.05$ m, the equation to describe the motion of the mass is given by

$$x(t) = -0.05\ \cos\ (5.7t)\quad \text{m}.$$

The period T is

$$T = \frac{2\pi}{\omega} = 1.1\ \text{sec},$$

and the frequency ν is $\nu = 1/T = 0.91/\text{sec}$.

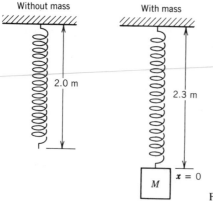

Without mass   With mass

2.0 m

2.3 m

$x = 0$

M

Fig. 1.5.   Example 2.

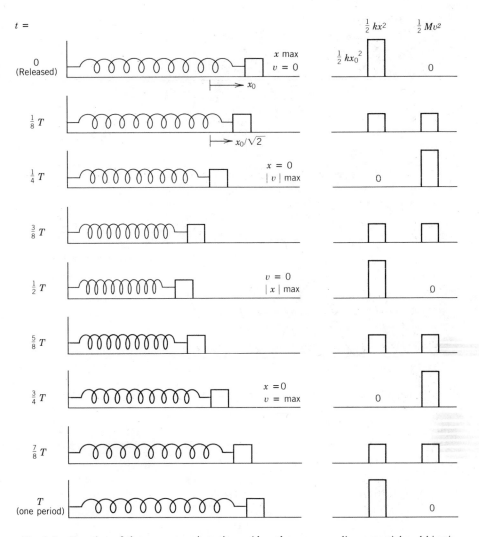

Fig. 1.6.   Location of the mass at various times. Also, the corresponding potential and kinetic energies are schematically shown.

That the total energy of the oscillating mass–spring system is conserved or constant can alternatively be shown directly from the equation of motion

$$M \frac{dv}{dt} = -kx. \tag{1.19}$$

Let us multiply this equation by the velocity v.

$$Mv \frac{dv}{dt} = -kvx. \tag{1.20}$$

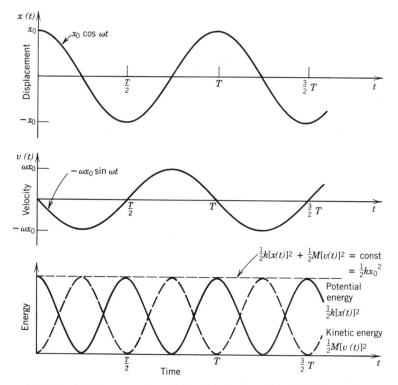

Fig. 1.7. Displacement $x(t)$, velocity $v(t)$ and potential and kinetic energies.

Since

$$\frac{d}{dt}v^2 = \frac{dv}{dt}\frac{dv^2}{dv} = 2v\frac{dv}{dt},$$

$$xv = x\frac{dx}{dt} = \frac{1}{2}\frac{d}{dt}x^2,$$

Eq. (1.20) becomes

$$\frac{d}{dt}(\tfrac{1}{2}Mv^2 + \tfrac{1}{2}kx^2) = 0, \tag{1.21}$$

which indeed states that the total energy is conserved,

$$\tfrac{1}{2}Mv^2 + \tfrac{1}{2}kx^2 = \text{const}.$$

1.4. Other Mechanical Oscillation Systems

Whenever there is a restoring force to act on a mass, oscillations are likely to occur. As we have seen, there must be two agents for mechanical oscillations to take place, one capable of storing potential energy (like the spring) and the other capable of storing kinetic energy (like the mass). In rotational devices (like the wheel balance in watches), restoring torque and rotational inertia replace restoring force and mass, respectively, but energy relations still hold.

Pendulum

A grandfather clock's accuracy is largely determined by the regularity of the pendulum oscillation frequency. You may already know that the pendulum frequency is totally determined by the length of the support l (Fig. 1.8) and does not depend on the mass.

The restoring force to act on the mass M in Fig. 1.8 is provided by the earth's gravity, which tends to make the mass stay at the equilibrium position, P, or the lowest position. The restoring force F is given by

$$F = Mg \sin \theta \quad \text{(toward } P\text{)} \tag{1.22}$$

and the equation of motion for the mass becomes

$$M \frac{dv}{dt} = -Mg \sin \theta \tag{1.23}$$

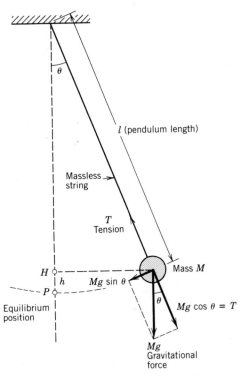

Fig. 1.8.   Pendulum having a mass M and a length l. $g = 9.8$ m/sec² is the gravitational acceleration.

(Do you see why the minus sign appears? See the discussion given in Section 2.) Since the velocity v is given by

$$v = l\frac{d\theta}{dt}, \tag{1.24}$$

Eq. (1.23) becomes

$$\frac{d^2\theta}{dt^2} = -\frac{g}{l}\sin\theta. \tag{1.25}$$

Although this equation looks simple, its solution cannot be expressed in terms of sine or cosine functions unless $|\theta|$ is much smaller than 1 rad. If $|\theta| \ll 1$ rad, $\sin\theta$ can be well approximated simply by θ (see Chapter 3), and Eq. (1.25) reduces to*

$$\frac{d^2\theta}{dt^2} = -\frac{g}{l}\theta. \tag{1.26}$$

This is mathematically identical to our previous equation, Eq. (1.7). We can immediately find the angular frequency as

$$\omega = \sqrt{\frac{g}{l}} \quad \text{(rad/sec).} \tag{1.27}$$

You should check that $\sqrt{g/l}$ indeed has the dimensions of 1/sec.

 Example 4. Find the length of the pendulum rod of a grandfather clock having an oscillation period of 2.0 sec. Assume that the mass of the rod is negligible compared with the mass to be attached.

 From $T = 2\pi/\omega = 2\pi\sqrt{l/g}$, we find $l = (T/2\pi)^2 g$. Substituting $T = 2$ sec and $g = 9.8$ m/sec^2, we find $l = 0.99$ m.

 Example 5. Assuming a solution of the form

$$\theta(t) = \theta_0 \cos\omega t$$

to Eq. (1.26), show that the total energy of the mass (potential and kinetic) is constant.

 The kinetic energy is

$$\text{K.E.} = \tfrac{1}{2}Mv^2 = \tfrac{1}{2}M\left(l\frac{d\theta}{dt}\right)^2$$

$$= \tfrac{1}{2}Ml^2\omega^2\theta_0^2\sin^2\omega t.$$

The potential energy is Mgh, where h is the height measured from the equilibrium position and is given by

$$h = OP - OH = l - l\cos\theta.$$

---

*This procedure is called *linearization* of Eq. (1.25), which is a nonlinear differential equation.

If θ is small, $\cos \theta$ can be approximated by (see Chapter 3)

$$\cos \theta \simeq 1 - \tfrac{1}{2}\theta^2.$$

Then

$$h \simeq \tfrac{1}{2}l\theta^2,$$

and the potential energy becomes

$$\text{P.E.} = Mgh = Mg\,\frac{l}{2}\,\theta^2 = \tfrac{1}{2}Mgl\theta_0^2 \cos^2 \omega t.$$

Recalling $\omega^2 = g/l$, we find

$$\text{K.E.} + \text{P.E.} = \tfrac{1}{2}Mgl\theta_0^2 \quad \text{(const)}.$$

Rotational Inertial Systems

The balance wheel (Fig. 1.9) in watches oscillates about its center. A spiral spring connected to the wheel balance provides a restoring torque rather than restoring force, and the rotational inertia of the wheel balance plays the role of mass inertia (translational inertia) in the mass–spring system.

Let the moment of inertia of the wheel balance be I ($kg \cdot m^2$) and the restoring torque provided by the spring be

$$\tau = -k_\tau \theta \quad \text{(N·m)}, \tag{1.28}$$

where k_τ is a constant (torsional constant, N·m) that plays the same role as the spring constant in the mass–spring system, and θ is the rotational angle of the wheel balance measured from the equilibrium (zero torsion) angular position.

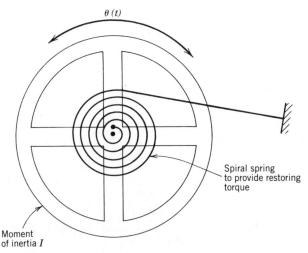

$\theta\,(t)$

Spiral spring
to provide restoring
torque

Moment
of inertia I

Fig. 1.9.   Wheel balance of watches connected to a spiral spring, an example of a linear oscillator.

Since the equation of motion for the rotational system is given by

$$I \frac{d^2\theta}{dt^2} = \text{torque, } \tau \tag{1.29}$$

where I is the moment of inertia, we find

$$I \frac{d^2\theta}{dt^2} = -k_\tau \theta. \tag{1.30}$$

This is again identical in mathematical form to Eq. (1.7).* The oscillation frequency is then given by

$$\omega = \sqrt{\frac{k_\tau}{I}}. \tag{1.31}$$

You should check that $\sqrt{k_\tau/I}$ indeed has the dimensions of frequency.

   Example 6.   A straight uniform stick having a length l (m) and a mass M (kg) is freely pivoted at one end as shown in Fig. 1.10. Find the frequency of oscillation about the pivot, assuming the angle θ is small.

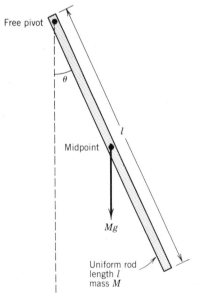

Free pivot

θ

l

Midpoint

Mg

Uniform rod
length l
mass M

Fig. 1.10.   Pivoted rod as a pendulum (physical pendulum).

*Compare Eq. (1.30) with Eq. (1.25). A wheel balance is a linear oscillator no matter how large θ is, as long as the spring stays within the elastic limit.

The restoring torque to act on the stick is

$$Mg \sin \theta \cdot \frac{l}{2} \simeq \tfrac{1}{2}Mgl\theta.$$

Therefore, $Mgl/2$ plays the role of restoring torque constant k_τ provided $|\theta| \ll 1$. (Recall $\sin \theta \simeq \theta$ if $|\theta| \ll 1$.) The moment of inertia about the end of the stick is

$$I = \int_0^l \frac{M}{l} x^2 \, dx = \tfrac{1}{3}Ml^2.$$

Then

$$\omega = \sqrt{\frac{k_\tau}{I}} = \sqrt{\frac{3g}{2l}}.$$

1.5. Electromagnetic Oscillation

We learned in classes on electricity and magnetism that an LC (inductance and capacitance) circuit oscillates with an angular frequency

$$\omega = \frac{1}{\sqrt{LC}}. \tag{1.32}$$

Although physical quantities we treat in electromagnetic oscillation are quite different from those in mechanical oscillation, the fundamental concept of oscillation mechanism—namely, the energy tossing mechanism—remains the same. Instead of kinetic and potential energies in the mass–spring system, we now have electric and magnetic energies stored in the capacitor and inductor, respectively.

Consider a capacitor charged to a charge q_0 (coulombs) suddenly connected to an inductor L (Fig. 1.11). The charge initially stored in the capacitor tends to flow toward the inductor and creates a current along the circuit. The voltage across the capacitor is

$$V_C = \frac{q}{C}$$

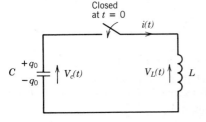

Fig. 1.11. LC tank circuit. q_0 is the initial charge on the capacitor.

and that across the inductor is

$$V_L = L \frac{di}{dt}.$$

Then Kirchhoff's voltage theorem requires

$$\frac{q}{C} = L \frac{di}{dt}. \tag{1.33}$$

Since we have chosen the direction of the current corresponding to a discharging capacitor, we have

$$i = -\frac{dq}{dt}. \tag{1.34}$$

Substituting Eq. (1.34) to Eq. (1.33), we find the following differential equation for the charge $q(t)$,

$$\frac{d^2q}{dt^2} = -\frac{1}{LC} q. \tag{1.35}$$

This is again mathematically identical to Eq. (1.7) for the mass–spring system, and we immediately find that the LC circuit would oscillate with the frequency

$$\omega = \frac{1}{\sqrt{LC}}.$$

Since the capacitor had an initial charge q_0, the equation to describe the charge at an arbitrary instant should be chosen as

$$q(t) = q_0 \cos \omega t. \tag{1.36}$$

Using Eq. (1.34), the current $i(t)$ becomes

$$i(t) = \omega q_0 \sin \omega t. \tag{1.37}$$

Therefore the electric energy stored in the capacitor is

$$U_E = \frac{1}{2} \frac{q^2}{C} = \frac{1}{2C} q_0^2 \cos^2 \omega t \tag{1.38}$$

and the magnetic energy stored in the inductor is

$$U_M = \tfrac{1}{2} L i^2 = \tfrac{1}{2} L \omega^2 q_0^2 \sin^2 \omega t. \tag{1.39}$$

Recalling $\omega^2 = 1/LC$, we find that the sum of the two energies is constant,

$$U_E + U_M = \frac{1}{2} \frac{q_0^2}{C},$$

and equal to the initial electric energy stored in the capacitor.

The capacitor and inductor exchange energy periodically as the mass and spring do, and we see that this energy-tossing mechanism is common to any kind of oscillation, mechanical or electromagnetic.

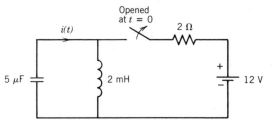

Fig. 1.12.   An example in which the initial current is not zero.

Example 7.   In the *LC* circuit shown in Fig. 1.12, the switch *S* is closed for a long time. Then the switch is opened at $t = 0$. Find the expressions for the current to flow in the *LC* circuit and the charge on the capacitor.

The initial current flowing through the inductor is

$$i_0 = \frac{12 \text{ V}}{2 \, \Omega} = 6.0 \text{ A}.$$

Then the current chosen clockwise is described by

$$i(t) = i_0 \cos \omega t = 6 \cos \omega t \quad \text{(amperes)},$$

where

$$\omega = \frac{1}{\sqrt{LC}} = \frac{1}{\sqrt{2 \times 10^{-3} \times 5 \times 10^{-6}}} = 10^4 \text{ rad/sec}.$$

The charge on the lower plate of the capacitor is given by

$$q(t) = \int_0^t i(t) \, dt = \int_0^t i_0 \cos \omega t \, dt$$

$$= \frac{i_0}{\omega} \sin \omega t$$

$$= 6 \times 10^{-4} \sin \omega t \text{ (coulomb)}.$$

Note that the initial condition in this example is different from that in Fig. 1.11.

1.6. Damped Oscillation

So far we have considered ideal cases in which energy dissipation can be completely neglected. For example, in the mass–spring system, we assumed that the floor on which the mass is placed is frictionless. Also, in the *LC* circuit, we neglected the resistance in the circuit. Both mechanical friction and electric resistance give rise to energy dissipation, and oscillation cannot continue forever, but should eventually be damped. Oscillation energy is converted into heat in an irreversible manner, or into radiation.

Consider now a capacitor C with a charge q_0 suddenly connected to an inductor L through a finite resistance R (Fig. 1.13). Using Kirchhoff's voltage theorem, we find

$$\frac{q}{C} = Ri + L\frac{di}{dt}. \tag{1.40}$$

Recalling

$$i = -\frac{dq}{dt},$$

we now have the following differential equation for the charge $q(t)$:

$$\frac{d^2q}{dt^2} + \frac{R}{L}\frac{dq}{dt} + \frac{1}{LC}q = 0. \tag{1.41}$$

In the limit of $R \to 0$ (zero resistance), we indeed recover Eq. (1.35).

Solving Eq. (1.41) is not straightforward because of the presence of the first-order derivative. However, in the absence of the inductance we know that the charge on the capacitor is exponentially damped,

$$q(t) = q_0 e^{-t/RC}, \tag{1.42}$$

where RC is the time constant. Therefore, we may expect that the solution to Eq. (1.41) is a combination of an oscillatory function and an exponential function, and we assume

$$q(t) = q_0 e^{-\gamma t} \cos \omega t, \tag{1.43}$$

where γ is the damping constant to be determined. The preceding form of solution, however, is valid only for the case of weak damping such that $\gamma \ll \omega$. The general case will be given as a problem of this chapter. Also in Chapter 13, the same problem will be solved by the method of Laplace transformation.

Notice that the solution for $q(t)$ given by Eq. (1.43) satisfies the initial condition

$$q(0) = q_0.$$

We now calculate dq/dt and d^2q/dt^2 (derive these):

$$\frac{dq}{dt} = q_0(-\gamma \cos \omega t - \omega \sin \omega t)e^{-\gamma t} \tag{1.44}$$

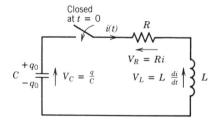

Fig. 1.13. *LCR* circuit. q_0 is the initial charge. An example of damped oscillation.

$$\frac{d^2q}{dt^2} = q_0[(\gamma^2 - \omega^2)\cos\omega t + 2\gamma\omega\sin\omega t]e^{-\gamma t}. \tag{1.45}$$

Substituting Eqs. (1.44) and (1.45) into Eq. (1.41) and eliminating the common factors q_0 and $e^{-\gamma t}$, we find

$$\left(\gamma^2 - \omega^2 - \frac{R}{L}\gamma + \frac{1}{LC}\right)\cos\omega t + \left(2\gamma\omega - \frac{R}{L}\omega\right)\sin\omega t = 0, \tag{1.46}$$

which must hold at any time. Then the coefficients of $\cos\omega t$ and $\sin\omega t$ must identically be zero,

$$\gamma^2 - \omega^2 - \frac{R}{L}\gamma + \frac{1}{LC} = 0 \tag{1.47}$$

$$2\gamma - \frac{R}{L} = 0. \tag{1.48}$$

From these we find

$$\gamma = \frac{R}{2L}, \qquad \omega = \frac{1}{\sqrt{LC}}, \tag{1.49}$$

where in Eq. (1.47) we have neglected terms containing γ, since we have assumed $\gamma \ll \omega$.

The function $q(t) = q_0 e^{-\gamma t}\cos\omega t$ is qualitatively shown in Fig. 1.14. The damped oscillation is confined between the two curves $\pm q_0 e^{-\gamma t}$, which are called envelopes. It should be emphasized again that the solution we have found is correct only for the case of small damping, $\gamma \ll \omega$, or, equivalently,

$$R \ll \sqrt{\frac{L}{C}}. \tag{1.50}$$

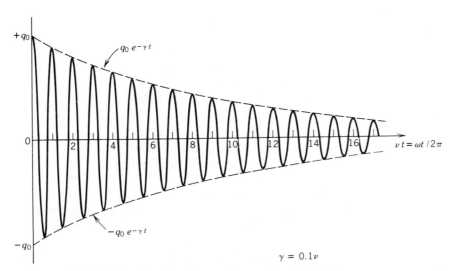

Fig. 1.14.　Behavior of the charge on the capacitor in Fig. 1.13.

Example 8. In the mass–spring oscillation system, assume there exists small but finite friction force between the mass and floor, which is proportional to the mass velocity,

$$F_{\text{friction}} = -fv = -f\frac{dx}{dt} \tag{1.51}$$

where f is a constant. Show that the differential equation for the displacement $x(t)$ is mathematically identical to Eq. (1.41) and find the condition for weakly damped oscillation.

The equation of motion for the mass is now given by

$$M\frac{d^2x}{dt^2} = -kx - f\frac{dx}{dt}$$

or

$$\frac{d^2x}{dt^2} + \frac{f}{M}\frac{dx}{dt} + \frac{k}{M}x = 0. \tag{1.52}$$

Comparing this with Eq. (1.41), we see that if the following substitution is made

$$x \to q, \quad M \to L, \quad f \to R, \quad k \to \frac{1}{C},$$

both equations are identical.

The condition for weakly damped oscillation

$$R \ll \sqrt{\frac{L}{C}}$$

can thus be translated as

$$f \ll \sqrt{kM}. \tag{1.53}$$

1.7. Forced Oscillation

In previous sections we found several oscillation frequencies appearing in both mechanical and electromagnetic systems. Those oscillation frequencies ($\omega = \sqrt{k/M}$, $1/\sqrt{LC}$, etc.) are also specifically called natural (or resonance) frequencies, since they appear when the oscillation systems are left alone, or isolated from external driving forces. Both mechanical and electromagnetic systems, however, can be forced to oscillate with frequencies other than the natural frequency.

A typical example of this forced oscillation is an ac circuit, in which an oscillating generator with an angular frequency ω is driving a current through L, C, R elements (Fig. 1.15). Even though there is a resistor R in the circuit, the current $i(t)$ does not damp, in contrast to the case we studied in Section 1.6, since the generator can continuously feed energy to compensate the amount of energy dissipated in the resistor.

The current flowing in the circuit will attain steady oscillation with the same frequency ω after the transient stage is over.* As we have studied in ac circuit theory, the amplitude of the current is given by

$$I_0 = \frac{V_0}{\sqrt{R^2 + (\omega L - 1/\omega C)^2}}. \tag{1.54}$$

This takes a maximum when

$$\omega L = \frac{1}{\omega C} \quad \text{or} \quad \omega = \frac{1}{\sqrt{LC}},$$

as shown in Fig. 1.16. The frequency determined from $1/\sqrt{LC}$ is thus appropriately called a resonance frequency, at which energy transfer from the generator to the resistor can be achieved most efficiently.

In mechanical oscillation systems, similar resonance phenomena can be found. When one pushes a swing, he naturally matches his pushing frequency with the natural or resonance frequency of the swing.

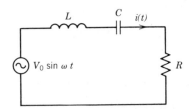

Fig. 1.15. *LCR* ac circuit. An example of forced oscillation.

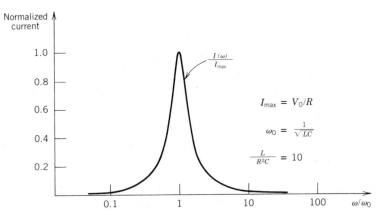

Fig. 1.16. Plot of Eq. (1.54) as a function of ω. If the frequency is plotted on a logarithmic scale, the graph becomes symmetric about the resonance frequency, $\omega_0 = 1/\sqrt{LC}$. The graph shown corresponds to the case $L/R^2C = 10$.

*The transient stage is discussed in Chapter 13 on Laplace transforms.

Problems

1. Calculate

$$\frac{d}{dx}\sin 5x, \quad \frac{d^2}{dx^2}\sin 5x, \quad \frac{d}{dx}\cos 3x, \quad \frac{d^2}{dx^2}\cos 3x.$$

2. If $|x| \ll 1$, a function $(1+x)^n$ can be approximated by

$$(1+x)^n \simeq 1+nx \quad \text{(binomial expansion)}.$$

   (a) Find the percent error caused by the approximation

   $$\sqrt{1+x} \simeq 1+\tfrac{1}{2}x \quad \text{for } x=0.1, 0.01.$$

   (b) Repeat (a) for

   $$(1+x)^{-1/2} \simeq 1-\tfrac{1}{2}x.$$

3. If $|\theta| \ll 1$ rad, $\sin\theta$ may be approximated by $\sin\theta \simeq \theta$. Calculate the percent error of this approximation for $\theta = 0.1$ rad, 0.01 rad.

4. Show that functions
   (a) $x = A\sin\omega t,$
   (b) $x = A\sin\omega t + B\cos\omega t,$
   (c) $x = A\cos(\omega t + \phi)$
   all satisfy Eq. (1.7), provided $\omega = \sqrt{k/M}$. A, B, and ϕ are constants.

5. In the configuration of Fig. 1.1, the mass (1.5 kg) is displaced 10 cm to the left and then released. Twenty oscillations are observed in 1 min. Find
   (a) The spring constant.
   (b) The equation describing the oscillation.
   (c) The energy associated with the oscillation.
   (*Answer:*   6.6 N/m, -10(cm) cos (2.1t), 3.3×10^{-2} J.)

6. A meter stick is freely pivoted about a horizontal axis at (a) the end of the stick or 100-cm mark and (b) the 75-cm mark. Find the oscillation frequencies in each case, assuming that the oscillation angle is small.
   (*Answer:*   0.61 Hz, 0.65 Hz.)

7. A thin circular hoop of radius a is hung over a sharp horizontal knife edge. Show that the hoop oscillates with an angular frequency $\omega = \sqrt{g/2a}$.

8. A marble thrown into a bowl executes oscillatory motion. Assuming that the inner surface of the bowl is parabolic ($y = ax^2$) and the marble has a mass m, find the oscillation frequency. Neglect friction and assume small oscillation amplitude.
   (*Answer:* $\omega = \sqrt{2ag}$. Note that a has dimensions of inverse length, m^{-1}.)

9. Place an object on a turntable of a record player. Observe the motion of the

object from the side. The motion is harmonic or sinusoidal of the form $x = x_0 \cos \omega t$. Prove this. If the turntable is revolving at $33\frac{1}{3}$ rpm, what is ω? What is v? What is the length of a pendulum to oscillate with the preceding frequency?

10. In the configuration of Fig. 1.11,
   (a) show that the current $i(t)$ is described by

$$i(t) = \frac{V_0}{\sqrt{L/C}} \sin \omega t,$$

   where $V_0 = q_0/C$ is the initial voltage on the capacitor.
   (b) The preceding expression indicates that the quantity $\sqrt{L/C}$ has the dimensions of ohms. Prove this. $\sqrt{L/C}$ is called the characteristic impedance.

11. A capacitor of $5\,\mu F$ charged to $1\,kV$ is discharged through an inductor of $2\,\mu H$. The total resistance in the circuit is $5\,m\Omega$.
   (a) Is this a weakly damped LCR circuit?
   (b) Find the time by which one-half the initial energy stored in the capacitor has been dissipated. The time is measured from the instant when discharge is started.

   (*Answer:* Yes, 0.28 msec.)

12. To solve Eq. (1.41) without the restriction of weak damping, assume

$$q(t) = e^{-\gamma t} (A \cos \omega t + B \sin \omega t),$$

   where ω, γ, A, and B are to be determined. From the initial condition, $q(0) = q_0$, we must have

$$A = q_0. \qquad \text{(i)}$$

   (a) Calculate dq/dt and d^2q/dt^2. Then substitute these into Eq. (1.41). You will obtain a relation like

$$f \cos \omega t + g \sin \omega t = 0, \qquad \text{(ii)}$$

   where f and g contain A, B, ω, γ. For Eq. (ii) to hold at any time, $f = g = 0$ must hold,

$$f = 0, \qquad \text{(iii)}$$

$$g = 0. \qquad \text{(iv)}$$

   (b) Another initial condition is that at $t = 0$, the current is zero, since the inductor behaves as if it were an infinitely large resistor right after the switch is closed. (Recall that inductors tend to resist any current variation.) Thus

$$i(0) = \frac{dq}{dt}\bigg|_{t=0} = 0. \qquad \text{(v)}$$

(c) Equations (i), (iii), (iv), and (v) constitute four simultaneous equations for four unknowns A, B, ω, γ. (We already have found A.) Solve these.

(d) Find the condition for ω to be real.

(*Answer:* $B = \gamma q_0/\omega$, $\gamma = R/2L$, $\omega = [(1/LC) - (R^2/4L^2)]^{1/2}$).

13. Prepare two eggs of approximately the same weight, one boiled and another raw. Make two pendulums using the eggs. (You can use Scotch tape or paper cups. Be careful with the raw egg.) Let the pendulums start oscillating. The pendulum with the raw egg would damp faster. Explain why. (Make sure that the pendulums both have the same length.)

14. Explain the function of shock absorbers installed on automobiles. What would happen without them?

15. Consider two cascaded spring–mass systems.

(a) Write down the equation of motion for *each* mass, assigning displacements $x_1(t)$ and $x_2(t)$ for the masses m_1 and m_2, respectively.

(b) Then eliminate $x_2(t)$ between the two equations to show that the differential equation for $x_1(t)$ is given by

$$m_1 m_2 \frac{d^4 x_1}{dt^4} + k(m_1 + 2m_2)\frac{d^2 x_1}{dt^2} + k^2 x_1 = 0.$$

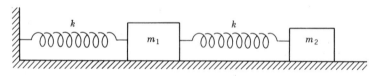

Fig. 1.17. Problem 15.

(c) Show that the oscillation frequency ω is given as solutions to

$$m_1 m_2 \omega^4 - k(m_1 + 2m_2)\omega^2 + k^2 = 0,$$

which allows two possible solutions for $|\omega|$.

16. Repeat Problem 15 for a two-mass, three-spring system whose both ends are clamped.

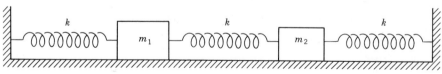

Fig. 1.18. Problem 16.

17. If $|\theta|$ is smaller than 1 rad, $\sin \theta$ can be approximated by

$$\sin \theta \simeq \theta - \tfrac{1}{6}\theta^3$$

and the pendulum equation [Eq. (1.25), p. 14] becomes

$$\frac{d^2\theta}{dt^2} = -\frac{g}{l}\,\theta(1 - \tfrac{1}{6}\theta^2).$$

This is still a nonlinear equation. However, it can tell us that as the oscillation amplitude of a pendulum increases, the oscillation frequency becomes smaller than the linear value, $\omega_0 = \sqrt{g/l}$. Explain (qualitatively) why this is so, referring to the preceding equation.

[*Hint:* The new frequency $\omega(\theta)$ is found approximately from

$$\omega^2(\theta) \simeq -\frac{1}{\theta}\frac{d^2\theta}{dt^2}\,.$$

Calculate the average of $1 - \tfrac{1}{6}\theta^2$, assuming $\theta = \theta_0 \sin \omega t$ to find

$$\omega^2(\theta_0) \simeq \frac{g}{l}\,[1 - \tfrac{1}{12}\theta_0^2].$$

A more exact analysis yields the correction to the frequency

$$\omega^2(\theta_0) \simeq \frac{g}{l}\,[1 - \tfrac{1}{8}\theta_0^2].$$

CHAPTER 2

Wave Motion

2.1. Introduction

The world is full of all kinds of waves. Sound waves, waves on the water surface, electromagnetic waves [radio and TV waves, microwaves, visible light, ultraviolet (invisible), and X rays], earthquake waves, and brain waves are only a few examples.

Waves are different from oscillations in the sense that waves propagate through a certain medium, or they move in space, while oscillations are localized. Consider sound waves in air. Our vocal chord is an oscillator and localized in our throat. But sound waves created by a vocal chord propagate through air, which is the medium for sound waves. Thus whenever we talk about waves, we have to have media for waves, and media must have spatial spread, large or small. For oscillations, time t was only the *independent variable*. If we choose time t, we can automatically find the instantaneous values of any oscillating physical quantities. For waves, however, *we have another independent variable, the spatial coordinate x.* This x should not be confused with that used in Chapter 1 to describe the displacement of the oscillating mass. There $x(t)$ was a *dependent variable*. In this chapter general properties of wave motion are outlined.

2.2. Creation of Waves on a String

Consider a mass M hanging from a ceiling through a spring (Fig. 2.1). This mass–spring system would start oscillating if we give it an initial pull or push. The oscillation would continue forever if all friction losses are negligibly small. Now connect a light string to the oscillating mass, at a certain instant. The end connected to the mass would start oscillating with the mass. At the same time a wavy structure starts propagating along the string, with a well-defined *spatial period*. We denote propagation velocity by c_w. The spatial period becomes shorter if the oscillation frequency of the mass increases (Fig. 2.2).

The figures shown in Fig. 2.1 depict snapshots taken at certain times. A point on the string just moves up and down, although the wavy structure itself is moving from left to right.

The perturbation on the rope created by the oscillating mass is one example

26

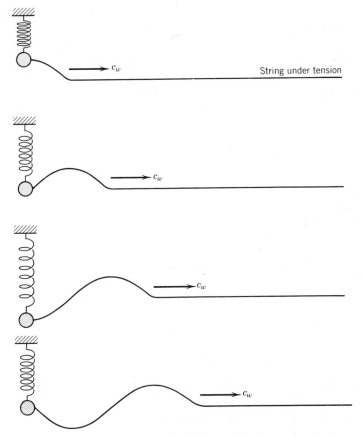

Fig. 2.1. When a string is suddenly connected to an oscillating mass–spring system, sinusoidal waves start propagating along the string.

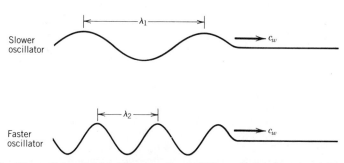

Fig. 2.2. The spatial period λ (m) becomes shorter if the oscillation frequency is increased.

of mechanical waves. The oscillating mass is the driver for the wave in this example. What is transferred from the oscillating mass to the wave is nothing but energy (and momentum as well, as we will see later). Thus it is expected that the oscillation amplitude of the mass will become smaller and smaller, since the mass–spring system is transferring its energy to the wave on the rope. In other words, the oscillating mass–spring system is doing work. Of course, the mass–spring system can be replaced by other driving mechanisms, such as our hands. Then the energy reserve can be large and we can create waves as long as we want.

The spatial period λ (m) is called the *wavelength*. The wavelength is related to the frequency v through

$$v\lambda = c_{\mathrm{w}} \tag{2.1}$$

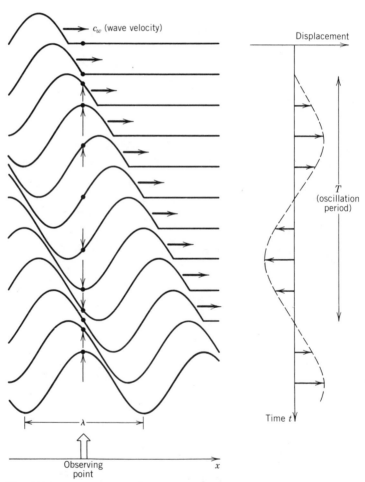

Fig. 2.3.   Sinusoidal waves seen by a stationary observer. Note that the observer detects a sinusoidal variation in time.

since in one second the oscillating mass creates v waves, which propagate with the velocity c_w. Thus an observer looking at a certain point on the string will see v oscillations passing by in one second, or the oscillation period is $1/v = T$ (Fig. 2.3).

If we know the propagation velocity c_w, we can immediately find the wavelength λ, which the wave with a given frequency v should have.

One of the major objectives of wave studies is to find this propagation velocity c_w for various kinds of waves, mechanical or electromagnetic. The propagation velocity is the fundamental quantity to characterize waves and is determined by the physical constants of the wave medium. You may know that the sound velocity in air is about 340 m/sec and the velocity of light in free space is 3.0×10^8 m/sec. Where do these numbers come from? How can we predict the sound velocity in water and the velocity of light in free space? These questions will be answered in subsequent chapters.

The waveform created on the string was sinusoidal (or harmonic) in the example. But this was simply due to the oscillatory (sinusoidal) motion of the mass. If a wave source oscillates sinusoidally, sinusoidal waves are created. Waves do not have to be sinusoidal, however. Suppose we hold the rope, and give a sudden jerk (Fig. 2.4). Then the waveform created is a pulse, rather than a continuous, sinusoidal wave. The pulse, however, propagates with the same speed as the sinusoidal wave.

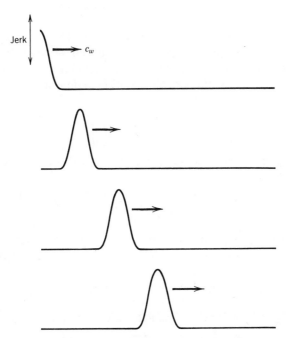

Fig. 2.4. Waves do not have to be sinusoidal. Here a single pulse is created by a sudden jerk.

Actually, waves we encounter in daily life are hardly simple sinusoidal waves. They all have complicated wave structures. The human voice can easily be demonstrated on an oscilloscope by detecting it with a microphone. As you will find, the waveform is rather complicated, hardly resembling a clear sinusoidal wave.

No matter how complicated the waveform is, there is a way to approximate the waveform in terms of many sinusoidal waves. This procedure is called the *Fourier analysis* and we will study this very powerful mathematical method in Chapter 13. Here all you have to remember is that any functions can be approximated by many sinusoidal functions. For this reason we can discuss waves in terms of simple sinusoidal functions, cos or sin, without much of a loss in generality.

2.3. Sinusoidal (Harmonic) Waves

Consider a sinusoidal wave with an amplitude A, frequency v cycles/sec, and wavelength λ (m) propagating in the positive x direction with a velocity c_w (m/sec). We want to find a mathematical expression that contains all the preceding information. We again consider snapshot pictures taken at equal time intervals, as shown in Fig. 2.5a. We choose the reference time $t = 0$ so that the snapshot taken at this time is described by the function

$$A \sin\left(\frac{2\pi}{\lambda} x\right), \qquad t = 0, \tag{2.2}$$

which is indeed periodic with a spatial period, λ (m). The next snapshot is taken at $t = \tau$, by which the whole wave pattern has moved in the positive x direction by a distance $c_w \tau$ (m). Since any function $f(x)$ shifted in the positive x direction by a distance a is given by $f(x - a)$, the equation to describe the wave pattern at $t = \tau$ is given by

$$A \sin\left[\frac{2\pi}{\lambda} (x - c_w \tau)\right]. \tag{2.3}$$

At $t = 2\tau$, the third snapshot is taken, and the equation for this wave pattern is given by

$$A \sin\left[\frac{2\pi}{\lambda} (x - 2c_w \tau)\right], \tag{2.4}$$

and so on. We can easily generalize this argument to the case of an arbitrary time t, and the general equation to describe the wave propagation is given by

$$A \sin\left[\frac{2\pi}{\lambda} (x - c_w t)\right]. \tag{2.5}$$

Here we introduce a function $f(x, t)$ and equate this to the preceding expression:

$$f(x, t) = A \sin\left[\frac{2\pi}{\lambda} (x - c_w t)\right]. \tag{2.6}$$

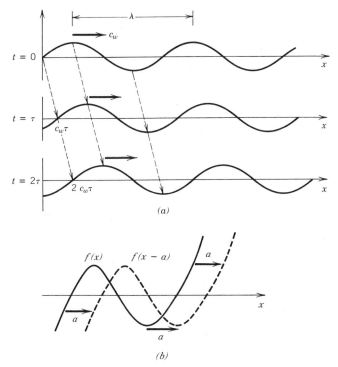

Fig. 2.5. (a) Snapshot pictures taken at successive times. The sinusoidal wavetrain at $t=0$ is shifted or translated by a distance $c_w t$. (b) If a function $f(x)$ is shifted in the positive x direction by a distance a, the function is described by $f(x-a)$. If the shift is in the negative x direction, $f(x+a)$ should be used.

The quantity A can be any physical quantity associated with any wave. For the case of the wave on the string, A can be chosen as the vertical displacement associated with the wave, and thus has the dimensions of distance m, for sound waves in air, A can be the displacement of air molecules about their equilibrium positions; for electromagnetic waves A can be electric and magnetic fields associated with the waves; and so on. All kinds of sinusoidal waves can be cast into the preceding general expression.

Let us fix the spatial coordinate $x=x_0$ (constant). This corresponds to observing the waves passing by as a function of time only. The observer must observe the point on the string at $x=x_0$ doing up and down motion in a sinusoidal manner. The frequency for this oscillation is equal to v. Mathematically, we have

$$f(x_0, t) = A \sin \left[\frac{2\pi}{\lambda} (x_0 - c_w t) \right],$$ (2.7)

which must have sinusoidal time dependence with the frequency v. Thus

$$\frac{2\pi}{\lambda} c_w = 2\pi v,$$

or

$$c_w = v\lambda.$$

This result is consistent with our previous qualitative argument.

Since $\omega = 2\pi v$, we may rewrite $f(x, t)$ as

$$f(x, t) = A \sin\left(\frac{2\pi}{\lambda} x - \omega t\right). \tag{2.8}$$

Furthermore, we may introduce a quantity defined by

$$k = \frac{2\pi}{\lambda}, \tag{2.9}$$

which has the dimensions of rad/m. The quantity k is called the *wavenumber*, although the actual number of waves in 1 m is given by $1/\lambda$. More appropriately, k should be called the angular wavenumber, just as we called ω the angular frequency. However, we follow tradition. You must be careful about the factor 2π whenever you use ω and k. This k should not be confused with the spring constant used in Chapter 1. Also, in electrical engineering, β is frequently used instead of k and is called the *phase constant*.

Using the angular frequency ω and the wavenumber k, Eq. (2.1) can be rewritten as

$$c_w = \frac{\omega}{k}. \tag{2.10}$$

You may wonder why we use ω and k instead of v and λ, particularly since v and λ alone can fully describe wave motion. The reason is more than just eliminating the factor 2π in Eq. (2.8). There is a lot more to it. The difference between ω and v is not significant. They just differ by a factor 2π. The use of k instead of λ (or $1/\lambda$) is more fundamental. We will see later that k is actually a vector directed in the direction of wave propagation.

Example 1. The velocity of *displacement waves* on a string is 40 m/sec. A driver oscillating with a frequency 15 Hz is connected to the one end of the string. Find the wavelength of the displacement wave created on the string.

Since the wavelength is given by

$$\lambda = \frac{c_w}{v},$$

we find

$$\lambda = \frac{40 \text{ m/sec}}{15/\text{sec}} = 2.67 \text{ m}.$$

Example 2. In the preceding example, determine A, k, and ω if the wave is written as

$$\xi(x, t) = A \sin(kx - \omega t).$$

Assume the driver oscillates with an amplitude of 1.0 cm.

The amplitude A is 1.0 cm. From $k = 2\pi/\lambda$, we find

$$k = \frac{2\pi}{2.67 \text{ m}} = 2.36 \text{ rad/m}$$

and from $\omega = 2\pi v$,

$$\omega = 2\pi \times 15 = 94.25 \text{ rad/sec.}$$

2.4. Wave Differential Equation, Partial Differentiation

In Chapter 1 we found the solution for oscillation from a differential equation of the form

$$\frac{d^2 f}{dt^2} + \omega^2 f = 0. \tag{2.11}$$

The solutions of this differential equation are indeed sinusoidal, as we have seen. Here we ask the following question: What differential equation can yield the wave equation, such as $A \sin (kx - \omega t)$?

Since $\sin (kx - \omega t)$ is periodic for both spatial coordinate x and time t, we expect that the differential equation contains both

$$\frac{d^2 f}{dt^2} \quad \text{and} \quad \frac{d^2 f}{dx^2}. \tag{2.12}$$

Let us then try to calculate these second derivatives for the particular function $f(x, t) = A \sin (kx - \omega t)$. Since

$$\frac{df}{dt} = \frac{d(kx - \omega t)}{dt} \frac{df}{d(kx - \omega t)} \tag{2.13}$$

we find

$$\frac{df}{dt} = -\omega A \cos (kx - \omega t). \tag{2.14}$$

Further differentiation yields

$$\frac{d^2 f}{dt^2} = -\omega^2 A \sin (kx - \omega t). \tag{2.15}$$

Similarly, for $d^2 f/dx^2$, we find

$$\frac{d^2 f}{dx^2} = -k^2 A \sin (kx - \omega t). \tag{2.16}$$

But we know that [see Eq. (2.10)]

$$\frac{\omega}{k} = c_w.$$

Then the function $f(x, t) = A \sin(kx - \omega t)$ satisfies the following differential equation:

$$\frac{d^2 f}{dt^2} = c_w^2 \frac{d^2 f}{dx^2} .$$ (2.17)

This may not be the only differential equation satisfied by the function $f(x, t)$, but it is at least one candidate.

Here we need some mathematics regarding the extension of differentiation. In the preceding calculation, we took for granted the following variable transformation:

$$\frac{df}{dt} = \frac{d}{dt}(kx - \omega t) \frac{df}{d(kx - \omega t)} ,$$ (2.18)

$$\frac{df}{dx} = \frac{d}{dx}(kx - \omega t) \frac{df}{d(kx - \omega t)} ,$$ (2.19)

with

$$\frac{d}{dt}(kx - \omega t) = -\omega,$$ (2.20)

$$\frac{d}{dx}(kx - \omega t) = k.$$ (2.21)

However,

$$\frac{d}{dt}(kx - \omega t) = -\omega$$

is valid (or meaningful) only if x does not depend on t. Since x and t are independent variables, we can choose in fact any values of x and t to calculate the preceding differentiation. However, it is meaningful only if we fix the value of x, and as long as the differentiation is concerned, the spatial coordinate x should be momentarily frozen, $x = $ constant. Physically, this corresponds to the situation in which we fix the observation point on the string and look at only the time variation of the wave motion at that particular point. Thus, more rigorously, we should have written Eq. (2.20) as

$$\frac{d}{dt}(kx - \omega t)\big|_{x = \text{const}} = -\omega.$$ (2.22)

Such a differentiation is called *partial differentiation*, and instead of the usual d/dt, we write

$$\frac{\partial}{\partial t}(kx - \omega t).$$ (2.23)

$\partial/\partial t$ automatically means that the spatial coordinate x is frozen, and then we carry out the differentiation with respect to time in a conventional manner.

Partial comes from the fact that the differentiation $\partial/\partial t$ can give us the time derivative only, and nothing about the spatial derivative. The spatial partial derivative can be written as

$$\frac{\partial}{\partial x}(kx - \omega t) = k, \tag{2.24}$$

in which we now freeze t, and vary only the spatial coordinate x. Physically, this corresponds to taking a snapshot picture at a given time.

The partial derivative of the function $f(x, t)$, which depends on both x and t, can be defined in exactly the same manner. For example, $\partial f/\partial t$ indicates that we carry out the usual differentiation with respect to t by freezing x, or assuming that x is a constant. Then for $f(x, t) = A \sin(kx - \omega t)$, we have

$$\frac{\partial f}{\partial t} = \frac{\partial}{\partial t}(kx - \omega t)\frac{d[A \sin(kx - \omega t)]}{d(kx - \omega t)}$$

$$= -\omega A \cos(kx - \omega t).$$

Notice that we did not write

$$\frac{\partial[A \sin(kx - \omega t)]}{\partial(kx - \omega t)},$$

since $\sin(kx - \omega t)$ contains the variables x and t, in the form $kx - \omega t$, which is now a single variable with respect to which we differentiate the function $A \sin(kx - \omega t)$. In this case we can define a new variable X by $X = kx - \omega t$, and the derivative becomes

$$\frac{d}{dX}(A \sin X).$$

This is simply an ordinary derivative with respect to X. Differentiation in this form is called *total differentiation*.

Using the symbols $\partial/\partial t$ and $\partial/\partial x$, we can rewrite Eq. (2.12) as

$$\frac{\partial^2 f}{\partial t^2} = c_w^2 \frac{\partial^2 f}{\partial x^2}, \tag{2.25}$$

where

$$\frac{\partial}{\partial t}\left(\frac{\partial f}{\partial t}\right) = \frac{\partial^2 f}{\partial t^2}, \qquad \frac{\partial}{\partial x}\left(\frac{\partial f}{\partial x}\right) = \frac{\partial^2 f}{\partial x^2}$$

just as in ordinary differentiation.

The partial differential equation we just found is called the wave (differential) equation, and we will encounter this many times in the following chapters. Most waves, mechanical and electromagnetic, can be cast into this form of differential equation. Whenever we end up with a differential equation

$$\frac{\partial^2 f}{\partial t^2} = \text{const}\,\frac{\partial^2 f}{\partial x^2}, \tag{2.26}$$

where the constant is *positive*, we can immediately find a propagation velocity from

$$c_w = \sqrt{\text{const.}} \tag{2.27}$$

For example, for electromagnetic waves in free space, the constant will become

$$\text{const} = \frac{1}{\varepsilon_0 \mu_0}, \tag{2.28}$$

where $\varepsilon_0 = 8.85 \times 10^{-12}$ F/m is the vacuum permittivity, and $\mu_0 = 4\pi \times 10^{-7}$ H/m is the vacuum permeability. $1/\sqrt{\varepsilon_0 \mu_0}$ gives 3.0×10^8 m/sec, the velocity of light in free space.

Example 3. A two-variable function is given by $f(x, y) = x^2 + y^2$. (a) Make a three-dimensional plot of $f(x, y)$. (b) Calculate $\partial f/\partial x$ and $\partial f/\partial y$. (c) Indicate how $\partial f/\partial x$ and $\partial f/\partial y$ can be graphically shown in the plot for $f(x, y)$.

(a)   See Fig. 2.6.

(b) $\partial f/\partial x = 2x$, $\partial f/\partial y = 2y$.

(c) $\partial f/\partial x$ indicates the slope of a straight line tangent to the parabola $f(x, y_0) = x^2 + y_0^2$ which is contained in the plane determined from $y = y_0$ (const) (Fig. 2.7). $\partial f/\partial y$ can be shown to have a similar meaning. It indicates the slope of a parabola, $f(x_0, y) = x_0^2 + y^2$ ($x_0 = $ const) in the plane $x = x_0$.

We can make a similar three-dimensional plot of the sinusoidal wave (see Fig. 2.8)

$$f(x, t) = A \sin (kx - \omega t)$$

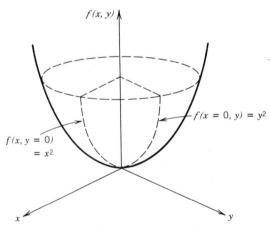

Fig. 2.6.   Example 3. Plot of $f(x, y) = x^2 + y^2$.

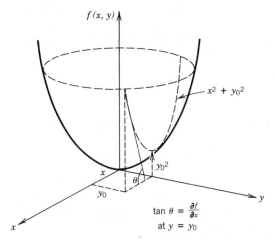

Fig. 2.7.   Example 3. $\partial f/\partial x$ indicates the slope of the tangent line to a parabola in the plane $y=$ const.

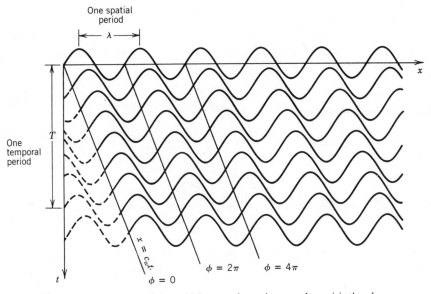

Fig. 2.8.   Propagation of sinusoidal waves shown in $x-t$ plane. ϕ is the phase.

on the plane of x and t. The spatial wave structure at $t=0$ progressively moves in the positive direction of x, as t increases. The trajectory of points (x, t) satisfying

$$\phi = kx - \omega t = \text{const} \tag{2.29}$$

is a straight line. For example, the line

$$kx = \omega t \qquad (\phi = 0)$$

or

$$\frac{x}{t} = \frac{\omega}{k},$$

starts off at the origin and its slope x/t indicates the propagation velocity ω/k.

Example 4.   Interpret the expression $f(x, t) = A \sin(kx + \omega t)$, where $k, \omega > 0$.

We rewrite $f(x, t) = A \sin[k(x + \omega t/k)]$. This indicates that the spatially sinusoidal wave $f(x, 0) = A \sin kx$ at $t = 0$ is shifted in the *negative* x direction by the distance $(\omega/k)t = c_w t$. Then the function expresses a sinusoidal wave propagating in the negative x direction with a velocity ω/k.

The quantity ϕ is called the phase and is measured in units of angle, radians or degrees. For a fixed time or the snapshot of the wave, the phase ϕ varies linearly with x,

$$\phi = kx - \omega t_0 \qquad (t_0 = \text{const})$$

However, since sinusoidal functions are periodic with the period of 2π, the two phases, ϕ and ϕ_1 differing by a multiple of 2π ($2\pi m$, $m = 1, 2, 3, \ldots$) yield exactly the same value for the wave quantity.

2.5. Nonsinusoidal Waves

The wave equation

$$\frac{\partial^2 f}{\partial t^2} = c_w^2 \frac{\partial^2 f}{\partial x^2}$$

actually allows any function $f(x, t)$ for its solution as long as f can be written as

$$f(x, t) = f(x - c_w t) \quad \text{or} \quad f(x + c_w t). \tag{2.30}$$

Solutions to the wave equation *do not have to be sinusoidal*, in contrast to the case of the oscillation we studied in Chapter 1. The proof is straightforward. Since

$$\frac{\partial f}{\partial t} = \pm c_w \frac{df}{dX}, \qquad X = x \pm c_w t$$

$$\frac{\partial^2 f}{\partial t^2} = c_w^2 \frac{d^2 f}{dX^2}$$

$$\frac{\partial f}{\partial x} = \frac{df}{dX} \quad \text{and} \quad \frac{\partial^2 f}{\partial x^2} = \frac{d^2 f}{dX^2}$$

we immediately find that

$$\frac{\partial^2 f}{\partial t^2} = c_w^2 \frac{\partial^2 f}{\partial x^2}.$$

This is a surprising feature of the wave equation. The function $f(x - c_w t)$ can take any form, sinusoidal, pulse, triangle, square, and so on. Let $f(x - c_w t)$ be an exponential function,

$$f(x - c_w t) = A e^{-(x - c_w t)^2 / a^2}. \tag{2.31}$$

where A is the amplitude and $a(m)$ determines the spatial width of the pulse (Fig. 2.9). At $t = 0$, we have an exponential function

$$f(x, 0) = A e^{-x^2 / a^2}. \tag{2.32}$$

This function is parallel-shifted or translated by a distance $c_w t$ after time t. That is, the exponential profile propagates with the velocity c_w, in the positive x direction. This exponential waveform can approximate the wave created on a string when we suddenly give a jerk to the end of the rope (Fig. 2.4).

No matter what waves are created, they all propagate with the same speed c_w, determined by the medium. As briefly discussed earlier, the waveforms of sounds we hear are extremely complicated, nothing like a clear sinusoidal wave. However, as long as the propagation velocity is concerned, any sound waveforms propagate at the same speed, about 340 m/sec at room temperature. Sound waves of a violin and those of a flute propagate with the same velocity. Imagine what would happen to an orchestra if this were not the case. Sounds from various instruments would be all mixed up, since they would reach your ears at different times!

You should be cautioned here that waves are not always described by the wave equation

$$\frac{\partial^2 f}{\partial t^2} = c_w^2 \frac{\partial^2 f}{\partial x^2},$$

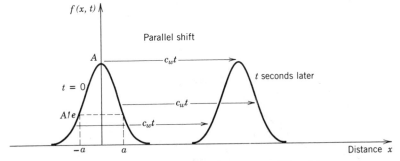

Fig. 2.9.   Exponential pulse wave shown at two different times.

which predicts the constancy of the propagation velocity independent of the waveform, or wave frequency. Sound waves and electromagnetic waves in free space can be well described by the preceding wave equation, but the waves on water surface, which is probably the most visible example of wave motion, cannot be described by this simple wave equation. The constancy of propagation velocity completely breaks down for water waves, and the propagation velocity depends on the waveform, or the wave frequency. The differential equation to describe the waves on water surface cannot be given by Eq. (2.25), but is more complicated. Another example is electromagnetic waves in matter, such as light waves in glass and water. As we will see, the velocity of light in glass is smaller than $c = 3.0 \times 10^8$ m/sec and is about $0.67c$ at the wavelengths region of visible light, $\lambda = 4 \times 10^{-7}$–$7 \times 10^{-7}$ m. More important, the velocity depends on the wavelength (or the wave frequency) even in that narrow wavelength region, and this in fact explains how a prism works (Chapter 7).

2.6. Phase and Group Velocities, Dispersion

At the end of Section 2.5 it was pointed out that certain waves cannot be described by the simple wave equation

$$\frac{\partial^2 f}{\partial t^2} = c_w^2 \frac{\partial^2 f}{\partial x^2} .$$

(2.33)

Waves described by this differential equation all have constant propagation velocities irrespective of wave frequencies or wavelengths,

$$c_w = \frac{\omega}{k} = \text{const.}$$

(2.34)

If we plot ω as a function of k, the graph is simply a straight line and its slope gives the propagation velocity (Fig. 2.10).

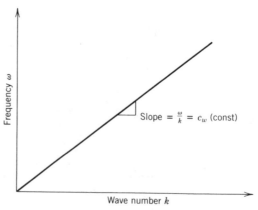

Fig. 2.10. ω–k diagram for a nondispersive wave, $\omega/k = c_w = \text{const.}$

Such waves with a constant propagation velocity are called nondispersive or dispersionless. If the propagation velocity depends on wave frequency, such waves are called dispersive. Dispersive waves do not have linear proportionality between ω and k. An example of dispersive waves is shown in Fig. 2.11. It describes how the wave frequency ω depends on the wavenumber k for the electromagnetic waves in a highly conductive medium, such as ionospheric plasma surrounding the earth. (This we will study in Chapter 9.) Note that the slope ω/k varies depending on the frequency ω, or wavelength, $\lambda = 2\pi/k$.

Nondispersive waves are described by Eq. (2.33). As we have seen in previous sections, a wave pattern initially created is simply translated as time goes on, keeping its waveform unchanged. If a pulse is created, the pulse propagates without deforming its shape (Fig. 2.12).

For dispersive waves this does not hold. The pulse initially created is severely deformed as it propagates. The pulse width Δx becomes larger and larger, and the pulse becomes widely spread or dispersed (Fig. 2.13). Now you see why we call such waves dispersive. The pulse, initially well defined and narrow, deforms its shape, and after a sufficiently long time it becomes difficult

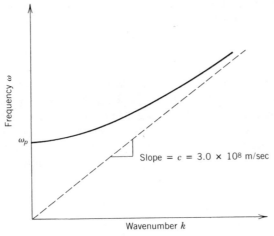

Fig. 2.11. ω–k relation for electromagnetic waves in a plasma. ω_p is called the plasma frequency and is determined by the density of free electrons. A similar ω–k relationship holds for electromagnetic waves in waveguides.

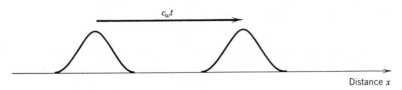

Fig. 2.12.   A pulse propagates undeformed in a nondispersive medium.

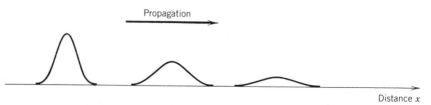

Fig. 2.13. In a dispersive medium, the pulse is strongly deformed, or dispersed.

to tell where the wave is located. Also, you can see that it is not easy to define the propagation velocity. By locating the peak at successive times, we can calculate how fast the peak is moving, but the portion of the pulse in front of the peak must propagate at a faster velocity than the peak, and the portion behind the peak must propagate at a slower velocity (Fig. 2.14). This spreading of pulse can be fully described once we find differential equations for dispersive waves.

Let us again consider a sinusoidal wave $f(x, t) = A \sin(kx - \omega t)$. As we have seen in Figs. 2.5 and 2.8, the collection of points all having the same phase,

$$\phi = kx - \omega t = \text{const}$$

forms a straight line on the x–t plane. The slope of the line gives the propagation velocity

$$\frac{\omega}{k} = c_w.$$

The velocity defined by ω/k is called the *phase velocity*, since the points of the same phase propagate with this velocity. In the ω–k diagram, the phase velocity is simply given by the slope of the straight line connecting the origin O and a certain point on the curve describing ω as a function of k, $\omega = \omega(k)$ (Fig. 2.15).

The velocity determined by the slope of a tangent line at a point on the curve, $d\omega/dk$, is called the *group velocity* and can be different from the phase velocity ω/k. For nondispersive waves,

$$\omega = c_w k, \quad c_w = \text{const}$$

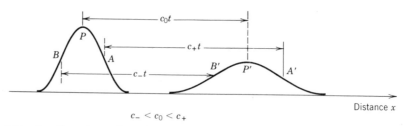

$$c_- < c_0 < c_+$$

Fig. 2.14. In a dispersive medium, the pulse is deformed as it propagates. The portion (A) in front of the peak propagates faster than the peak, and the portion (B) behind the peak more slowly than the peak.

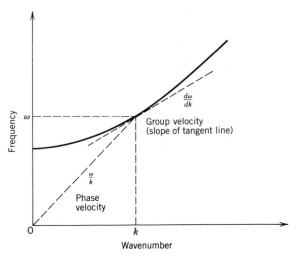

Fig. 2.15.   In dispersive waves, the group velocity $d\omega/dk$ is different from the phase velocity ω/k.

and the phase and group velocities are the same. In the example shown in Fig. 2.15, the group velocity is slower than the phase velocity, as you can easily see. Thus we may redefine dispersive and nondispersive waves as follows:

$$\text{Dispersive waves} \quad \frac{d\omega}{dk} \neq \frac{\omega}{k} \tag{2.35}$$

$$\text{Nondispersive waves} \quad \frac{d\omega}{dk} = \frac{\omega}{k} = \text{const.} \tag{2.36}$$

2.7. Superposition of Two Waves, Beats

Where does the name *group velocity* come from? Group of what? To answer this question, let us consider two sinusoidal waves of equal amplitude A, but different frequencies, ω_1 and ω_2, both propagating in the positive x direction. The two waves thus have wavenumbers k_1 and k_2, respectively, which can be found if the relationship between ω and k is known. The sum of the two waves then becomes

$$f(x, t) = A[\sin (k_1 x - \omega_1 t) + \sin (k_2 x - \omega_2 t)]. \tag{2.37}$$

We know that for any α and β (see Appendix B),

$$\sin \alpha + \sin \beta = 2 \sin \frac{\alpha + \beta}{2} \cos \frac{\alpha - \beta}{2}. \tag{2.38}$$

Then

$$f(x, t) = 2A \sin \left[\frac{(k_1 + k_2)x - (\omega_1 + \omega_2)t}{2} \right]$$
$$\times \cos \left[\frac{(k_1 - k_2)x - (\omega_1 - \omega_2)t}{2} \right]. \tag{2.39}$$

If ω_1 and ω_2 are exactly equal, then k_1 and k_2 are too, and we simply have

$$f(x, t) = 2A \sin (k_1 x - \omega_1 t), \tag{2.40}$$

or the amplitude is simply doubled, as it should be. Now let us consider a case in which ω_1 and ω_2 are slightly different, so that

$$\omega_1 = \omega_2 + \Delta\omega, \quad \Delta\omega \text{ small.} \tag{2.41}$$

Similarly, we let

$$k_1 = k_2 + \Delta k, \quad \Delta k \text{ small.} \tag{2.42}$$

Then

$$f(x, t) = 2A \sin (k_1 x - \omega_1 t) \cos \left[\frac{\Delta k}{2} x - \frac{\Delta\omega}{2} t \right] \tag{2.43}$$

since

$$\frac{\omega_1 + \omega_2}{2} = \frac{2\omega_1 + \Delta\omega}{2} \simeq \omega_1$$

and

$$\frac{k_1 + k_2}{2} = \frac{2k_1 + \Delta k}{2} \simeq k_1.$$

Let us see what this new waveform looks like. For this we take a snapshot at $t = 0$,

$$f(x, 0) = 2A \sin k_1 x \cos \frac{\Delta k}{2} x. \tag{2.44}$$

Since $\Delta k \ll k$, the wavelength associated with $\Delta k/2$

$$\lambda = \frac{2\pi}{\Delta k/2}, \tag{2.45}$$

is much longer than that corresponding to k_1,

$$\lambda_1 = \frac{2\pi}{k_1}. \tag{2.46}$$

Thus the function that is a product of two sinusoidal functions is shown in Fig. 2.16.

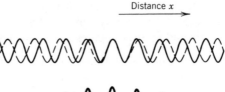

Distance x

$$\lambda_1 \simeq \frac{2\pi}{k_1}$$

$\lambda/2$
Beat wavelength

$$\frac{k_1}{k_2} = 1.125, \ \Delta k = 0.125 k_1$$

Fig. 2.16. When two waves (solid and broken lines in the top figure) with slightly different wavelengths (and thus frequencies) are added, the amplitude is modulated and clumps are formed.

The fine ripples of the short wavelength propagate with the phase velocity

$$c_{ph} = \frac{\omega_1}{k_1} .$$

But the envelope determined by the factor

$$\cos\left(\frac{\Delta k}{2}x - \frac{\Delta \omega}{2}t\right)$$

propagates with the velocity

$$\frac{\Delta \omega/2}{\Delta k/2} = \frac{\Delta \omega}{\Delta k} .$$

By making $\Delta \omega$ and Δk sufficiently small, $\Delta \omega/\Delta k$ approaches the group velocity

$$c_g = \frac{d\omega}{dk} . \tag{2.47}$$

The clumps formed by several short waves may appropriately be called groups of waves, and these clumps propagate with the group velocity, which can be different from the phase velocity for dispersive waves.

We will see in later chapters that any waves must carry energy and momentum. It will be shown that *energy is transferred with the group velocity, rather than the phase velocity*, and in this respect, the group velocity is a more fundamental quantity.

Example 5. The dispersion relation of electromagnetic waves in the ionospheric plasma is given by

$$\omega^2 = \omega_p^2 + c^2 k^2 ,$$

where ω_p is a constant (called the plasma frequency) and c is the speed of light in free space, 3.0×10^8 m/sec. Assuming $\omega_p = 1.0 \times 10^8$ rad/sec, calculate the phase and group velocities at a frequency of $v = 20$ MHz (see Fig. 2.17).

Since $\omega = 2\pi v = 1.26 \times 10^8$ rad/sec, we find

$$k = \frac{1}{c}\sqrt{\omega^2 - \omega_p^2} = 0.254 \text{ rad/m}.$$

Then the phase velocity is

$$c_{ph} = \frac{\omega}{k} = \frac{1.26 \times 10^8}{0.254} = 4.97 \times 10^8 \text{ m/sec}.$$

The group velocity is given by

$$c_g = \frac{d\omega}{dk} = \frac{d}{dk}\sqrt{\omega_p^2 + c^2 k^2} = \frac{c^2 k}{\sqrt{\omega_p^2 + c^2 k^2}} = \frac{c^2}{\omega/k}.$$

Then

$$c_g = \frac{(3.0 \times 10^9)^2}{4.97 \times 10^8} = 1.81 \times 10^8 \text{ m/sec}.$$

In the preceding example the phase velocity is actually larger than the speed of light c. This, however, is not in contradiction to Einstein's relativity theory, which claims that nothing can travel faster than the speed of light. As briefly mentioned earlier, the energy of waves travels with the group velocity rather than the phase velocity, and as long as the group velocity does not exceed c, no contradiction to the relativity theory should arise.

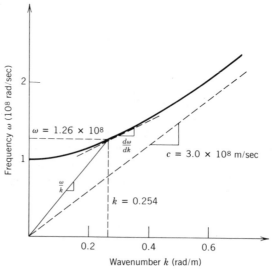

Fig. 2.17. Example 5. ω–k relation. The phase velocity ω/k is larger than the speed of light, but the group velocity $d\omega/dk$ is not.

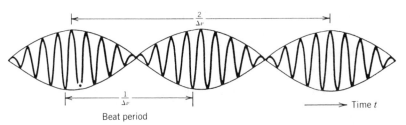

Fig. 2.18. Beats caused by the superposition of two waves of slightly different frequencies.

The superposition of two waves of slightly different frequencies yields an important phenomenon called beats (Fig. 2.18). Let us fix the spatial coordinate x in Eq. (2.43), say at $x=0$. This corresponds to an observer standing at $x=0$ and seeing the waves passing by. He will observe a waveform given by

$$f(0, t) = -2A \sin \omega_1 t \cos \frac{\Delta \omega}{2} t, \tag{2.48}$$

which indicates that the amplitudes of high-frequency (ω_1) oscillations are modulated by the slowly varying ($\Delta \omega \ll \omega_1$) sinusoidal function, cos ($\Delta \omega t/2$). The clumps appear every $2\pi/\Delta \omega = 1/\Delta v$ sec. Thus in the case of sound waves, for example, one hears the sound intensity going up and down with a frequency

$$\Delta v = \frac{\Delta \omega}{2\pi} = |v_1 - v_2|. \tag{2.49}$$

This intensity modulation is called the beats. A piano tuner uses this beat phenomenon when he tunes a piano using tuning forks.

Example 6. When a certain note of a piano is sounded with a tuning fork of a frequency 580 Hz, 5 beats are heard every second. Find the frequency of the note.

Let the frequency be v. Then

$$|580 - v| = 5, \quad v = 575 \text{ Hz or } 585 \text{ Hz}.$$

The beat frequency alone cannot determine which is the note frequency. However, in practice the piano tuner can tell which (piano or fork) has a higher frequency by sounding one immediately after another. Beats are frequently used in tuning musical instruments. Zero beat indicates that two sound sources have the same frequency.

Problems

1. A displacement wave on a string is described by $0.02 \sin [2\pi(0.5x - 10t)]$ (m), where x is in meters and t in seconds. Find

   (a) The propagation velocity.

(b) Wavelength λ and wavenumber k.

(c) Frequency ν and angular frequency ω.

(d) Direction of propagation.

(e) Amplitude of the wave.

(*Answer:* 20 m/sec, 2m, π rad/m, 10 Hz, 20π rad/sec, positive x, 2 cm.)

2. Find the expression for the wave in Problem 1 propagating in the negative x direction.

3. Repeat Problem 1 for a wave given by 0.03 sin $(5x - 150t)$ (m).

4. Plot $f(x, t) = 0.1$ sin $[2\pi(0.5x - 20t)]$ as a function of x at $t = $(a) 0 sec, (b) 0.0125 sec, (c) 0.025 sec, (d) 0.0375 sec, and (e) 0.05 sec. Convince yourself that the wave pattern progresses in the positive x direction as time goes on.

5. The speed of electromagnetic waves in free space is 3.0×10^8 m/sec. CBC Regina uses a frequency of 540 kHz (Hz, hertz=cycles/second). Find the wavelength.

    (*Answer:* 556 m.)

6. Find the wavelength of wavy groove structure on an LP record ($33\frac{1}{3}$ rev/min) for a sound wave with a frequency 1 kHz. Consider two radial positions on the record, $r = $(a) 15 cm and (b) 10 cm.

    (*Answer:* 0.52 *mm*, 0.35 *mm*.)

7. Calculate

$$\frac{\partial f}{\partial x}, \quad \frac{\partial f}{\partial y}, \quad \frac{\partial^2 f}{\partial x^2}, \quad \frac{\partial^2 f}{\partial y^2}, \quad \frac{\partial^2 f}{\partial x \partial y}$$

    for functions

    (a) $f(x, y) = \sin(x - 2y)$

    (b) $f(x, y) = x^2 + xy + y^2$.

8. (a) Show that $f(x, t) = 0.1$ sin $[2\pi(0.5x - 20t)]$ satisfies the wave equation

$$\frac{\partial^2 f}{\partial t^2} = 1600 \frac{\partial^2 f}{\partial x^2}.$$

    (b) What is the propagation velocity?

9. (a) Plot $f(x, t) = 0.5\, e^{-(x - 5t)^2}$ as functions of $x(m)$ at $t = 0$, 0.5 sec, 1.0 sec. Convince yourself that the preceding expression describes a pulse propagating in the positive x direction.

    (b) Find the differential equation the preceding expression should satisfy.

10. A certain wave has $\omega - k$ relationship (or dispersion relation) given by

$$\omega = 10^3 k - 3 \times 10^{-5} k^3.$$

(a) Plot ω against k for $0 \leqslant k \leqslant 3 \times 10^3$ rad/m.

(b) Is the wave dispersive or nondispersive?

(c) Find the phase and group velocities at $k = 1 \times 10^3$ rad/m.

(*Answer:* 970 m/sec, 910 m/sec.)

11. (a) Add two sinusoidal waves

$$2 \sin (5x - 1500t), \quad 2 \sin (5.1x - 1530t).$$

(b) What is the beat frequency?

(c) What is the beat wavelength?

(*Answer:* 4.8 Hz, 62.8 m.)

12. The phase velocity of surface waves in deep water is given by

$$\frac{\omega}{k} = \sqrt{\frac{g}{k} + \frac{T_s}{\rho} k},$$

where T_s is the surface tension (7.3×10^{-2} N/m) and ρ is the water mass density.

(a) Find the expression for the group velocity.

(b) When a stone is thrown into a pond, waves of many wavelength components start propagating. What would be the slowest wavelength? Note: A stone thrown into water gives an impulse-like disturbance, which contains many wavelength components.

CHAPTER 3

Some Mathematics

Taylor expansion of mathematical functions is used frequently in the following chapters. Since you may not be too familiar with this powerful mathematical technique, it would be appropriate to devote one chapter to this subject and related mathematical formulas.

Taylor's theorem goes as follows: A function $f(x)$ can be expanded in terms of a power series of x as

$$f(x) = f(a) + f'(a)(x-a) + \frac{f''(a)}{2!}(x-a)^2 + \frac{f'''(a)}{3!}(x-a)^3$$

$$+ \cdots + \frac{f^n(a)}{n!}(x-a)^n \ldots, \tag{3.1}$$

where a is an arbitrary value of x and $f^{(n)}(a)$ is the nth derivative of $f(x)$ evaluated at $x=a$. The proof is straightforward. Let $f(x)$ be expanded in a power series as

$$f(x) = A_0 + A_1(x-a) + A_2(x-a)^2 + \cdots = \sum_{m=0}^{\infty} A_m(x-a)^m$$

where A_n's are constant. A_0 can be immediately found as $f(a)$ by letting $x=a$. To determine A_n, we differentiate $f(x)$ n times,

$$\frac{d^n}{dx^n} f(x) = \sum_{m \geqslant n} A_m m(m-1)(m-2) \cdots (m-(n-1))(x-a)^{m-n}$$

and let $x=a$. Then only the term $m=n$ remains nonzero, and

$$\frac{d^n}{dx^n} f(x)\big|_{x=a} = n(n-1) \cdots 3 \cdot 2 \cdot 1 \, A_n = n! \, A_n$$

or

$$A_n = \frac{1}{n!} f^{(n)}(a),$$

which gives Eq. (3.1).

As an example, let $f(x) = \sin x$. Since

$$\frac{d}{dx}\sin x = \cos x$$

$$\frac{d^2}{dx^2}\sin x = -\sin x$$

$$\frac{d^3}{dx^3}\sin x = -\cos x$$

$$\vdots$$

we find

$$\sin x = \sin a + (\cos a)(x-a)$$
$$-\frac{\sin a}{2!}(x-a)^2 - \frac{\cos a}{3!}(x-a)^3$$
$$+\frac{\sin a}{4!}(x-a)^{\,4} + \frac{\cos a}{5!}(x-a)^5 \dots \tag{3.2}$$

If we choose $a=0$, $\sin 0 = 0$, and $\cos 0 = 1$, and we find

$$\sin x = x - \frac{1}{3!}x^3 + \frac{1}{5!}x^5 - \frac{1}{7!}x^7 + \cdots. \tag{3.3}$$

This is the series expansion of $\sin x$ about the origin $x=0$. If x is small (x is measured in radians here; small x means $|x| \ll 1$ rad), $\sin x$ can be approximated by $\sin x \simeq x$. (This has been used for analyzing pendulum oscillation in Chapter 1.)

You should prove the following power series expansions by yourself.

$$\cos x = 1 - \frac{1}{2!}x^2 + \frac{1}{4!}x^4 - \frac{1}{6!}x^6 + \cdots \tag{3.4}$$

$$\tan x = x + \frac{1}{3}x^3 + \frac{2}{15}x^5 + \cdots \tag{3.5}$$

$$e^x = 1 + x + \frac{1}{2!}x^2 + \frac{1}{3!}x^3 + \frac{1}{4!}x^4 + \cdots. \tag{3.6}$$

A particularly important case is the binomial expansion

$$(1+x)^n = 1 + nx + \frac{n(n-1)}{2!}x^2 + \frac{n(n-1)(n-2)}{3!}x^3 + \cdots, \tag{3.7}$$

where n is an arbitrary number (not necessarily an integer). For this we let $f(x) = (1+x)^n$ in Eq. (3.1), and $a=0$. Then

$$f'(0) = n$$

$$f''(0) = n(n-1)$$

$$f'''(0) = n(n-1)(n-2)$$

$$\vdots$$

and we readily obtain Eq. (3.7). We have also already used this binomial expansion in Chapter 1; n can be any number—positive, negative, or even complex. For example, let $n=\frac{1}{2}$. Then

$$(1+x)^{1/2}=1+\frac{1}{2}x-\frac{1}{8}x^2+\frac{3}{48}x^3-\cdots. \tag{3.8}$$

For $n=-1$, we find

$$\frac{1}{1+x}=1-x+x^2-x^3+x^4-\cdots. \tag{3.9}$$

If x is small, we may approximate $(1+x)^n$ by

$$(1+x)^n\simeq1+nx. \tag{3.10}$$

Let x be replaced by $x+a$ in Eq. (3.1). Then we find another form of the Taylor expansion

$$f(x+a)=f(a)+f'(a)x+\frac{f''(a)}{2!}x^2+\cdots. \tag{3.11}$$

Thus if x is small, we can approximate $f(x+a)$ by

$$f(x+a)\simeq f(a)+f'(a)x, \tag{3.12}$$

retaining up to only the first-order term in the power series. This approximation will be frequently used in the following chapters. (See Problem 6.)

Using Eqs. (3.3), (3.4), and (3.6) we can prove the following important relationship:

$$e^{i\theta}=\cos\theta+i\sin\theta, \tag{3.13}$$

where $i=\sqrt{-1}$. To prove this, let $x=i\theta$ in Eq. (3.6), where θ is a real number (positive or negative). Then

$$
\begin{aligned}
e^{i\theta}&=1+i\theta+\frac{1}{2!}(i\theta)^2+\frac{1}{3!}(i\theta)^3+\frac{1}{4!}(i\theta)^4+\cdots\\
&=1+i\theta-\frac{1}{2!}\theta^2-i\frac{1}{3!}\theta^3+\frac{1}{4!}\theta^4\\
&\quad+i\frac{1}{5!}\theta^5-\frac{1}{6!}\theta^6+\cdots\\
&=1-\frac{1}{2!}\theta^2+\frac{1}{4!}\theta^4-\frac{1}{6!}\theta^6\cdots\\
&\quad+i[\theta-\frac{1}{3!}\theta^3+\frac{1}{5!}\theta^5-\cdots].
\end{aligned}
\tag{3.14}
$$

Recalling Eqs. (3.3) and (3.4), we see that Eq. (3.13) indeed holds. Using this

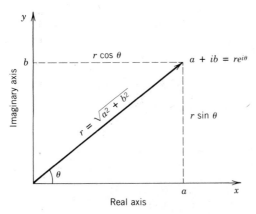

Fig. 3.1.   Polar representation of the complex number $a+ib=re^{i\theta}$.

formula, we can express any complex number $a+ib$ (a, b real) in the form of (Fig. 3.1)

$$re^{i\theta}, \qquad (3.15)$$

where

$$r=\sqrt{a^2+b^2} \qquad (3.16)$$

$$\theta=\tan^{-1}\left(\frac{b}{a}\right) \qquad (3.17)$$

since

$$re^{i\theta}=\sqrt{a^2+b^2}\,(\cos\theta+i\sin\theta)$$

and

$$\cos\theta=\frac{a}{\sqrt{a^2+b^2}}$$

$$\sin\theta=\frac{b}{\sqrt{a^2+b^2}}.$$

Equation (3.15) is called the polar representation of complex numbers.

Example 1.   Write $1\pm i\sqrt{3}$ in the form of polar representation (Fig. 3.2.).
Since $a=1$, $b=\pm\sqrt{3}$, we find $1\pm i\sqrt{3}=\sqrt{1+3}\,e^{\pm i\theta}$, where

$$\theta=\tan^{-1}(\sqrt{3}/1)=\pi/3 \text{ rad.}$$

Then

$$1\pm i\sqrt{3}=2e^{\pm i(\pi/3)}.$$

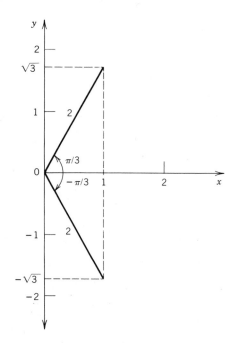

Fig. 3.2. $1 \pm i\sqrt{3}$ on the complex plane.

Example 2. Assume a solution of the form $f = Ae^{i(kx-\omega t)}$ to the wave equation

$$\frac{\partial^2 f}{\partial t^2} = c_w^2 \frac{\partial^2 f}{\partial x^2}.$$

Show that f given previously satisfies the wave equation provided $\omega^2 = c_w^2 k^2$.

Since

$$\frac{\partial}{\partial x} Ae^{i(kx-\omega t)} = Aike^{i(kx-\omega t)}$$

$$\frac{\partial^2}{\partial x^2} Ae^{i(kx-\omega t)} = A(ik)^2 e^{i(kx-\omega t)}$$

$$\frac{\partial}{\partial t} Ae^{i(kx-\omega t)} = A(-i\omega)e^{i(kx-\omega t)}$$

$$\frac{\partial^2}{\partial t^2} Ae^{i(kx-\omega t)} = A(-i\omega)^2 e^{i(kx-\omega t)},$$

we find

$$-A\omega^2 e^{i(kx-\omega t)} = -c_w^2 Ak^2 e^{i(kx-\omega t)}$$

or

$$\omega^2 = c_w^2 k^2.$$

Example 2 indicates that the partial derivative $\partial f/\partial t$ can be replaced by $-i\omega f$, if $f(x, t)$ is assumed to be

$$A e^{i(kx - \omega t)}.$$

Similarly, $\partial f/\partial x$ can be replaced by $ik f$. When we want a solution of the form $A \cos(kx - \omega t)$, we can take the real part of $A e^{i(kx - \omega t)}$. For $A \sin(kx - \omega t)$, we can take the imaginary part of $A e^{i(kx - \omega t)}$. This so-called *operator method* greatly simplifies mathematical analyses of oscillation and wave phenomena. We will study this subject in more detail in Chapter 13.

Problems

1. Prove Eqs. (3.4), (3.5), and (3.6).

2. What is the power series expansion of e^{-x}?

3. Show that

$$\cos x = \frac{e^{ix} + e^{-ix}}{2}, \qquad \sin x = \frac{e^{ix} - e^{-ix}}{2i}.$$

4. Electromagnetic waves in the ionosphere are described by the following differential equation,

$$\frac{\partial^2 E}{\partial t^2} + \omega_p^2 E = c^2 \frac{\partial^2 E}{\partial x^2},$$

where ω_p (called the plasma frequency) is a constant, $c = 3.0 \times 10^8$ m/sec is the speed of light in vacuum, and $E(x, t)$ is the electric field associated with the waves. Using the operator method, show that the dispersion relation for the wave is given by

$$\omega^2 = \omega_p^2 + c^2 k^2.$$

5. Show that the wave differential equation to be satisfied by the dispersion relation given in Problem 10 (Chapter 2) is

$$\frac{\partial f}{\partial t} = -10^3 \frac{\partial f}{\partial x} - 3 \times 10^{-5} \frac{\partial^3 f}{\partial x^3}.$$

6. Show that if Δx is small, $f(x + \Delta x)$ can be approximated by

$$f(x) + f'(x) \Delta x.$$

7. Show that the differential equation to yield the dispersion relation of water wave $\omega^2 = gk$ can be given by either

$$\frac{\partial^2 \xi}{\partial t^2} = ig \frac{\partial \xi}{\partial x} \quad (i = \sqrt{-1})$$

or

$$\frac{\partial^4 \xi}{\partial t^4} + g^2 \frac{\partial^2 \xi}{\partial x^2} = 0.$$

8. Calculate $\sqrt{i}$.

(*Answer:* $\pm (1+i)/\sqrt{2}$).

CHAPTER 4

Mechanical Waves

4.1. Introduction

Waves can be classified into two major categories, mechanical waves and electromagnetic waves. Mechanical waves are those that can be created and propagated in elastic material media. Sound waves in gases, liquids, and solids, and waves on an elastic string are typical examples. All mechanical waves can be described fully by Newton's equation of motion once we can find appropriate forces to act on a small volume (or segment) of media. Electromagnetic waves, on the other hand, do not require any material media. Vacuum is an excellent medium for electromagnetic waves (light, radio and TV waves, etc.). However, electromagnetic waves can propagate in material media, too, depending on wave frequencies. Maxwell's equation are needed to describe electromagnetic waves. Electromagnetic waves in material media need both Maxwell's equations and Newton's equation of motion.

In this chapter general aspects of mechanical waves are discussed. One important requirement for a medium to accommodate mechanical waves is that the medium be elastic. If it is compressed or expanded by a force, it should be able to restore its original shape when the force is removed. Sound waves can be propagated in a hard soil, but not in soft clay. Waves can be created on a rope under a tension, but not on a rope without tension.

Another essential fact is that any material media (air, water, solids) have a mass density. Air is light but has a finite mass density of 1.29 kg/m^3 at 0°C, 1 atmospheric pressure, which determines, together with the pressure, the velocity of sound waves in air.

Elasticity and inertia (mass) are two major physical qualities that determine the propagation velocity of mechanical waves. You should recall that we have already encountered similar requirements for mechanical oscillations. For a mass–spring system, the spring provides the elasticity or restoring force and determines the potential energy, and the mass provides the inertia of the system and determines the kinetic energy. Thus elasticity and inertia (mass) are physical qualities common to oscillating systems and medium for mechanical waves.

4.2. Mass–Spring Transmission Line

Here we attempt to construct or model mechanical waves and their media using many units of mass–spring combination (Fig. 4.1). A single unit (one spring and one mass) cannot accomodate waves since it has no spatial spread. For waves we need a medium spatially spread, and thus many mass–spring units connected in series.

Suppose that each unit contains a spring with a spring constant k and a mass m and has a length Δx, which is also assumed to be the natural length of each spring. Thus if we do not apply any force to disturb the system, the masses are located at equilibrium positions, $x = \Delta x, 2\Delta x, 3\Delta x, \ldots$. Let us give the left end a sudden push displacing it by a distance ξ_0. The first spring is then compressed and tends to push the first mass. Because of its inertia it takes the mass a finite time before it moves to the right and then pushes the second spring. If the mass finally moves the distance ξ_0, the first spring is relaxed back to its natural length. Since the first mass now moves the distance ξ_0, the second spring repeats what the first spring did but does this after some delay. The sequence repeats on and on along the system (Fig. 4.2).

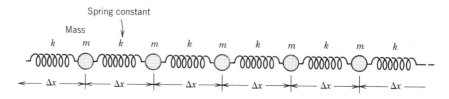

Fig. 4.1. Mechanical transmission line composed of mass–spring units.

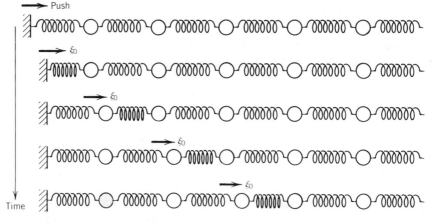

Fig. 4.2. When a sudden push is given to the end of the transmission line, the disturbance starts propagating. ξ_0 indicates the displacement of masses from their original positions.

The time delay depends on how strong the spring is and how heavy the mass is. If the spring is strong (larger k), we expect the spring can restore its original length quicker, and the time delay will be shorter, and if the mass (m) is larger, the time delay will be larger. Since the delay time is a measure of wave velocity, we expect that the expression for the wave velocity contains a factor k/m.

In the preceding example, we can already see some of the fundamental properties of mechanical waves. First of all, the model transmission line has both elasticity (provided by the springs) and a mass density $m/\Delta x$ (kg/m). Therefore it is expected to be a medium for mechanical waves. Second, we can see that what propagates with the disturbance (or wave) is energy. By squeezing the first spring, we have provided it with an initial potential energy $\frac{1}{2}k\xi_0^2$. The spring then transfers this energy to the first mass, which now acquires a kinetic energy. Since the mass then tends to squeeze the second spring, the energy is transferred to the second spring, and so on. Notice that neither springs nor masses move *along* with the disturbance. They stay more or less at original positions that are slightly displaced by a distance ξ_0 after the disturbance has passed. The wave velocity thus has nothing to do with the material velocity, and in fact the two velocities are independent as long as the displacement ξ is sufficiently small (linear wave medium).

This far we can go without using mathematics. But we wish to find out exactly how the propagation velocity is related to the factor k/m. We can do this with our newly acquired knowledge of Taylor expansion.

4.3. Derivation of Wave Equation

Let us pick up one unit of the mass–spring system located at a distance x from the left end. When a wave is created, each mass deviates from its equilibrium position. We denote the displacement of the mass originally (in equilibrium) located at x by $\xi(x)$, that of the mass at $x+\Delta x$ by $\xi(x+\Delta x)$ and that of the mass at $x-\Delta x$ by $\xi(x-\Delta x)$. (Fig. 4.3) The spring to the left of the mass at x suffers a

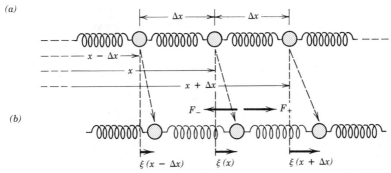

Fig. 4.3. A portion of the transmission located at x is shown (*a*) without wave (equilibrium) and (*b*) with wave (perturbed).

net length change given by

$$\Delta x + \xi(x) - [\Delta x + \xi(x - \Delta x)] = \xi(x) - \xi(x - \Delta x) \tag{4.1}$$

and exerts a force on the mass given by

$$F_- = k[\xi(x) - \xi(x - \Delta x)] \tag{4.2}$$

directed to the left. The spring to the right of the mass at x similarly exerts a force

$$F_+ = k[\xi(x + \Delta x) - \xi(x)] \tag{4.3}$$

directed to the right. Thus the net force to act on the mass at x is

$$F = F_+ - F_- = k[\xi(x + \Delta x) + \xi(x - \Delta x) - 2\xi(x)]. \tag{4.4}$$

You should be careful about the direction of two forces, which is introduced only for mathematical formality. Actual direction of each force of course depends on the magnitude of displacements $\xi(x - \Delta x)$, $\xi(x)$, and $\xi(x + \Delta x)$. Consider the special case $\xi(x - \Delta x) = 0$, $\xi(x + \Delta x) = 0$, shown in Fig. 4.4. F_- is given by $k\xi(x)$, a positive force directed to the left. F_+ is given by $-k\xi(x)$, a negative force directed to the right, which is equivalent to a positive force directed to the left. In Fig. 4.4 it can be clearly seen that both springs tend to push the mass to the left, as intuitively expected.

The new location of the mass originally located at x is given by

$$x + \xi(x). \tag{4.5}$$

But x is simply the location of the mass in equilibrium and is an independent variable. Hence our equation of motion for the mass becomes

$$m\frac{d^2\xi(x)}{dt^2} = k[\xi(x + \Delta x) + \xi(x - \Delta x) - 2\xi(x)]. \tag{4.6}$$

However, the displacement ξ is a function of two independent variables, t and x, and we should write, using the partial derivative,

$$m\frac{\partial^2\xi(x, t)}{\partial t^2} = k[\xi(x + \Delta x, t) + \xi(x - \Delta x, t) - 2\xi(x, t)]. \tag{4.7}$$

We are getting closer to the wave equation in Chapter 2. To finish up, let us expand $\xi(x + \Delta x, t)$ and $\xi(x - \Delta x, t)$ in terms of the power series of Δx using

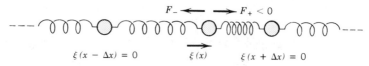

Fig. 4.4.   Sign convention of forces exerted on a mass by springs. Actual force can be opposite to the formal force.

Taylor expansion. Assuming that Δx is small, we find

$$\xi(x+\Delta x,\ t)\simeq \xi(x,\ t)+\frac{\partial \xi}{\partial x}\Delta x+\frac{1}{2}\frac{\partial^2 \xi}{\partial x^2}(\Delta x)^2 \tag{4.8}$$

$$\xi(x-\Delta x,\ t)\simeq \xi(x,\ t)-\frac{\partial \xi}{\partial x}\Delta x+\frac{1}{2}\frac{\partial^2 \xi}{\partial x^2}(\Delta x)^2. \tag{4.9}$$

Then the right-hand side of Eq. (4.7) is simplified as

$$k\frac{\partial^2 \xi}{\partial x^2}(\Delta x)^2,$$

and the equation becomes

$$m\frac{\partial^2 \xi}{\partial t^2}=k(\Delta x)^2\frac{\partial^2 \xi}{\partial x^2} \tag{4.10}$$

or

$$\frac{\partial^2 \xi}{\partial t^2}=\frac{k}{m}(\Delta x)^2\frac{\partial^2 \xi}{\partial x^2}. \tag{4.11}$$

This is exactly of the form of the wave equation we had before [Eq. (2.26)]. The propagation velocity c_w is determined as

$$c_w=\Delta x\sqrt{\frac{k}{m}}, \tag{4.12}$$

which indeed contains the factor k/m, as we expected.
    We may rewrite the velocity as

$$c_w=\sqrt{\frac{k\,\Delta x}{m/\Delta x}}. \tag{4.13}$$

$k\,\Delta x$ is called the elastic modulus of the spring and has the dimensions of force, Newtons. We denote this by $K=k\,\Delta x$. This elastic modulus is actually more convenient than the spring constant k, since it is a normalized constant. Let a spring have a spring constant k and a natural length l. To elongate (or compress) the spring by a length Δl, we have to apply a force

$$F=k\,\Delta l,$$

which can be rewritten as

$$F=kl\frac{\Delta l}{l}=K\frac{\Delta l}{l}.$$

In this form, the strain $\Delta l/l$ is a normalized quantity, and the elastic modulus K is a constant determined from the spring material and its shape, but independent of the length.

Using the elastic modulus K, the velocity now becomes

$$c_w = \sqrt{\frac{K}{m/\Delta x}}. \tag{4.14}$$

Since $m/\Delta x$ is the mass per unit length or the linear mass density (kg/m), denoted by ρ_l, we find

$$c_w = \sqrt{\frac{\text{elastic modulus}}{\text{mass density}}} = \sqrt{\frac{K}{\rho_l}}. \tag{4.15}$$

In fact, all mechanical waves we are going to study have a similar expression for their propagation velocities. *The propagation velocities of mechanical waves are determined by the elasticity and the mass density of media.*

Example 1.   A spring of a total mass of 0.5 kg and a natural length of 1.5 m is elongated by 5 cm when it is stretched by a force of 20 N. Find the velocity of mechanical waves along the spring.

Since

$$F = k\,\Delta l = K\frac{\Delta l}{l},$$

we find

$$K = \frac{Fl}{\Delta l} = \frac{20\,\text{N} \times 1.5\,\text{m}}{0.05\,\text{m}}$$

$$= 600\,\text{N}.$$

The linear mass density is

$$\rho_l = \frac{0.5\,\text{kg}}{1.5\,\text{m}} = 0.33\,\text{kg/m}.$$

Then

$$c_w = \sqrt{\frac{K}{\rho_l}} = 42.4\,\text{m/sec}.$$

Notice that in the preceding example, the mass density is that of the spring itself, and you may wonder why we can use the wave equation, Eq. (4.11). which was actually derived from a model having discrete springs (massless) and masses. The answer to this question is that Δx, the length occupied by the unit mass–spring combination, can be chosen as small as we want, since the velocity is determined by K, the elastic modulus, and the mass density ρ_l. Both quantities remain finite no matter how small Δx is chosen. Thus although what we started from was a *discrete* medium, taking a proper limit ($\Delta x \to 0$), we can go into a *continuous* medium, such as the spring we used in the example. The spring has a mass uniformly distributed along its natural length. A similar technique will be used in Chapter 9 on electromagnetic waves.

4.4. Energy Carried by Waves

In Chapter 2 it was briefly pointed out that any waves should carry energy and momentum with them. Let us see if this is the case for the waves on the mass–spring transmission line. We again look at the unit located at x. The displacement of the mass at this location was given by $\xi(x, t)$. Thus the velocity of the mass becomes

$$v(x, t) = \frac{\partial \xi}{\partial t} \quad \text{(velocity wave)*} \tag{4.16}$$

and the kinetic energy of the mass is given by

$$\text{K.E.} = \frac{1}{2}mv^2 = \frac{1}{2}m\left(\frac{\partial \xi}{\partial t}\right)^2 \quad \text{(J).} \tag{4.17}$$

The potential energy stored in the spring to the right of the mass is

$$\text{P.E.} = \tfrac{1}{2}k[\xi(x+\Delta x) - \xi(x)]^2. \tag{4.18}$$

Using the Taylor expansion for $\xi(x+\Delta x)$, we find

$$\text{P.E.} = \tfrac{1}{2}k(\Delta x)^2 \left(\frac{\partial \xi}{\partial x}\right)^2. \tag{4.19}$$

However, since

$$\frac{\partial \xi}{\partial t} = -c_w \frac{d\xi}{dX}, \quad X = x - c_w t$$

and

$$\frac{\partial \xi}{\partial x} = \frac{d\xi}{dX},$$

we find

$$\text{K.E.} = \tfrac{1}{2}mc_w^2 \left(\frac{d\xi}{dX}\right)^2 \tag{4.20}$$

$$\text{P.E.} = \tfrac{1}{2}k(\Delta x)^2 \left(\frac{d\xi}{dX}\right)^2. \tag{4.21}$$

Recalling

$$c_w^2 = \frac{k(\Delta x)^2}{m},$$

we conclude *in traveling mechanical waves, the potential energy (because of elasticity) and the kinetic energy (because of mass motion) are the same* everywhere and anytime.

---

*Not to be confused with the wave velocity c_w!

The total energy is then

$$\text{Wave energy} = 2 \times \tfrac{1}{2}mc_w^2 \left(\frac{d\xi}{dX}\right)^2$$

$$= mc_w^2 \left(\frac{d\xi}{dX}\right)^2. \tag{4.22}$$

We may define the energy per unit length of the transmission line by simply dividing this by Δx.

$$\text{Wave energy density} = \rho_l c_w^2 \left(\frac{d\xi}{dX}\right)^2 \quad \text{(J/m)}. \tag{4.23}$$

To illustrate an example, let us consider a sinusoidal wave,

$$\xi(x, t) = \xi_0 \sin (kx - \omega t) \tag{4.24}$$

where

$$\frac{\omega}{k} = c_w = \sqrt{\frac{K}{\rho_l}} \quad (k = \text{wavenumber, not spring constant}) \tag{4.25}$$

Then

$$\frac{d}{dX} \xi_0 \sin [k(x - c_w t)] = \xi_0 \frac{d}{dX} \sin kX$$

$$= k\xi_0 \cos [k(x - c_w t)], \tag{4.26}$$

and we find

$$\text{Wave energy density} = \rho_l c_w^2 k^2 \xi_0^2 \cos^2 (kx - \omega t)$$

$$= \rho_l \omega^2 \xi_0^2 \cos^2 (kx - \omega t). \tag{4.27}$$

This function is plotted in Fig. 4.5 at $t=0$ (snapshot). These energy clumps propagate with the wave velocity c_w, since Eq. (4.27) also satisfies the wave equation. The average value of the energy density is just one half of the peak value (Fig. 4.5).

$$\text{Average wave energy density} = \tfrac{1}{2}\rho_l \omega^2 \xi_0^2 \quad \text{(J/m)}. \tag{4.28}$$

Thus the energy carried by the wave in 1 sec. is

$$\text{Rate of energy transfer} = \tfrac{1}{2}c_w \rho_l \omega^2 \xi_0^2 \quad \text{(J/sec.)}. \tag{4.29}$$

The concept of the average power resembles that in ac circuit theory. If a sinusoidal current $I_0 \sin \omega t (A)$ is flowing through a resistor $R(\Omega)$, the instantaneous power is $RI_0^2 \sin^2 \omega t$ (watts). The time averaged power is defined by

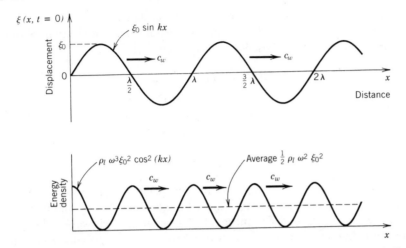

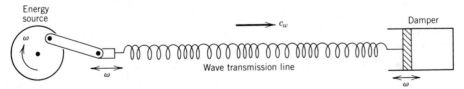

Fig. 4.5. Displacement $\xi(x, 0)$ (top) and wave energy density (bottom). The broken line indicates the average wave energy density $\frac{1}{2}\rho_l\omega^2\xi_0^2$.

$$\frac{1}{T}\int_0^T RI_0^2 \sin^2 \omega t\, dt = \frac{1}{T}\int_0^T RI_0^2 \frac{1-\cos 2\omega t}{2}\, dt$$

$$= \tfrac{1}{2}RI_0^2$$

where T is the period of the oscillation, $T = 2\pi/\omega$ (sec). The factor $\frac{1}{2}$ always appears for the average value of energy (or power) associated with oscillating physical quantities. You may recall the rms (root-mean-square) value of the current in the preceding example is $I_0/\sqrt{2}$, which comes from

$$\tfrac{1}{2}RI_0^2 \equiv RI_{rms}^2.$$

The oscillator connected to the transmission line creates waves on the line (Fig. 4.6). To create the waves the oscillator has to give energy to the waves. The waves now propagate down toward the damper, where the wave energy is converted into heat. How much energy can be transferred in one second is totally determined by the amplitude of displacement waves, and the oscillator frequency, in a given wave medium.

Fig. 4.6. Energy can be carried by waves, from one place to another. The damper dissipates wave energy into heat.

Example 2. A mechanical oscillator with a frequency of 30 Hz is connected to the spring in Example 1. The oscillator creates a wave with an amplitude of 1.5 cm. Find the power the oscillator has to deliver.

$$\text{Power} = \tfrac{1}{2} c_w \rho_l \omega^2 \xi_0^2$$

$$= \tfrac{1}{2} \times 42.4 \text{ m/sec} \times 0.33 \text{ kg/m}$$

$$\times \left(2\pi \times 30 \, \frac{1}{\text{sec}} \right)^2 \times (1.5 \times 10^{-2} \text{ m})^2$$

$$= 55.9 \text{ J/sec} = 55.9 \text{ W}.$$

As Eqs. (4.20) and (4.21) indicate, the kinetic and potential energies associated with mechanical waves must be the same. This reminds us of the energy relations we had for mechanical oscillations. However, there is an important difference between the two cases. In waves, potential and kinetic energies are the same *at any time and at any location,* while in oscillation, although the amplitudes of both energies are the same, they are mutually exclusive, tossing the energy back and forth. Later, in Chapter 6, we shall see that a wave medium of a finite spatial extent can become essentially an oscillation system.

4.5. Momentum Carried by Waves

The momentum carried by waves can be similarly calculated. You would conclude prematurely that since the momentum of the mass at x is

$$mv = m \frac{\partial \xi}{\partial t} = m \frac{\partial}{\partial t} [\xi_0 \sin (kx - \omega t)]$$

$$= -m\omega\xi_0 \cos (kx - \omega t), \qquad (4.30)$$

which oscillates and thus whose average value is zero, the sinusoidal wave cannot carry net momentum. This, however, is a wrong argument. As a mass moving with a constant velocity has both a kinetic energy $mv^2/2$ and a momentum mv, any waves (mechanical and electromagnetic) should carry both energy *and* momentum.

To see this let us consider the wave on a spring with an elastic modulus K and a linear mass density ρ_l. The quantity we have to watch carefully is the mass density ρ_l. If we stretch the spring, the mass density decreases (although the total mass m is unchanged), and if we compress it, the mass density increases. In the presence of waves any point on the spring suffers elongation and compression, and the local mass density certainly varies! Thus the momentum density (in this case, the momentum per unit length) should be written as

$$(\rho_l + \Delta\rho_l)v, \qquad (4.31)$$

where ρ_l is the mass density in the absence of the wave and a constant, $\Delta\rho_l$ is the change in the mass density due to the wave, and v is the velocity of the local point on the spring,

$$v = \frac{\partial \xi}{\partial t}. \tag{4.32}$$

The mass density change $\Delta \rho_l$ can be found as follows. Let us pick up a portion on the spring occupying a length Δx in the absence of wave (Fig. 4.7). The total mass of this section is

$$\Delta x \, \rho_l \text{ (const)}. \tag{4.33}$$

Suppose the point originally at x suffers a displacement $\xi(x)$ and that at $x + \Delta x$ suffers a displacement $\xi(x + \Delta x)$. The original length Δx now becomes

$$\Delta x + \xi(x + \Delta x) - \xi(x). \tag{4.34}$$

Using the Taylor expansion for $\xi(x + \Delta x)$, [Eq. (3.12)]

$$\xi(x + \Delta x) \simeq \xi(x) + \frac{\partial \xi}{\partial x} \Delta x, \tag{4.35}$$

we see that the mass $\Delta x \rho_l$ is now distributed over a length

$$\Delta x + \frac{\partial \xi}{\partial x} \Delta x. \tag{4.36}$$

However, the total mass should be unchanged and equal to $\rho_l \Delta x$. Therefore we must have

$$(\rho_l + \Delta \rho_l)\left(\Delta x + \frac{\partial \xi}{\partial x} \Delta x\right) = \rho_l \Delta x, \tag{4.37}$$

or, after expanding LHS, we find

$$\Delta \rho_l = -\rho_l \frac{\partial \xi / \partial x}{1 + \partial \xi / \partial x}. \tag{4.38}$$

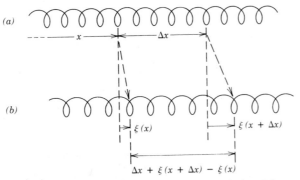

Fig. 4.7.   (a) Spring in equilibrium. (b) In the presence of a wave the length becomes $\Delta x + \xi(x + \Delta x) - \xi(x)$, which causes the change in the mass density.

If $|\partial\xi/\partial x| \ll 1$, which we assume, we can neglect $\partial\xi/\partial x$, compared with 1 in the denominator, and finally obtain

$$\Delta\rho_l = -\rho_l \frac{\partial\xi}{\partial x}. \tag{4.39}$$

(This appropriately defines the "density wave.")
Substituting this into Eq. (4.31), we find

$$\text{Momentum density} = \rho_l \frac{\partial\xi}{\partial t} - \rho_l \frac{\partial\xi}{\partial t}\frac{\partial\xi}{\partial x}.$$

Assume a sinusoidal wave

$$\xi(x, t) = \xi_0 \sin(kx - \omega t). \tag{4.40}$$

Then

$$\text{Momentum density} = -\rho_l \omega \xi_0 \cos(kx - \omega t)$$
$$+ \rho_l \omega k \xi_0^2 \cos^2(kx - \omega t). \tag{4.41}$$

As we did for the energy density, we take the average over the spatial coordinate to find

$$\text{Average momentum density} = \tfrac{1}{2}\rho_l \omega k \xi_0^2 = \frac{k}{\omega}\tfrac{1}{2}\rho_l\omega^2\xi_0^2$$
$$= \tfrac{1}{2}\rho_l\omega^2\xi_0^2/c_w. \tag{4.42}$$

Comparing this with Eq. (4.28), we find the following important conclusion:

$$\text{Average momentum density} = \frac{\text{average energy density}}{\text{wave velocity }(c_w)}. \tag{4.43}$$

Although we derived this conclusion for a particular mechanical wave, it holds quite generally, even for electromagnetic waves. Remember that whenever energy is transferred, momentum must be transferred as well, and there exists a simple relation between them.

Example 3. Find the rate of momentum transfer (the momentum transferred in one second) in Example 2.

The energy transfer rate is 55.9 J/sec. Thus the momentum transfer rate is

$$\frac{55.9 \text{ J/sec}}{42.4 \text{ m/sec}} = 1.32 \text{ J/m} = 1.32 \text{ N}.$$

Note that the momentum transfer rate is equivalent to a force N. in this one-dimensional case.

Example 4. A giant laser pulse with a power of 500 MW (megawatts) and a duration of 10 nsec is completely reflected by a mirror. Find the momentum gained by the mirror. The velocity of light is 3.0×10^8 m/sec.

The momentum transfer rate is $500 \text{ MW}/3 \times 10^8 \text{ m/sec} = 1.67 \text{ N}$. Since the laser beam is reflected, the laser beam suffers a momentum change twice as much as it would when completely absorbed by a black object. Thus the momentum gained by the mirror is

$$\Delta p = 1.67 \text{ N} \times 10^{-8} \text{ sec} \times 2$$

$$= 3.33 \times 10^{-8} \text{ N} \cdot \text{sec}.$$

Example 4 will be studied again later. Here this example is given only to illustrate that electromagnetic waves can indeed carry momentum with them. The force exerted by electromagnetic waves is called the *radiation pressure*. When you enjoy sunlight on a hot summer day, you should realize that sunlight is actually pushing you, although the force is extremely small. The radiation pressure can become substantial for powerful laser beams, as the preceding example indicates. Sound waves in air also exert a net force on our eardrums.

4.6. Transverse Waves on a String

In Section 2.2 we briefly discussed the waves on a string. To create waves on a string the string must be under a tension, as we often experience in string musical instruments. This tension provides the elasticity that is a fundamental requirement for creating mechanical waves and plays exactly the same role as the spring elastic modulus, $K(\text{N})$. In fact, the velocity of the waves on a string under a tension $T(\text{N})$ has a form very similar to Eq. (4.15) and is given by

$$c_w = \sqrt{\frac{T}{\rho_l}}, \tag{4.44}$$

where ρ_l (kg/m) is the linear mass density of the string.

However, there is a fundamental difference between the waves on a string and those in a spring. The difference is in the direction of motion of mass. In the case of the spring [Fig. 4.8a], segments of the spring move in the same direction (parallel or antiparallel) as the wave velocity and create mass compression and rarefaction. Such waves are called *longitudinal waves*. Sound waves in solids, liquids, and gases are all longitudinal waves and will be studied in the following chapter. On the other hand, the waves on a string are associated with the mass motion perpendicular to the wave velocity, as clearly seen from Figs. 2.1 and 2.2. Such waves are called *transverse* waves, since mass motion is in transverse directions relative to the wave velocity. In general, waves that can be described by vector quantities perpendicular to the wave velocity are called transverse waves. In the case of the string, the string displacement ξ is normal to the string, or the direction of propagation. Most *electromagnetic waves are transverse waves*, as we will study later.

Let us derive the expression for the wave propagation velocity on a string [Eq. (4.44)]. As before, we pick up a small segment on the string having a length

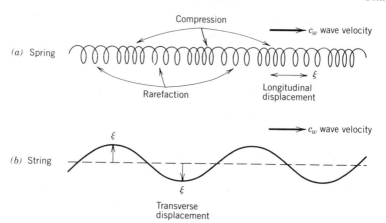

Fig. 4.8. Longitudinal and transverse waves. In longitudinal waves the displacement ξ is in the same direction as the wave velocity, $\mathbf{c}_w$. In transverse waves, the displacement ξ is normal to $\mathbf{c}_w$.

Δx (Fig. 4.9). The segment has a mass given by

$$\rho_l \Delta x \quad \text{(kg)}. \tag{4.45}$$

Our purpose here is to derive an equation of motion for this mass. We also assign the displacements for each end of the segment $\xi(x)$ and $\xi(x+\Delta x)$, which are now perpendicular to the x axis, the direction of wave propagation. Since the string is under a tension force T (N), we find the net vertical force

$$F = F_+ - F_- = T \sin \theta_2 - T \sin \theta_1, \tag{4.46}$$

where θ_2 and θ_1 are the angles of tangent lines at A and B, respectively,

$$\tan \theta_1 = \frac{\partial \xi}{\partial x} \quad \text{at } x$$

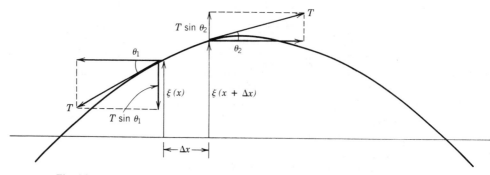

Fig. 4.9. A segment Δx on the string suffers a displacement $\xi(x, t)$ normal to the string.

and

$$\tan \theta_2 = \frac{\partial \xi}{\partial x} \quad \text{at } x + \Delta x.$$

If the displacement $\xi(x, t)$ is small, so are the angles, and we may approximate $\sin \theta$ by θ and $\tan \theta$ by θ (see Chapter 3). Therefore

$$F = T(\theta_2 - \theta_1)$$

$$= T \left(\frac{\partial \xi}{\partial x} \Big|_{x + \Delta x} - \frac{\partial \xi}{\partial x} \Big|_{x} \right) \tag{4.47}$$

where $\partial \xi / \partial x|_{x + \Delta x}$ indicates the value of $\partial \xi / \partial x$ evaluated at $x + \Delta x$. Using Taylor expansion [Eq. (3.12)], we find (put $f = \partial \xi / \partial x$)

$$\frac{\partial \xi}{\partial x} \Big|_{x + \Delta x} - \frac{\partial \xi}{\partial x} \Big|_{x} \simeq \Delta x \frac{\partial^2 \xi}{\partial x^2} \tag{4.48}$$

and then

$$F = \Delta x T \frac{\partial^2 \xi}{\partial x^2}. \tag{4.49}$$

This is the force to act on the segment with the mass $\rho_l \Delta x$. Therefore Newton's equation of motion gives

$$\rho_l \Delta x \frac{\partial^2 \xi}{\partial t^2} = \Delta x T \frac{\partial^2 \xi}{\partial x^2}$$

or

$$\frac{\partial^2 \xi}{\partial t^2} = \frac{T}{\rho_l} \frac{\partial^2 \xi}{\partial x^2}. \tag{4.50}$$

This equation is identical to Eq. (4.15) except K is replaced by T, both having the same dimensions, N. Equation (4.50) immediately yields the velocity of transverse waves on the string, Eq. (4.44).

Example 5.   A sinusoidal transverse wavetrain is moving along a string having a mass density of 20 g/m. The string is under a tension of 40 N. The amplitude of the wave is 5 mm, and the wave frequency is 80 cycles/sec. (a) Write down an expression for the displacement wave $\xi(x, t)$. (b) For the expression assumed in (a), find the expression for the velocity wave. (c) Calculate the average energy density, power, and momentum transfer rate.

(a)   The wave velocity is

$$c_w = \sqrt{T/\rho_l} = \sqrt{40 \text{ N}/0.02 \text{ kg/m}} = 44.7 \text{ m/sec.}$$

Since the frequency is 80 cycles/sec and the amplitude is 5 mm, we may write

$$\xi(x, t) = 5 \times 10^{-3} \text{ (m) } \sin\left(2\pi\frac{x}{\lambda} - 2\pi vt\right)$$

$$= 5 \times 10^{-3} \text{ (m) } \sin(11.25x - 503t)$$

where $\lambda = c_w/v = 0.56$ m is the wavelength.

(b) Since the velocity wave is given by

$$v(x, t) = \partial\xi/\partial t$$

we find

$$v(x, t) = 5 \times 10^{-3} \times (-503) \cos(11.25x - 503t) \text{ m/sec}$$

$$= -2.5 \text{ (m/sec) } \cos(11.25x - 503t).$$

(c) It can be shown that the expressions we obtained before for energy, power, and momentum transfer rate in the longitudinal waves in a spring are also applicable for the transverse waves on a string (Problem 6). Then from Eqs. (4.28), (4.29), and (4.42), we find

$$\text{Average energy density} = \tfrac{1}{2}\rho_l\omega^2\xi_0^2$$

$$= 6.3 \times 10^{-2} \text{ J/m}$$

$$\text{Average power} = c_w \times \text{energy density}$$

$$= 2.8 \text{ W.}$$

$$\text{Average momentum transfer rate} = \text{power}/c_w$$

$$= 6.3 \times 10^{-2} \text{ N.}$$

You may wonder how a momentum can be transferred *along* the string even though the displacement ξ is *transverse* to the string. In the case of spring, spring elements (or mass) move along the spring, and we could calculate the net momentum transfer rate rather naturally. Actually, string elements, too, do have small but finite parallel displacement. In Fig. 4.9 ξ indicates only the *transverse* component of the total displacement. For the purpose of wave propagation velocity, only the transverse displacement is required. However, as we discovered in finding the momentum transfer rate for waves on a spring, we have to be careful in retaining small, higher-order quantities to calculate the net momentum.

Let us consider a string segment Δx that experiences a transverse displacement ξ, as shown in Fig. 4.10. The instantaneous velocity of the segment is *perpendicualr to the displaced string*. The velocity component parallel to the unperturbed string is then given by

$$v_\parallel = -\theta v \simeq -\frac{\partial\xi}{\partial x}\frac{\partial\xi}{\partial t}, \tag{4.51}$$

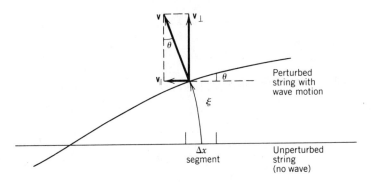

Fig. 4.10. There exists a small but finite parallel velocity component associated with waves on a string. This is responsible for the momentum transfer along the string.

where we have substituted $\theta \simeq \partial\xi/\partial x$ and $v = \partial\xi/\partial t$, and assumed a small angle θ so that $\sin\theta \simeq \theta$. Then the parallel momentum of the segment Δx is

$$dp = -\Delta x \rho_l \frac{\partial\xi}{\partial x}\frac{\partial\xi}{\partial t}.$$

Assuming a harmonic displacement $\xi(x, t) = \xi_0 \sin(kx - \omega t)$, and averaging over x, we obtain the RMS momentum density

$$\tfrac{1}{2}\rho_l(\omega\xi_0)^2 \frac{k}{\omega} \; (\text{N} \cdot \text{sec/m}) \tag{4.52}$$

[Steps of derivation are left as an exercire (Problem 6).] This is identical to what we found for the momentum density in a spring and consistent with the general relationship between energy density and momentum density, Eq. (4.43).

Problems

1. When a helical spring of mass 0.1 kg and natural length 2 m is stretched by a force of 30 N, an elongation of 10 cm results. Find the velocity of longitudinal waves along the spring, assuming that the spring is at its natural length.

   (*Answer:* 109.5 m/sec.)

2. A device called a *wave demonstrator* consists of rods [moment of inertia of each I (kg·m^2)] and springs [torsional constant of each τ (N·m)] alternatively connected with a spacing Δx as shown in Fig. 4.11.

   (a) Derive a wave differential equation for the angular displacement $\theta(x, t)$.

   (b) What is the propagation velocity?

   (*Answer:* (b) $c_w = \sqrt{\tau(\Delta x)^2/I}$.)

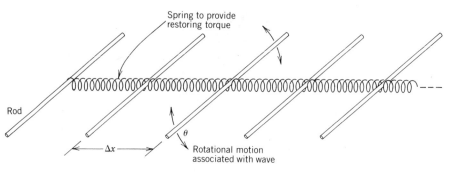

Fig. 4.11.   Problem 2.

3. A driver attached to one end of a long spring of mass density 0.5 kg/m and elastic modulus 300 N creates sinusoidal waves of displacement amplitude 2 cm and frequency 40 cycles/sec. Neglecting wave reflection from the other end, find (a) the power (average, or RMS), (b) the average rate of momentum transfer.

    (*Answer:* 155 W, 6.3 N·sec/sec.)

4. Consider a pulse

$$\xi(x, t) = \xi_0 e^{-(x-c_wt)^2/a^2}$$

    propagating with a velocity c_w along a spring of mass density ρ_l (kg/m) and elastic modulus K (N). Calculate

    (a) kinetic energy
    (b) potential energy
    (c) momentum

    associated with the pulse.

    (*Answer:* K.E. = P.E. = $(\sqrt{\pi/2}\rho_l)(\xi_0 c_w)^2/2a$, momentum = $\sqrt{\pi/2}\rho_l\xi_0^2 c_w/a$).

5. A sinusoidal wavetrain is moving along a spring. The amplitude of the displacement wave is 0.5 cm and the propagation velocity is 25 m/sec.

    (a) Write down an expression to describe the displacement wave, $\xi(x, t)$.
    (b) For the expression assumed in (a), find the expression for the velocity wave.

6. Show that the expressions for the average energy density, the average power, and the average momentum transfer rate associated with sinusoidal transverse waves on a string are identical to those derived for longitudinal waves in a spring.

7. A steel wire of radius 0.5 mm is subject to a tension of 10 N. Steel has a

volume mass density of 7800 kg/m^3. Find the velocity of transverse waves on the wire.

(*Answer:* 40.4 m/sec.)

8. A steel wire having a radius of 0.4 mm hangs from the ceiling. When a mass of 5.0 kg is hung from the free end, what is the velocity of transverse waves on the wire?

(*Answer:* 112 m/sec.)

9. Show that if the tension is varied by a small amount ΔT, the change in the velocity of transverse waves on a string is given approximately by

$$\Delta c_w = \frac{1}{2} \frac{\Delta T}{T} c_w,$$

where T and c_w are the original tension and velocity, respectively.

10. Show that if the mass density is varied by a small amount $\Delta \rho_l$, the change in the velocity of transverse waves on a string is given approximately by

$$\Delta c_w = -\frac{1}{2} \frac{\Delta \rho_l}{\rho_l} c_w,$$

where ρ_l is the original mass density.

11. A mechanical oscillator connected to the end of a stretched string creates transverse displacement of the end given by

$$\xi = 0.01 \text{ (m) sin } (20t).$$

The tension in the string is 10 N and the string has a linear mass density of 20 g/m. Find (a) the velocity of transverse waves, (b) the frequency v, (c) the wavelength, (d) the average power delivered by the oscillator.

(*Answer:* 22.4 m/sec, 3.2 Hz, 7.0 m, 9 mW.)

12. Assuming that about 20% of the power of an incandescent lamp is converted into light power (the rest is wasted as heat), estimate the radiation pressure at a distance 2 m away from a 100-W light bulb. The velocity of light is 3.0×10^8 m/sec and the bulb radiates light isotropically (i.e., in every spherical direction).

(*Answer:* 1.3×10^{-9} N/m^2 for black absorber, 2.6×10^{-9} N/m^2 for a mirror.)

13. It is possible to find an exact dispersion relation for a longitudinal wave on the periodic mass–spring transmission line as shown in Fig. 4.1 without the restriction that the wavelength be much longer than Δx. In other words, the equation of motion, Eq. (4.7), can be solved exactly. Let $\xi(x, t) =$

$\xi_0 \sin(kx - \omega t)$ in

$$m \frac{\partial^2 \xi}{\partial t^2} = k_s[\xi(x + \Delta x) + \xi(x - \Delta x) - 2\xi(x)],$$

where k is the wavenumber and k_s is the spring constant.

(a)   Calculate $\xi(x + \Delta x) - \xi(x)$ and $\xi(x - \Delta x) - \xi(x)$ using

$$\sin \alpha - \sin \beta = 2 \cos \frac{\alpha + \beta}{2} \sin \frac{\alpha - \beta}{2}.$$

(b)   Show that the equation of motion yields the following dispersion relation:

$$\omega^2 = \frac{k_s}{m} 4 \sin^2 \left(\frac{\Delta x \, k}{2} \right).$$

(c)   Check that in the long wavelength limit $\Delta x k \ll 1$, we recover

$$\omega = \sqrt{k_s/m} \, \Delta x \, k.$$

(d)   Plot ω as a function k for $0 \leqslant k \leqslant \pi/\Delta x$.

(e)   Is the wave dispersive or nondispersive? Calculate the phase and group velocities.

[*Note:* Equation (4.7) is called a second-order *difference* equation, which is closely related to the differential equation. Under appropriate limits (such as long wavelength limit), a difference equation reduces to a differential equation. Since in atomic or molecular scales, each molecule must be treated as a discrete particle, the difference equation is more fundamental than the differential equation, which holds only if the medium of concern can be regarded as a continuous medium. Another important application area of difference equation can be found in numerical (computational) integration of differential equations. In this scheme a differential equation is translated into a difference equation, which computers can handle with ease.]

CHAPTER 5

Sound Waves in Solids, Liquids, and Gases

5.1. Introduction

Longitudinal waves in an elastic body (or medium) are generally called sound waves. The most familiar sound waves are those in air. However, even in solids or liquids, sound waves can be propagated.

Sound waves are associated with compressional and rarefactional motion of molecules in the direction of wave propagation, in a manner similar to the longitudinal waves along a spring. In solids, transverse waves can also exist. Earthquakes generally produce both longitudinal waves and transverse waves, the latter propagating slower than the former. When we are hit by an earthquake, we first feel horizontal movements corresponding to the longitudinal waves, and some time later, tumbling vertical movement corresponding to the transverse waves.

In this chapter we study general properties of sound waves (longitudinal waves) in solids, liquids, and gases.

5.2. Sound Velocity Along a Solid Rod

In the previous chapter we learned that propagation velocities of mechanical waves are in general given by

$$c_w = \sqrt{\frac{\text{elastic modulus}}{\text{mass density}}}.$$ (5.1)

As we learned in mechanics, the elastic modulus is a constant that relates the stress to the strain as

$$\text{stress} = \text{elastic modulus} \times \text{strain}.$$ (5.2)

For the case of a continuous spring having a mass density uniformly distributed, we had

$$F = K \frac{\Delta l}{l},$$ (5.3)

where the stress is the force itself (N) and the strain is $\Delta l/l$ (dimensionless).

77

The proportionality between the stress and the strain is known as Hooke's law.

Hooke's law for an elastic body (rather than the spring) takes a slightly different form. Consider a solid rod with a natural length l (m) and a cross section A (m^2) (Fig. 5.1). If a tension F(N) is applied along the rod, the rod will be elongated by a length Δl, and we may write

$$F = \text{const}\, \frac{\Delta l}{l} \tag{5.4}$$

However, if the cross section is increased, a larger force must be applied to get the same deformation, Δl. Therefore the constant in the preceding equation is still size dependent and is not a real material constant. If we divide Eq. (5.4) by the cross section A and write

$$\frac{F}{A} = Y\, \frac{\Delta l}{l}, \tag{5.5}$$

the inconvenience can be removed. The constant Y(N/m^2) is called the Young's modulus and is a material constant. Equation (5.5) may be regarded as the microscopic form of the Hooke's law.

The stress in this case is thus given by the force per unit area N/m^2, and the elastic modulus Y has the same dimension.

The mass density of the rod should be the volume mass density, ρ_v (kg/m^3), rather than the linear mass density, as in the case of the spring and string. The propagation velocity of longitudinal waves in the rod is thus given by

$$c_w = \sqrt{\frac{\text{elastic modulus}}{\text{mass density}}} = \sqrt{\frac{Y}{\rho_v}}. \tag{5.6}$$

It should be remembered that we are dealing with a rod, rather than the unbounded volume of solid. The expression for the wave velocity, Eq. (5.6), is

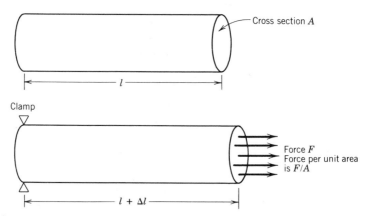

Fig. 5.1.   Relative elongation of a solid rod $\Delta l/l$ is proportional to the force per unit area F/A (Hooke's law).

valid only for a rod, along which pure longitudinal waves are propagated. The velocity of longitudinal, spherical waves in an unbounded solid contains an additional factor in the elastic modulus (shear modulus) and is larger than that given by Eq. (5.6). Liquids and gases, on the other hand, cannot support shear stress (and thus transverse waves), and such complication does not occur. Transverse waves on a string discussed in the previous chapter may be alternatively called shear waves.

Any longitudinal mechanical waves can appropriately be called sound waves. The most familiar form of sound waves is of course that in air, in which air molecules move back and forth in the same direction as the velocity of the sound waves. The molecules in the rod do the same thing. They suffer displacements in the same direction as the wave velocity. Whenever molecules are displaced, there occurs density increase or decrease (corresponding to compression and rarefaction, respectively) just as in the longitudinal waves in the spring.

The energy and momentum densities associated with the sound waves in solids can similarly be found, if we replace ρ_l (linear mass density) by ρ_v (volume mass density) in Eqs. (4.28) and (4.43),

$$\text{Energy density} = \tfrac{1}{2}\rho_v\omega^2\xi_0^2 \quad (\text{J/m}^3) \tag{5.7}$$

$$\text{Momentum density} = \frac{\tfrac{1}{2}\rho_v\omega^2\xi_0^2}{c_w} \quad (\text{N·sec/m}^3) \tag{5.8}$$

both for a sinusoidal wave with a displacement amplitude ξ_0 (m). The power flow needs some explanation. If the rod has a cross section A (m^2), the wave energy per unit length along the rod is (see Fig. 5.2)

$$\tfrac{1}{2}\rho_v\omega^2\xi_0^2\, A \quad (\text{J/m}). \tag{5.9}$$

In one second, the energy of an amount

$$\tfrac{1}{2}\rho_v\omega^2\xi_0^2\, A\cdot c_w \quad (\text{J}) \tag{5.10}$$

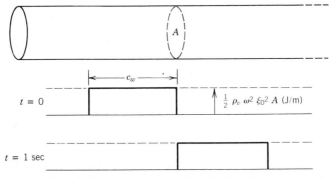

Fig. 5.2. In 1 sec, $\tfrac{1}{2}c_w\rho_v\omega^2\xi_0^2 A$ (J) energy is transferred through the cross section of the rod.

goes through the cross section A (m^2) at an arbitrary location. Hence, the power density (energy flow rate across a unit area in unit time) is

$$\tfrac{1}{2}\rho_v\omega^2\xi_0^2\, c_w \quad (\text{W/m}^2). \tag{5.11}$$

This quantity is called the *intensity* of the wave.

Example 1. Steel has the Young's modulus of 2×10^{11} N/m^2 and the volume mass density 7800 kg/m^3. Assuming a sinusoidal longitudinal wave of displacement amplitude of 1.0×10^{-6} mm and a frequency $v = 5$ kHz, in a steel rod, find (a) the wave velocity, and (b) the wave intensity.

(a) $c_w = \sqrt{\dfrac{Y}{\rho_v}} = \sqrt{\dfrac{2 \times 10^{11} \text{ N/m}^2}{7.8 \times 10^3 \text{ kg/m}^3}} = 5.1 \times 10^3 \text{ m/sec}.$

(b) From Eq. (5.11),

Intensity $= \tfrac{1}{2}\rho_v\omega^2\xi_0^2 c_w$

$\quad = \tfrac{1}{2} \times 7800 \text{ kg/m}^3 \times (2\pi \times 5 \times 10^3 \text{ rad/sec})^2 \times (10^{-9} \text{ m})^2 \times 5.1 \times 10^3 \text{ m/sec}$

$\quad = 1.96 \times 10^{-2} \text{ W/m}^2.$

In the table, Young's moduli, densities, and sound velocities of typical materials at room temperature are shown.

Material	$Y(\times 10^{10}$ N/m$^2)$	ρ_v (kg/m^3)	c_w ($\times 10^3$ m/sec)
Aluminum	6.9	2,700	5.0
Cast iron	19	7,200	5.1
Copper	11	8,900	3.5
Lead	1.6	11,340	1.2
Steel	20	7,800	5.1
Glass	5.4	2300	5.0
Brass (70% Cu, 30% Zn)	10.5	8,600	3.5

5.3. Rigorous Derivation of Sound Velocity Along a Solid Rod

In finding the propagation velocity of sound waves in solid rods, we have just used the general expression obtained in Chapter 4,

$$c_w = \sqrt{\dfrac{\text{elastic modulus}}{\text{mass density}}}. \tag{5.12}$$

Here, for redundancy, we try to derive Eq. (5.6) directly from the Hooke's law and the equation of motion. The procedure, however, is almost exactly the same as we did in Chapter 4.

Consider a long, uniform rod with a cross section A (m²) having Young's modulus Y (N/m²) and a mass density ρ_v (kg/m³). We pick up a thin slice of thickness Δx, located at a distance x (m) from one end of the rod (Fig. 5.3).

When a sound wave is excited along the rod, the thin slice will move about its original location. At the same time, the slice is deformed since the forces to act on the cross sections A_x and $A_{x+\Delta x}$ will be different. Let the displacements of the cross sections of A_x and $A_{x+\Delta x}$ be $\xi(x)$ and $\xi(x+\Delta x)$, respectively (Fig. 5.4). Then the net deformation is

$$\xi(x+\Delta x) - \xi(x) \simeq \frac{\partial \xi}{\partial x} \Delta x \tag{5.13}$$

where we have used the Taylor expansion for $\xi(x+\Delta x)$,

$$\xi(x+\Delta x) \simeq \xi(x) + \frac{\partial \xi}{\partial x} \Delta x. \tag{5.14}$$

Then Hooke's law,

$$\text{Stress} = Y \times \text{strain} \tag{5.15}$$

becomes

$$\frac{F}{A} = Y \frac{\Delta x \, (\partial \xi/\partial x)}{\Delta x} = Y \frac{\partial \xi}{\partial x}. \tag{5.16}$$

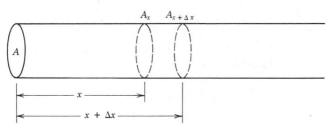

Fig. 5.3.   Solid rod in equilibrium. The small element having a volume of $A \, \Delta x$ experiences no net force.

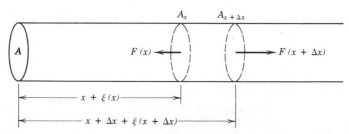

Fig. 5.4.   With the presence of a wave, the volume $A \, \Delta x$ is displaced. At the same time, the volume is deformed.

Notice that Δx is the original thickness of the slice, and corresponds to l in Eq. (5.5), and $\Delta x \, (\partial\xi/\partial x)$ corresponds to Δl. Next, the equation of motion,

$$\text{Mass} \times \text{acceleration} = \text{net force}, \tag{5.17}$$

can be written down as

$$\rho_v \Delta x A \frac{\partial^2 \xi}{\partial t^2} = F(x+\Delta x) - F(x) \simeq \Delta x \frac{\partial F}{\partial x}, \tag{5.18}$$

where we have Taylor-expanded $F(x+\Delta x)$. Substituting Eq. (5.16) into Eq. (5.18), we find

$$\frac{\partial^2 \xi}{\partial t^2} = \frac{Y}{\rho_v} \frac{\partial^2 \xi}{\partial x^2}, \tag{5.19}$$

and the propagation velocity can immediately be found as

$$c_w = \sqrt{\frac{Y}{\rho_v}}. \tag{5.20}$$

Notice that any physical quantities associated with the sound wave should obey the same equation. If we differentiate Eq. (5.19) with respect to x, we get

$$\frac{\partial^3 \xi}{\partial x \, \partial t^2} = \frac{Y}{\rho_v} \frac{\partial^3 \xi}{\partial x^3}. \tag{5.21}$$

But

$$\frac{F}{A} = Y \frac{\partial \xi}{\partial x} \quad \text{(pressure or force wave)}.$$

Hence we find that the force $F(x, t)$ obeys the same differential equation. Similarly, the velocity defined by

$$v = \frac{\partial \xi}{\partial t} \quad \text{(velocity wave)}$$

should obey the same equation, too. However, you should realize that the force F and the velocity v are not of the same form as the displacement ξ. For example, if we assume a sinusoidal wave given by $\xi(x, t) = \xi_0 \sin{(kx - \omega t)}$, the force becomes

$$F(x, t) = AY \frac{\partial \xi}{\partial x} = AYk\xi_0 \cos{(kx - \omega t)}, \tag{5.22}$$

which is 90° out of phase with respect to the displacement ξ (Fig. 5.5). The velocity is also 90° out of phase with respect to ξ. Since the work is given by force $\times$ displacement, we can easily see that the rod is not doing any work, or gaining any energy, on average. The average of the function

$$\sin{(kx - \omega t)} \cdot \cos{(kx - \omega t)} \tag{5.23}$$

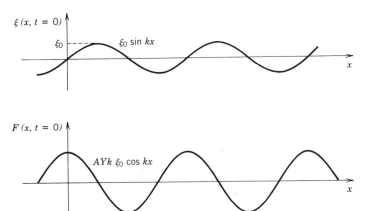

Fig. 5.5. The force wave is 90° out of phase relative to the displacement wave. Both waves are shown at $t=0$.

is exactly zero. The rod is transferring energy from left to right, but not dissipating or creating energy at all. The rod can accommodate energy that is moving or propagating in the rod, and only acts as a medium for the wave.

In fact, good wave media all have this nondissipation property. By *good* it is meant here that media do not dissipate energy. Otherwise, the wave energy is gradually eaten up or dissipated by the medium, and by the time the wave propagates down to the other end, little energy is left. In practice, however, any waves in material media suffer damping. In our studies, however, we assume that dissipation is small, except for one case in the chapter on electromagnetic waves in metals (skin effects). In other words, media we study are all *reactive* media. The concept of reactive media will become clear in chapters on electromagnetic waves.

5.4. Sound Waves in Liquids

Sound waves require a compressive (and thus rarefactive) medium. Sound waves in solids are possible because solids are elastically compressive. Water is compressive, too, and there is a relationship between the force F and the change in water volume ΔV just as in Hooke's law for solids.

Consider a liquid in a rigid cylinder occupying a volume V (m³) in the absence of compression force (Fig. 5.6). If the cylinder has a cross-sectional area A (m²), the stress due to the force F (N) is F/A (N/m²), and the strain is $-\Delta l/l$, where $-\Delta l$ is essentially the change in the volume $-\Delta V$, since $V=Al$. The stress–strain relation for liquids is written as

$$\frac{F}{A} = -M_B \frac{\Delta l}{l} = -M_B \frac{\Delta V}{V} \qquad (5.24)$$

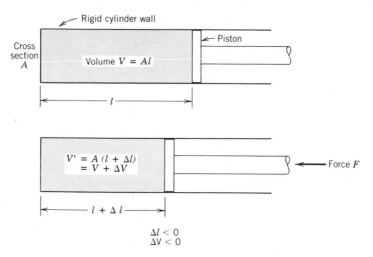

Fig. 5.6. Water (or liquids) can be compressed by an external force.

where M_B (N/m²) plays the role of Young's modulus in solids and is called the bulk modulus of liquids. Notice that for a compressive force as shown in Fig. 5.6, the volume change ΔV is negative, $\Delta V < 0$. The bulk modulus is a material constant and has the same dimension as Young's modulus. Since the mass density is inversely proportional to the volume

$$\rho_v V = \text{const} \tag{5.25}$$

the decrease in the volume causes an increase in the mass density, and we may write Eq. (5.24) in an alternative form,

$$\frac{F}{A} = M_B \frac{\Delta \rho_v}{\rho_v}, \qquad \Delta \rho_v > 0 \text{ if } \Delta V < 0. \tag{5.26}$$

Once we find the elastic modulus (M_B in this case), the velocity of longitudinal (sound) waves is immediately found as

$$c_w = \sqrt{\frac{M_B}{\rho_v}} \tag{5.27}$$

in analogy to Eq. (5.6). Since liquids cannot support shear stress, this expression is not subject to the geometrical constraint (rod) as the sound waves in solids.

Example 2. Water has a mass density of 10^3 kg/m² and a bulk modulus of 2.1×10^9 N/m². Find the velocity of sound waves in water.

$$c_w = \sqrt{\frac{M_B}{\rho_v}} = \sqrt{\frac{2.1 \times 10^9 \text{ N/m}^2}{10^3 \text{ kg/m}^3}}$$

$$= 1.45 \times 10^3 \text{ m/sec.}$$

Example 3.   Derive Eq. (5.27) directly from Newton's equation of motion for a small elementary volume in a liquid. Follow the procedure employed in Section 5.3 for the sound waves in solids.

We consider a pipe filled with a liquid in which a sound wave is excited. A small element with a thickness Δx has a mass $\rho_V A \Delta x$, where ρ_v is the unperturbed liquid mass density, and A is the cross section of the pipe. Let the displacements of surfaces S_1 and S_2 be $\xi(x)$ and $\xi(x+\Delta x)$, respectively (Fig. 5.7). Then the net volume change from the original volume $A \Delta x$ is

$$A[\Delta x + \xi(x+\Delta x) - \xi(x)] - A \Delta x$$
$$= A[\xi(x+\Delta x) - \xi(x)]$$
$$\simeq A \frac{\partial \xi}{\partial x} \Delta x,$$

where we have used Taylor expansion for the quantity $\xi(x+\Delta x)$,

$$\xi(x+\Delta x) \simeq \xi(x) + \frac{\partial \xi}{\partial x} \Delta x.$$

This change in the volume must be caused by the internal force F acting on the surface of the element. From Eq. (5.24), we then find

$$\frac{F}{A} = -M_B \frac{1}{A \Delta x} A \Delta x \frac{\partial \xi}{\partial x} = -M_B \frac{\partial \xi}{\partial x}. \tag{A}$$

However, the force to act on the surface S_1 and that on S_2 must be different in order to cause the displacement of the whole volume, $A \Delta x$. The net force directed to the right is

$$F(x) - F(x+\Delta x) \simeq -\Delta x \frac{\partial F}{\partial x}, \tag{B}$$

where we have again Taylor-expanded $F(x+\Delta x)$. Then the equation of motion for the segment can be written as

$$\rho_v A \Delta x \frac{\partial^2 \xi}{\partial t^2} = -\Delta x \frac{\partial F}{\partial x}. \tag{C}$$

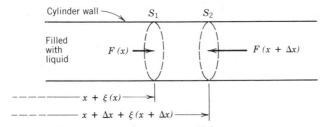

Fig. 5.7.   Displacement of two surfaces S_1 and S_2 originally separated by Δx.

Substituting the force in Eq. (A) into Eq. (C), we finally obtain

$$\frac{\partial^2 \xi}{\partial t^2} = \frac{M_B}{\rho_v} \frac{\partial^2 \xi}{\partial x^2}, \tag{D}$$

which immediately yields the sound velocity, Eq. (5.27).

5.5. Sound Waves in Gases

Sound waves in air are probably the most familiar wave phenomena. Human ears can detect sound waves with frequencies ranging from about 20 Hz to 20 KHz (*audio* frequency range). Some animals (dogs, bats) apparently can detect sound waves with higher frequencies (*ultrasonic* frequencies). Frequencies below the audible limit are called *infrasonic* frequencies. Quakes are usually accompanied by infrasonic waves in air beside physical waves in the ground. Acoustics is one physicoengineering branch devoted to the studies of sound waves and their applications.

Sound waves can be created by physical objects oscillating or vibrating. When we speak, our vocal chords vibrate to create compressive and rarefactive motion of molecules in air. At room temperature, $\simeq 20°C$, these compressive and rarefactive air molecule patterns propagate at a speed of 340 m/sec. Since sound waves in air (or gases, in general) are still one of the mechanical waves, we can find the sound velocity from the general formula

$$c_w = \sqrt{\frac{\text{elastic modulus}}{\text{mass density}}}.$$

Air, for example, has a mass density

$$\rho_v = 1.29 \text{ kg/m}^3$$

at 0°C and the atmospheric pressure. The elastic modulus of gases is defined in the same manner as that for liquids.

As before (see Fig. 5.8), we consider a cylinder with a cross-sectional area $A \text{ (m}^2)$ and a length l filled with a gas having a pressure P. If the piston is pushed by an external force, F (N), the pressure rises (ΔP, the change in the pressure, is positive), but the volume decreases by $A \, \Delta l$. If we recall that gas pressure P and the volume occupied by the gas V are related through *the equation of state*

$$PV^\gamma = \text{const}, \tag{5.28}$$

where γ is the ratio of specific heats, we find [after differentiating Eq. (5.28)]

$$\Delta P V^\gamma + P\gamma V^{\gamma-1} \Delta V = 0. \tag{5.29}$$

Therefore

$$\Delta P = -\gamma P \frac{\Delta V}{V}. \tag{5.30}$$

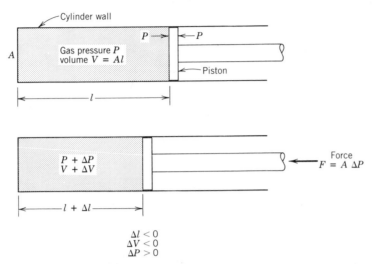

$$\Delta l < 0$$
$$\Delta V < 0$$
$$\Delta P > 0$$

Fig. 5.8. When a gas is compressed, the pressure increases, which is due to the increase in both molecule density and temperature.

Since $V = Al$, $\Delta V = A\,\Delta l$, we finally obtain

$$\Delta P = -\gamma P \frac{\Delta l}{l}. \tag{5.31}$$

This is identical to Eq. (5.24) (stress–strain relation for solids) provided we replace F/A (which has the dimensions of pressure, N/m^2) by ΔP, and M_B by γP. Therefore we may define the bulk modulus of a gas having a pressure P and a ratio of specific heats γ by

$$M_B(\text{gas}) = \gamma P \quad (\text{N/m}^2). \tag{5.32}$$

Denoting the volume mass density of the gas by ρ_v (kg/m^2), we find the sound velocity is given by

$$c_w = \sqrt{\frac{\gamma P}{\rho_v}} \quad (\text{m/sec}). \tag{5.33}$$

The origin of the additional factor γ (compared with the sound velocity in liquids) stems from the fact that gases tend to be heated when compressed. The pressure of a gas is given by

$$P = nk_BT \quad (\text{N/m}^2), \tag{5.34}$$

where n is the number density of gas molecules (m^{-3}), k_B is the Boltzmann constant, 1.38×10^{-23} J/K, and T is the absolute temperature (K). If the gas is compressed, the molecule density obviously increases. At the same time, the gas is heated if the gas is *thermally insulated* from external agents. Then the total pressure increase is given by

$$\Delta P = k_B(\Delta n T + n\,\Delta T)$$

$$= P\left(\frac{\Delta n}{n} + \frac{\Delta T}{T}\right). \tag{5.35}$$

Recalling

$$\frac{\Delta n}{n} = -\frac{\Delta V}{V} \tag{5.36}$$

Eq. (5.34) becomes

$$\Delta P = -P\frac{\Delta V}{V} + P\frac{\Delta T}{T}, \tag{5.37}$$

which is obviously larger than the pressure change due to the density (or volume) change alone. The coefficient γ is defined from

$$-P\frac{\Delta V}{V} + P\frac{\Delta T}{T} \equiv -\gamma P\frac{\Delta V}{V} \tag{5.38}$$

and is always larger than or equal to 1. Its numerical value depends on what molecules the gas contains. For gases with *monatomic molecules*, such as helium (He) and argon (Ar), $\gamma = 5/3$. For gases with *diatomic molecules* such as oxygen (O_2), nitrogen (N_2), and air (mixture of oxygen and nitrogen gases), $\gamma = 7/5$. γ is given by $\gamma = (f+2)/f$, where f is the number of degrees of freedom of molecular energy partition. For monatomic gases, $f = 3$ (or $\gamma = 5/3$) corresponding to three possible kinetic energies, $\frac{1}{2}mv_x^2, \frac{1}{2}mv_y^2, \frac{1}{2}mv_z^2$ in x, y, z coordinates. In diatomic gas additional two degrees of freedom come in (Fig. 5.9). These are

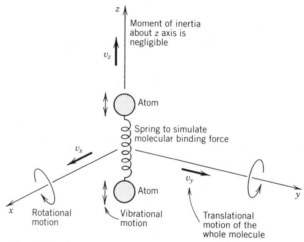

Fig. 5.9.   Translational, rotational and vibrational energies of a diatomic molecule.

associated with rotational energies about two axes of bound atoms as shown. Rotation about z axis has negligible energy because of the small moment of inertia. You may wonder why one should not include the vibrational energy along the z axis. Only quantum mechanics can answer this question. As the energy of the electron in a hydrogen atom is not arbitrary but quantized (or discrete),* so is the energy of this vibration in the diatomic molecule. That is, the vibration energy cannot increase with temperature, and this energy does not contribute to the number of freedoms. (Bound vibration is not *free* in a sense.) A brief introduction to quantum mechanics is given in Chapter 14. It is interesting that quantum effects appear even in sound waves, which have been an important physical subject since Newton's era.

Example 4. Find the velocity of sound in (a) air and (b) helium gas at the atmospheric pressure and a temperature of 0°C.

(a) Since air is largely composed of nitrogen and oxygen gases both having diatomic molecules, we may choose $\gamma = 7/5$. The pressure P is 1.0×10^5 N/m^2 and the mass density ρ_v is 1.29 kg/m^2 at 0°C, 1 atm pressure. Then

$$c_w = \sqrt{\frac{\gamma P}{\rho_v}} = \sqrt{\frac{1.4 \times 1.0 \times 10^5 \text{ N/m}^2}{1.29 \text{ kg/m}^3}} = 330 \text{ m/sec.}$$

(b) Since helium gas is monatomic, we choose $\gamma = 5/3$. The mass density ρ_v can be found as follows. One mole of a gas occupies 22.4 L under the atmospheric pressure at 0°C and contains 6×10^{23} (Avogadro's number) molecules. Since helium has an atomic weight of 4.0,† a helium atom has a mass of $4 \times 1.67 \times 10^{-27}$ kg, where 1.67×10^{-27} kg is the proton mass. Then the mass density of the helium gas becomes

$$6 \times 10^{23} \times 4 \times 1.67 \times 10^{-27} \text{ kg} \times \frac{10^3 \text{ L}}{22.4 \text{ L}} \text{ m}^{-3} = 0.18 \text{ kg/m}^3.$$

The velocity of sound is

$$c_w = \sqrt{\frac{\gamma P}{\rho_v}} = \sqrt{\frac{5/3 \times 1.0 \times 10^5 \text{ N/m}^2}{0.18 \text{ kg/m}^3}}$$
$$= 960 \text{ m/sec.}$$

The expression for the velocity of sound waves in gases [Eq. (5.33)] can be rewritten in terms of microscopic quantities. Since the pressure P is $nk_B T$ [Eq. (5.34)] and the mass density is

$$\rho_v = nm \quad (\text{kg/m}^3), \tag{5.39}$$

---

*See Chapter 14.

†Recall that the He nucleus has two protons and two neutrons.

where m is the mass of one molecule, Eq. (5.38) becomes

$$c_w = \sqrt{\frac{\gamma k_B T}{m}}. \tag{5.40}$$

Notice that the velocity of sound in gases is actually independent of the gas density n and determined by the gas temperature and the molecular mass. If you have studied kinetic theory of gases, you may recall that the sound velocity given by Eq. (5.40) is very similar to the *random molecular velocity*,

$$\sqrt{\frac{3 k_B T}{m}}. \tag{5.41}$$

The only difference is in the numerical factors, γ and 3. The close resemblance is actually due to the propagation mechanism of sound waves or the origin of the bulk modulus (elasticity) of gases.

If we further multiply both numerator and denominator in Eq. (5.4) by the Avogadro number, $N = 6 \times 10^{23}$/mole, we obtain another way to express the sound velocity.

$$c_w = \sqrt{\frac{N k_B T}{N m}}. \tag{5.42}$$

However, $N k_B = 8.3$ J/K $\equiv R$ is known as the gas constant, and $NM = M_{mol}$ is the mass of one mole. Then, using these, we have

$$c_w = \sqrt{\frac{\gamma R T}{M_{mol}}}. \tag{5.43}$$

Example 5.   Find the velocity of sound in air at 20° C. One mole of air has a mass of 29 g.

Substituting $\gamma = 7/5$, $R = 8.3$ J/K·mol, $T = 273 + 20 = 293$K, and $M_{mol} = 0.029$ kg, we find

$$c_w = \sqrt{\frac{7/5 \times 8.3 \text{ J/K} \times 293 \text{ K}}{0.029 \text{ kg}}}$$

$$= 343 \text{ m/sec}.$$

5.6. Intensity of Sound Waves in Gases

In Section 5.2 we have derived expressions for average energy density and power density for sinusoidal sound waves in solids. Those expressions can also be applied for sound waves in gases. The quantity of practical importance is the power density (W/m^2), which indicates how much energy (J) passes through a unit area (1 m^2) per unit time (sec). The power density is alternatively called the intensity of sound waves.

If a sinusoidal displacement wave train is described by

$$\xi(x, t) = \xi_0 \sin(kx - \omega t),$$

the intensity I is given by [Eq. (5.11)]

$$I = \tfrac{1}{2}\rho_v \omega^2 \xi_0^2 c_w \quad (\text{W/m}^2) \tag{5.44}$$

The human ear is a very sensitive organ. At the same time the human ear is very flexible and can stand a tremendously wide intensity range. The lower limit of audible intensity is of the order of 10^{-12} W/m² and the maximum safety limit is of the order of 1 W/m². The ratio between these two values is 10^{12}! The intensity of ordinary conversation is of the order of 10^{-6} W/m; street traffic, 10^{-5} W/m²; jet plane, 10^{-2} W/m².

Example 6. Sinusoidal sound waves in air have an intensity 1.0×10^{-6} W/m² and a frequency of 2 kHz. Assuming a room temperature (20°C) and 1 atm pressure, find the amplitude ξ_0 of the displacement waves. What is the amplitude of the velocity waves? of the pressure waves?

At 20°C the velocity of sound is 343 m/sec (Example 5). In Eq. (5.44), $I = 1.0 \times 10^{-6}$ W/m, $\rho_v = 1.29$ kg/m³ $\times (273°\text{K}/293°\text{K}) = 1.20$ kg/m³, $\omega = 2\pi \times 2 \times 10^3$ rad/sec, and $c_w = 343$ m/sec. Then

$$\xi_0 = \frac{1}{\omega}\sqrt{\frac{2I}{\rho_v c_w}} = \frac{1}{2\pi \times 2 \times 10^3 \text{ rad/sec}}\sqrt{\frac{2 \times 10^{-6} \text{ W/m}^2}{1.20 \text{ kg/m}^3 \times 343 \text{ m/sec}}}$$

$$= 5.55 \times 10^{-9} \text{ m}.$$

The amplitude of the velocity wave is $\omega \xi_0$ since

$$v = \frac{\partial \xi}{\partial t} = -\omega \xi_0 \cos(kx - \omega t).$$

Then

$$\omega \xi_0 = 2\pi \times 2 \times 10^3 \times 5.55 \times 10^{-9} \text{ m/sec}$$

$$= 7.0 \times 10^{-5} \text{ m/sec}.$$

The amplitude of the pressure wave can be found from

$$\Delta P = -\gamma P \frac{\partial \xi}{\partial x}$$

in analogy to the sound waves in liquids, Eq. (A), Example 3. Then

$$\Delta P = -\gamma P k \xi_0 \cos(kx - \omega t),$$

and the amplitude of the pressure wave is

$$\gamma P k \xi_0,$$

where the wavenumber k can be found from the dispersion relation

$$\frac{\omega}{k}=c_w \quad \text{or} \quad k=\frac{\omega}{c_w}.$$

Substituting $\gamma=7/5$, $P=1.013 \times 10^5$ N/m^2, $\xi_0=5.55 \times 10^{-9}$ m, and $k=2\pi \times 2 \times 10^3/343=36.6$ rad/m, we find

$$\gamma P\xi_0=2.85 \times 10^{-2} \text{ N/m}^2.$$

As this example indicates, the air molecules hardly move (they move only 5.6×10^{-9} m!) for the intensity of 10^{-6} W/m^2, which is the typical intensity of conversation. The pressure perturbation relative to the equilibrium pressure is only

$$\frac{2.85 \times 10^{-2}}{1.0 \times 10^5}=2.85 \times 10^{-7}.$$

We indeed realize how sensitive our ears are.

Because the audible intensity range is so wide (10^{12}!), A. G. Bell (the inventor of the telephone, who was also very sympathetic to those with hearing difficulties) introduced a logarithmic expression to indicate sound intensity. The decibel (*deci*="ten", *bel* after "Bell"), dB, is defined as

$$dB = 10 \log_{10} (I/I_0) \tag{5.45}$$

where I_0 is the standard sound intensity chosen as 1.0×10^{-12} W/m^2, which is about the threshold of human hearing. For example, the intensity of 1.0×10^{-6} W/m^2 is equivalent to 60 dB, since

$$10 \log_{10} (10^{-6}/10^{-12})=10 \log_{10} 10^6=60.$$

Apparently, human feeling for sound intensity (loudness of sound) is in logarithmic form, not in linear form. In this respect the introduction of decibels is rather natural. The decibel representation is also used in electrical engineering to express a power relative to a standard power. 3 dB, which often appears in electrical engineering, indicates a difference by a factor of 2 between two powers. (Recall that $\log_{10} 2 \simeq 0.3$.)

Problems

1. Ice has a density of 920 kg/m^3 and Young's modulus of 1×10^{10} N/m^2 at 0°C. Estimate the speed of sound along an ice rod.

   (*Answer:* 3.3×10^3 m/sec.)

2. Longitudinal earthquake waves typically have a velocity of 5×10^3 m/sec. Assuming the average earth density is 1500 kg/m^3, estimate the elastic modulus of the earth.

   (*Answer:* 3.8×10^{10} N/m^2.)

3. A long steel rod having a diameter of 5 cm is forced to transmit longitudinal waves by a mechanical oscillator connected to the end. The amplitude of the displacement waves is 10^{-5} m and the frequency is 400 Hz. Find

   (a) The expression to describe the displacement wave.

   (b) The average energy per unit length of the rod.

   (c) The average power transfer through a cross section of the rod.

   (d) The power delivered by the oscillator.

   (*Answer:* 4.8×10^{-3} J/m, 25 W, 25 W.)

4. In Problem 3, find the expression for

   (a) The velocity wave.

   (b) The force wave.

   (c) The energy wave.

   Assume that the displacement wave is described by $\xi_0 \sin(kx - \omega t)$.

5. Compute the velocity of sound in

   (a) A hydrogen (H_2) gas.

   (b) An argon (Ar) gas. Both gases are at $0°C$.

   The atomic mass of hydrogen atom is 1.0 and that of argon is 40. Watch the value of γ.

   (*Answer:* 1260 m/sec, 307 m/sec.)

6. Show that the change in the sound velocity Δc_w caused by a small change in the temperature ΔT is given by

$$\Delta c_w = \frac{1}{2} \frac{\Delta T}{T} c_w$$

   where $c_w = \sqrt{\gamma RT/M_{mol}}$ is the original velocity.

7. An observer detects the sound in air caused by an explosion on a lake 2 sec after he detects the sound in water caused by the same explosion. How far is the explosion point from the observer? Assume the second velocity in water is 1500 m/sec and that in air is 340 m/sec.

   (*Answer:* 880 m.)

8. A sinusoidal sound wave in air ($20°C$, 1 atm pressure) has an intensity of 1×10^{-7} W/m². What is

   (a) The amplitude of the displacement wave?

   (b) The amplitude of the pressure wave?

   (c) What is the intensity expressed in dB?

   Assume $v = 400$ Hz.

   (*Answer:* 8.8×10^{-9} m, 9.0×10^{-3} N/m², 50 dB.)

9. The air molecule displacement associated with a harmonic (sinusoidal)

sound wave train in air (20°C, 1 atm pressure) is described by

$$\xi(x,\ t)=1.0\times 10^{-8}\ (\text{m})\ \sin\ (kx-\omega t)$$

where $\omega=2\pi\times 10^3$ rad/sec.

(a) What is the value of k, the wavenumber?

(b) How intense is the wave? Answer in terms of W/m² and dB.

(c) What is the amplitude of the pressure wave? How does this compare with the equilibrium pressure?

(*Answer:* 18 rad/m, 8.1×10^{-7} W/m², 59 dB, 2.56×10^{-2} N/m².)

10. Two sound waves, one in air and one in water, have the same frequency and the same intensity. What is the ratio between molecular displacement amplitudes? Assume $T=20°C$.

(*Answer:* $\xi_{air}/\xi_{water}=57.2$.)

11. The displacement wave of a harmonic sound wave in air is given by

$$\xi(x,\ t)=\xi_0\ \sin\ (kx-\omega t),\qquad \frac{\omega}{k}=c_s.$$

Show that the ratio between the pressure wave $-\gamma P$ $(\partial\xi/\partial x)$ and the velocity wave $\partial\xi/\partial t$ is constant and given by $\sqrt{\gamma P\rho_v}$, where ρ_v is the mass density. This quantity is called the "impedance" for sound wave. [See Eq. (6.28).]

CHAPTER 6

Wave Reflection and Standing Waves

6.1. Introduction

When a mass hits a hard wall, the mass is reflected by the wall. This reflection phenomena can alternatively be interpreted as the reflection of energy and momentum associated with the mass. If the wall is soft, the collision is inelastic, and the wall completely absorbs the energy and momentum of the mass. No reflection occurs in this case.

As we have seen, waves carry energy and momentum, and whenever waves encounter an obstacle, they are reflected by the obstacle. Echoes are caused by the reflection of sound waves. Radars utilize the reflection of electromagnetic (micro) waves by metal objects such as airplanes.

Wave reflection gives rise to an important wave phenomenon called standing waves, which play essential roles in most musical instruments. As the name indicates, standing waves do not propagate and therefore are not associated with energy and momentum transfer. They are essentially oscillators with a spatial spread and can create waves in a surrounding medium. Steel strings in a piano, for example, oscillate with respective frequencies determined by the length, tension, and mass of each string. Each string can create sound waves in air with a particular frequency.

6.2. Reflection at a Fixed Boundary, Standing Waves

Suppose a transverse pulse is propagating along a string from left to right toward the end, which is rigidly clamped (Fig. 6.1). When the pulse hits the end, it exerts a force on the end vertically upward. However, the end is clamped and cannot move. Therefore the wall should exert a force on the string vertically downward. This new force in turn creates a pulse that is propagating from right to left with opposite polarity. We then conclude

At a fixed boundary, the displacement ξ stays zero and the reflected wave changes its polarity.

Notice that the magnitude of the displacement $|\xi|$ remains unchanged and the

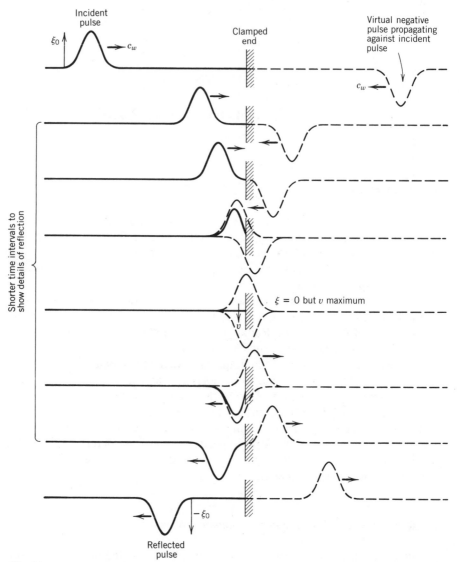

Fig. 6.1.   Wave reflection at a fixed boundary. $\xi = 0$ at the boundary, and the reflected wave changes its sign. A virtual wave, which is an inverted mirror image of the wave on the string, is convenient to analyze wave reflection.

pulse is completely reflected, keeping the same shape. A boundary to cause the polarity change in the reflected wave (relative to the original wave) is called a *hard* boundary. The rigid, clamped boundary in the preceding case is one example.

Let us next consider a sinusoidal wavetrain propagating toward the clamped end. We choose $x = 0$ at the end and describe the incident wave by

$$\xi_+(x,\, t)=\xi_0 \sin (kx-\omega t), \qquad \frac{\omega}{k}=c_w, \tag{6.1}$$

where + indicates a positive-going wave. Note that x is negative in the region to the left of the end, $x=0$, but the preceding expression still describes the wavetrain propagating to the right or toward the end. When the wavetrain is reflected, the direction of propagation should be reversed. Also the polarity of the wave must be reversed. Therefore the reflected wavetrain should be described by

$$\xi_-(x,\, t)=-\xi_0 \sin (-kx-\omega t). \tag{6.2}$$

Since

$$\sin(-\theta)=-\sin \theta,$$

ξ_- becomes

$$\xi_-(x,\, t)=\xi_0 \sin (kx+\omega t). \tag{6.3}$$

The total displacement is the sum of ξ_+ and ξ_-,

$$\xi(x,\, t)=\xi_0[\sin (kx-\omega t)+\sin (kx+\omega t)]. \tag{6.4}$$

Recalling (see Appendix B)

$$\sin \alpha + \sin \beta = 2 \sin \left(\frac{\alpha+\beta}{2}\right) \cos \left(\frac{\alpha-\beta}{2}\right),$$

we find

$$\xi(x,\, t)=2\xi_0 \sin kx \cos \omega t. \tag{6.5}$$

[Notice that $\cos(-\theta)=\cos \theta$.] Equation (6.5) is *not* of the form of a traveling wave since it does not contain a factor, $X=kx-\omega t$ or $kx+\omega t$. Although we started from two propagating (traveling) waves going in opposite directions, we end up with something that does not propagate!

In Fig. 6.2 the snapshots of Eq. (6.5) at several instances are shown. We can clearly see that the sinusoidal wave patterns are not moving along the x axis. Rather they simply oscillate up and down, and a picture with a long exposure time (superposition of many snapshots) indicates the formation of clumps with a *spatial period* $\lambda/2$, where λ is the wavelength of the original waves. Such wave patterns are called *standing waves*, in contrast to the traveling waves we have so far been studying.

Standing waves are formed when two sinusoidal waves of the same frequency (and thus the same wavelength) propagating in opposite directions are superposed.

Although they are not propagating, standing waves must be created by traveling waves going opposite directions. Since standing waves are not propagating, *no energy can be carried by the standing waves*. Rather, energy is confined between "nodes," where the displacement ξ is zero. The boundary is one of the

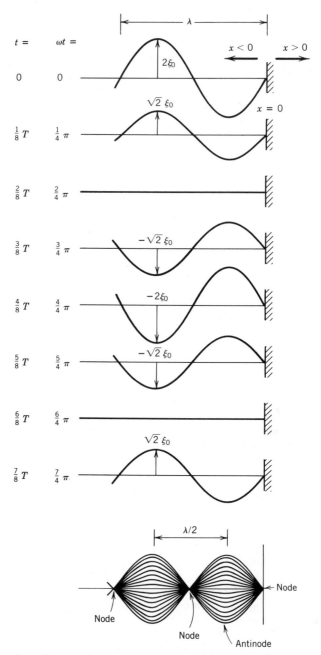

Fig. 6.2. Snapshots of the standing wave, Eq. (6.5), and the picture taken with a long exposure time. Standing waves are essentially oscillators with a spatial spread.

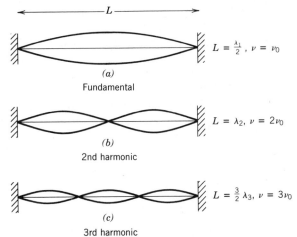

Fig. 6.3. Allowed standing waves on a string clamped at ends a distance L apart.

nodes, as can be seen in Fig. 6.3. The absence of energy flow in the standing waves is understandable since the positive-going wavetrain ξ_+ and the negative going wavetrain ξ_- each carry the same amount of energy in opposite directions, and the net energy flow must be zero.

String musical instruments (piano, guitar, violin, etc.) all utilize this standing-wave phenomena. A string with a given length L clamped at both ends allows standing waves with *discrete* wavelengths, starting with the longest wavelength, $\lambda_1 = 2L$. This determines the frequency of the oscillation through $\lambda\nu = c_w$, where c_w is the velocity of the transverse waves on the string, $c_w = \sqrt{T/\rho_l}$, with T the tension (N) and ρ_l the linear mass density (kg/m). This lowest frequency is called the fundamental frequency. We denote this by ν_0.

The next possible wavelength is $\lambda_2 = L$ (Fig. 6.3b). The frequency for this mode is $2\nu_0$ and called the *second harmonic*. The third possible wavelength is $\lambda_3 = 2L/3$, the third harmonic is $3\nu_0$, and so on.

Example 1.   A piano string having mass 15 g and length 1 m (clamped at both ends) is used for a note, $\nu = 220$ Hz (fundamental mode). Find the tension to be applied to the string.

The wavelength of the fundamental mode is $\lambda = 2 \times 1 = 2m$. Then

$$c_w = \lambda\nu$$

$$= 2 \text{ m} \times 220 \text{ sec}^{-1}$$

$$= 440 \text{ m/sec}$$

From $c_w = \sqrt{T/\rho_l}$ [see Eq. (4.44)], we find

$$T = c_w^2 \rho_l = (440 \text{ m/sec})^2 \times 0.015 \text{ kg/m}$$

$$= 2.9 \times 10^3 \text{ N}.$$

6.3. Reflection at a Free Boundary

At a free boundary, the material composing a wave medium is free to move in contrast to the case of the fixed boundary. In Fig. 6.4 we again show a pulse-shaped transverse wave traveling on a string toward the end that is now free to move. When the pulse hits the free end, it is reflected but with *the same polarity*, in contrast to the case of the fixed end. To see why, let us recall that the vertical restoring force to act on a segment in the string was given by [Eq. (4.46)]

$$F = T \frac{\partial \xi}{\partial x}. \tag{6.6}$$

Let the displacements of incident wave and reflected waves be described by

$$\xi_+ = \xi_0 \sin (kx - \omega t) \tag{6.7}$$

$$\xi_- = A \sin (-kx - \omega t) = -A \sin (kx + \omega t), \tag{6.8}$$

where A is an amplitude to be determined. Since the total displacement is

$$\xi = \xi_+ + \xi_- = \xi_0 \sin (kx - \omega t) - A \sin (kx + \omega t), \tag{6.9}$$

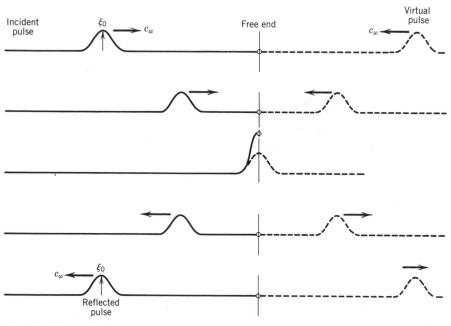

Fig. 6.4. Wave reflection at a free boundary. The amplitude at the boundary is doubled, and the reflected wave has the same polarity as the incident wave. The virtual wave in this case is an erect mirror image.

the vertical force to act on the end of the string is

$$T\frac{\partial \xi}{\partial x} = T[\xi_0 k \cos (kx - \omega t) - Ak \cos (kx + \omega t)], \qquad (6.10)$$

which is to be evaluated at the end, $x = 0$. Then

$$T\frac{\partial \xi}{\partial x}\bigg|_{x=0} = Tk(\xi_0 - A) \cos \omega t. \qquad (6.11)$$

The force is a restoring force and would tend to push the string back to the equilibrium position. However, at the free end this force is obviously absent, $T(\partial \xi/\partial x) = 0$, at any time. From this we must conclude that

$$A = \xi_0, \qquad (6.12)$$

and the reflected wave should have the same polarity as the incident wave. (You should apply this argument to the case of fixed boundary and show that $A = -\xi_0$ results in this case.) Hence we conclude

At a free boundary, the restoring force is zero, or $\partial \xi/\partial x = 0$, and the reflected wave has the same polarity as the incident wave. The amplitude at the free boundary is twice as large as that of the incident wave.

Free boundaries are often called *soft* boundaries.

To speak of a free boundary on a string has little practical meaning since it would be almost impossible to make the end move without friction if the string were subject to a tension. However, in wind musical instruments (including a pipe organ), the reflection of sound waves at free boundaries plays an essential role.

If $A = \xi_0$ in Eq. (6.9), we find the expression for standing waves created by the free boundary,

$$\xi = 2\xi_0 \cos kx \sin (-\omega t)$$
$$= -2\xi_0 \cos kx \sin \omega t. \qquad (6.13)$$

The amplitude of the standing waves at $x = 0$ is indeed $2\xi_0$, twice as large as that of the original incident wave. The long exposure picture is shown in Fig. 6.5.

The pipes of a pipe organ must have at least one open end. Otherwise standing sound waves created in a pipe cannot come out and we hear nothing! A pipe of length L having a closed end can create a standing wave whose wavelength is

$$\lambda = 4L \qquad (6.14)$$

and frequency

$$v_1 = \frac{c_w}{4L} \qquad (6.15)$$

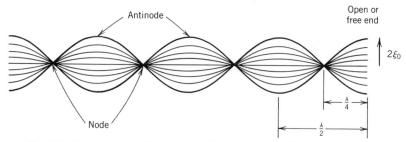

Fig. 6.5.   Long exposure picture of standing waves created by a free boundary.

as shown in Fig. 6.6a. Note that the closed end acts as a fixed boundary, and the open end as a free boundary. At the closed end, the molecule displacement must be zero since the wall is rigidly fixed. At the open end, air molecules are free to move. The amplitude becomes the maximum at the open end, which acts as a speaker to radiate sound waves into air.

A pipe having two open ends has its fundamental wavelength (Fig. 6.6b)

$$\lambda_1 \doteq 2L \qquad\qquad (6.16)$$

and a corresponding fundamental frequency

$$v_0 = \frac{c_w}{2L}. \qquad\qquad (6.17)$$

You should find higher resonance frequencies in both cases.

Example 2.   Find the length of a closed pipe to create sound waves of frequency 40 Hz. What is the next resonance frequency? Assume $T = 20°C$.

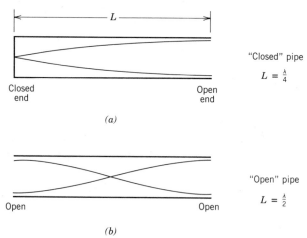

Fig. 6.6.   Standing waves in an organ pipe: (*a*) One end closed and (*b*) both ends open. Only fundamental modes are shown.

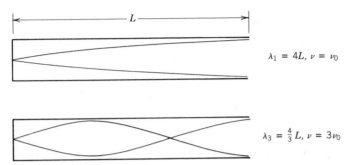

Fig. 6.7.   Fundamental and third harmonic modes in a "closed" organ pipe.

At 20°C, the sound velocity is $c_w = 343$ m/sec (see Chapter 5). For the closed pipe, the fundamental wavelength is $\lambda = 4L$, with L the length of the pipe. Since $c_w = \lambda v$, we find

$$L = \frac{c_w}{4v} = \frac{343 \text{ m/sec}}{40/\text{sec}} = 8.6 \text{ m.}$$

The next resonance wavelength is $4L/3$ as shown in Fig. 6.7. Then the frequency is 3 times higher than the fundamental. The next resonance frequency is 3×40 Hz $= 120$ Hz.

6.4. Theory of Wave Reflection, Mechanical Impedance

In previous sections we have considered two ideal cases of wave reflection, reflection at a fixed boundary and reflection at a free boundary. In both cases, reflection is *complete* in a sense that all energy associated with an incident wave is reflected. In practice, however, complete reflection rarely happens. Consider sound waves incident on a concrete wall. Although most energy is reflected by the wall because concrete is much "harder" than air, we know that sound waves can exist in solids such as concrete, and some energy of the sound wave can penetrate into the concrete wall. Then the reflection is not 100%. A similar (probably more important) phenomena can be readily observed in optics. Light incident on glass is reflected, as we all know. In optical devices such as cameras it is desirable to avoid light reflection at the lens surface (see Chapter 11).

Wave reflection takes place whenever waves in one medium try to enter another medium. In the preceding examples, air is one medium of sound waves, and the concrete wall is another medium. The sound velocity in air is quite different from that in a solid. In the case of a light incident on a glass surface, too, the velocity of light in air and that in glass are different (the velocity of light in glass is about 67% of that in air or vacuum) and air and glass are different media for light waves. If a wave medium is cut off as in the two ideal cases considered in the previous sections, wave reflection is complete, since beyond the cutoff point, nothing can accommodate waves.

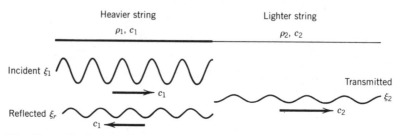

Fig. 6.8. Wave reflection at a joint of two strings having different mass densities is not complete. Some energy can cross the boundary.

To illustrate *incomplete reflection*, let us consider two strings with different mass densities, ρ_1 (kg/m) and ρ_2 (kg/m), respectively, both subject to a common tension, T (N) (Fig. 6.8). The velocities of transverse waves on each string are

$$c_1 = \sqrt{\frac{T}{\rho_1}}, \qquad c_2 = \sqrt{\frac{T}{\rho_2}}, \tag{6.18}$$

respectively (Chapter 4). We saw (Example 5 and Problem 6, Chapter 4) that the energy flow associated with a sinusoidal wavetrain with amplitude ξ_0 and frequency $\omega = 2\pi \nu$, along a string having a mass density ρ_l (kg/m), is given by

$$\tfrac{1}{2}\rho_l c_w \omega^2 \xi_0^2 \quad \text{(W)}, \tag{6.19}$$

where c_w is the velocity of transverse waves. We assume that the incident wave has an amplitude ξ_1, the reflected wave, ξ_r, and the transmitted wave, ξ_2, respectively. The energy flow associated with the incident wave is

$$\tfrac{1}{2}\rho_1 c_1 \omega^2 \xi_1^2 \; (\rightarrow) \text{ incident waves}, \tag{6.20}$$

that with the reflected wave

$$\tfrac{1}{2}\rho_1 c_1 \omega^2 \xi_r^2 \; (\leftarrow) \text{ reflected wave}, \tag{6.21}$$

and that with the transmitted wave

$$\tfrac{1}{2}\rho_2 c_2 \omega^2 \xi_2^2 \; (\rightarrow) \text{ transmitted wave}, \tag{6.22}$$

where arrows indicate the direction of energy flow. Therefore, from the energy conservation principle, we must have

$$\tfrac{1}{2}\rho_1 c_1 \omega^2 \xi_1^2 - \tfrac{1}{2}\rho_1 c_1 \omega^2 \xi_r^2 = \tfrac{1}{2}\rho_2 c_2 \omega^2 \xi_2^2, \tag{6.23}$$

or

$$\rho_1 c_1 (\xi_1^2 - \xi_r^2) = \rho_2 c_2 \xi_2^2. \tag{6.24}$$

This result is of desired form. For example, the case $\xi_2 = 0$ indicates complete reflection, and yields $\xi_1^2 = \xi_r^2$, or the reflected wave must have the same amplitude as the incident wave. (Notice $\xi_r = \pm \xi_1$, the plus sign for the complete reflection at the free boundary and the minus sign at the fixed boundary.) Second, if there is no reflected wave, $\xi_r = 0$, and all the incident energy is transmitted through the boundary.

To find the amplitudes ξ_r and ξ_2 in terms of ξ_1 (the given amplitude of the incident wave) we notice that

$$\xi_2 = \xi_1 + \xi_r, \tag{6.25}$$

which results from the requirement that the displacement at the boundary be continuous. For example, at the fixed boundary, $\xi_2 = 0$, and we indeed recover $\xi_r = -\xi_1$, the complete reflection with the polarity change in the reflected wave. At the free boundary, the displacement at the boundary was twice as large as the incident displacement. Equation (6.25) indeed yields $\xi_2 = 2\xi_1$ with $\xi_r = \xi_1$, the complete reflection with the same polarity.

Equations (6.24) and (6.25) are two simultaneous equations for two unknowns, ξ_r and ξ_2. Solving these, we find

$$\xi_r = \frac{c_1\rho_1 - c_2\rho_2}{c_1\rho_1 + c_2\rho_2}\xi_1 = \frac{\sqrt{\rho_1 T} - \sqrt{\rho_2 T}}{\sqrt{\rho_1 T} + \sqrt{\rho_2 T}}\xi_1 \tag{6.26}$$

and

$$\xi_2 = \frac{2\sqrt{\rho_1 T}}{\sqrt{\rho_1 T} + \sqrt{\rho_2 T}}. \tag{6.27}$$

Notice that the tension T could have been canceled between numerator and denominator, but we did not do this. The reason is that the quantity

$$Z_m = \sqrt{\text{mass density} \times \text{elastic modulus}} \tag{6.28}$$

has an important physical meaning and for the sound waves in solids, $\sqrt{\rho_v Y}$ can play the same role. In the case of two strings the elastic moduli (tension) are common, but this is generally not the case.

The quantity defined by Eq. (6.28) is called *mechanical impedance*. The concept of impedance was introduced in electrical engineering for electromagnetic waves, but can be applied to mechanical waves as well. Of course, the *mechanical impedance does not have the dimensions of ohms* as electromagnetic impedance.

For sound waves in a gas, Eq. (6.28) takes the following form

$$Z_m = \sqrt{\rho_v \gamma P}$$

where ρ_v is the volume mass density and γP is the elastic modulus. As we have seen already (Problem 11, Chapter 5), the impedance is given as the ratio between the pressure (or force) wave and the velocity wave. If one wishes to make an analogy between mechanical waves and electromagnetic waves, the following correspondence may be established

Mechanical		Electromagnetic
Force wave	$\leftrightarrow$	Voltage wave
Velocity wave	$\leftrightarrow$	Current wave
$Z_m = \dfrac{\text{force wave}}{\text{velocity wave}}$	$\leftrightarrow$	$Z = \dfrac{\text{voltage wave}}{\text{current wave}}$

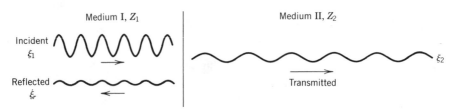

Fig. 6.9. Wave reflection occurs whenever waves try to penetrate into a medium having a different mechanical impedance. If $Z_1 = Z_2$, no wave reflection occurs. If $Z_2 = 0$ or ∞, the reflection is complete.

The impedance Z_m is a quantity characterized by wave media. Using Z_m, we can rewrite Eq. (6.26) as

$$\xi_r = \frac{Z_1 - Z_2}{Z_1 + Z_2}\, \xi_1, \tag{6.29}$$

where Z_1 is the impedance of medium 1 and Z_2 of medium 2. From Eq. (6.29) we can draw an important conclusion:

The condition for the absence of wave reflection at a boundary between two media is that the impedance of the media be the same.

The fraction of reflected wave energy is given by (Fig. 6.9)

$$\left(\frac{\xi_r}{\xi_1}\right)^2 = \left(\frac{Z_1 - Z_2}{Z_1 + Z_2}\right)^2. \tag{6.30}$$

Example 3. Sound waves in air are incident normal to water surface. Calculate how much (in %) energy is reflected at the water surface. Water has a bulk modulus of 2.1×10^9 N/m². Assume that the temperature of air is 20°C and the pressure is 1 atm.

The water impedance for sound waves is

$$Z_{\text{water}} = \sqrt{\rho M_B} = \sqrt{10^3 \text{ kg/m}^3 \times 2.1 \times 10^9 \text{ N/m}^2}$$
$$= 1.45 \times 10^6 \text{ kg/m}^2 \cdot \text{sec}.$$

The air impedance for sound is

$$Z_{\text{air}} = \sqrt{\rho \gamma P} = \sqrt{1.29 \text{ kg/m}^3 \times 7/5 \times 1.0 \times 10^5 \text{ N/m}^2}$$
$$= 4.3 \times 10^2 \text{ kg/m}^2 \cdot \text{sec}.$$

Then the energy reflection coefficient is

$$\left(\frac{Z_{\text{air}} - Z_{\text{water}}}{Z_{\text{air}} + Z_{\text{water}}}\right)^2 = 0.9988 = 99.88\%.$$

This indicates practically all energy is reflected at the water surface. Only 0.1% penetrates into water.

Example 4. A pulse of amplitude 1 cm is propagating along a string toward a boundary where the string is connected to another string having a mass density four times larger. Both strings are subject to a common tension. (a) Find the amplitudes of reflected and transmitted pulses. (b) Find the energy reflection coefficient. (c) Sketch qualitatively reflected and transmitted pulses after the incident pulse reached the boundary.

(a) Since the elastic modulus (tension in this case) is common, the mechanical impedance is proportional to the square root of the mass density $Z_1/Z_2 = 1/2$. From Eq. (6.29), we find

$$\xi_r = \frac{Z_1/Z_2 - 1}{Z_1/Z_2 + 1}\, \xi_1 = \frac{-0.5}{1.5}\, \xi_1 = -\tfrac{1}{3}\xi_1 = -0.33 \text{ cm}$$

and from Eq. (6.27),

$$\xi_2 = \frac{2}{1 + Z_2/Z_1}\, \xi_1 = \tfrac{2}{3}\xi_1 = 0.67 \text{ cm}.$$

The reflected pulse has opposite polarity relative to the incident pulse as expected since the second string is heavier and thus "harder."

(b) The energy reflection coefficient, Eq. (6.30), becomes

$$\left(\frac{Z_1/Z_2 - 1}{Z_1/Z_2 + 1}\right)^2 = \left(-\frac{1}{3}\right)^2 = \frac{1}{9} = 11\%.$$

(c) See Fig. 6.10.

The opposite case in which a pulse is propagating on the heavier string toward a lighter string is given as a problem. You will find that the energy

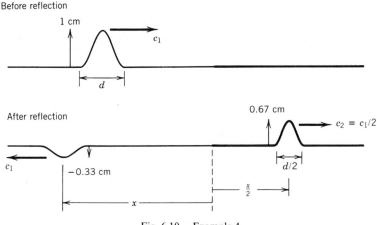

Fig. 6.10.   Example 4.

reflection coefficient remains unchanged since

$$\left(\frac{Z_1-Z_2}{Z_1+Z_2}\right)^2 = \left(\frac{Z_2-Z_1}{Z_2+Z_1}\right)^2.$$

To summarize, we have seen that waves are reflected at a boundary where a discontinuity in (mechanical and electromagnetic) impedance exists. The same propagation velocity in two media does not necessarily ensure the absence of reflection at the boundary between the two media. The impedance must be the same. The subject treated in this section may be too advanced for you at this stage, but is an extremely important concept of any wave phenomena. We will study the reflection of electromagnetic waves in later chapters using the same concept, the impedance for electromagnetic waves.

Problems

1. A certain violin string is 50 cm long between its fixed ends and has a mass of 0.5 g. It sounds the A note (440 Hz) without fingering.
   (a) What is the tension to be applied?
   (b) Where must one put one's finger to play a C note (528 Hz)?
   (*Answer*: 194 N, 8.3 cm.)

2. A string 2 m long has a mass of 500 g. If the string is fixed at the ends with a tension of 20 N, what is the fundamental frequency of oscillation?
   (*Answer*: 7.1 Hz.)

3. What is the fundamental resonance frequency of a 50-cm-long pipe with one end closed? What are the higher resonance frequencies? Take $c_s = 340$ m/sec.
   (*Answer*: 170 Hz, 510 Hz, 850 Hz, ...)

4. What are the resonance frequencies of a well 20 m deep? Take $c_s = 340$ m/sec.
   (*Answer*: 4.25 Hz, 12.75 Hz, 21.25 Hz, ...)

5. When wave reflection is not complete, incomplete standing waves are formed. Let the incident harmonic wave be $\xi_0 \sin(kx - \omega t)$ and the reflected wave be $-\Gamma\xi_0 \sin(kx + \omega t)$, where Γ is the reflection coefficient defined by

$$\Gamma = \frac{\text{reflected wave at } x=0}{\text{incident wave at } x=0}.$$

   If $|\Gamma| = 1$, we have complete reflection and thus complete standing waves.
   (a) Where does the amplitude maximum occur? Answer in terms of the distance from the discontinuity, $x=0$.
   (b) Where does the amplitude minimum occur?

(c) Show that the ratio between the amplitude maximum and minimum is given by

$$\frac{1+|\Gamma|}{1-|\Gamma|}.$$

This quantity is called the *standing-wave ratio* and is a measure of how much reflection takes place.

6. A pulse 1 cm high on a string having a mass density of 50 g/m is propagating toward the boundary where the string is connected to another with a mass density 20 g/m. Both strings are under the same tension.

(a) Is the boundary a hard or soft boundary for the incident pulse?

(b) Find the amplitude (height) of the reflected pulse.

(c) Find the amplitude of the transmitted pulse.

(*Answer:* Soft, 0.23 cm, 1.23 cm.)

7. The transverse wave on a string discussed so far is a one-dimensional wave having only one spatial variable, x. Generalization to multidimensional waves can be done by suitable changes in the elastic moduli and mass densities. As an example, let us consider transverse waves on a rectangular membrane whose four edges are rigidly clamped like a membrane of a drum. The membrane has a *surface* mass density $\rho_s\,(\text{kg/m}^2)$ and is subject to a uniform, isotropic surface tension $T_s\,(\text{N/m})$.

(a) The wave differential equation for the vertical displacement $\xi(x, y, t)$ is given by

$$\frac{\partial^2 \xi}{\partial t^2} = \frac{T_s}{\rho_s}\left(\frac{\partial^2 \xi}{\partial x^2} + \frac{\partial^2 \xi}{\partial y^2}\right).$$

What is the velocity of the wave on the membrane? (You should attempt to derive the wave equation following the same procedure used for the waves on a string.)

(b) The standing waves on a string whose ends are clamped were described by

$$\xi(x, t) = \xi_0 \sin\left(\frac{m\pi}{L}x\right)\sin \omega t,$$

where m is a nonzero integer and L is the length of the string. Show that for the standing waves on the membrane the appropriate expression for the standing waves is

$$\xi(x, y, t) = \xi_0 \sin\left(\frac{m\pi}{a}x\right)\sin\left(\frac{n\pi}{b}y\right)\sin \omega t,$$

where m and n are nonzero integers. Explain why $\cos\,[(m\pi/a)x]$, $\cos\,[(n\pi/b)y]$ are not allowed for the clamped membrane.

(c) Show that the resonance frequencies of the membrane are given by

$$\omega = \sqrt{\frac{T_s}{\rho_s}} \, \pi \left[\left(\frac{m}{a}\right)^2 + \left(\frac{n}{b}\right)^2 \right]^{1/2} \quad \text{(rad/sec)}.$$

Are the higher harmonics integer multiples of the lowest-order resonance frequency,

$$\omega(m=1, n=1) = \sqrt{\frac{T_s}{\rho_s}} \, \pi \left(\frac{1}{a^2} + \frac{1}{b^2} \right)^{1/2} ?$$

[*Note:* The harmonics are not always of integer multiples of the fundamental in contrast to the case of strings. The ratio $\omega(m, n)/\omega(m=1, n=1)$ may not even be a rational number. This explains why sounds created by a drum are rather uncomfortable. Analysis of a circular membrane reveals the same conclusion. In this case higher harmonics are always an irrational number $\times$ the fundamental. We have to know about Bessel functions to fully understand waves on a circular membrane.]

8. The three-dimensional sound wave is described by

$$\frac{\partial^2 \xi}{\partial t^2} = c_s^2 \left(\frac{\partial^2 \xi}{\partial x^2} + \frac{\partial^2 \xi}{\partial y^2} + \frac{\partial^2 \xi}{\partial z^2} \right), \quad c_s = \text{sound velocity}.$$

Following the procedure outlined for Problem 7, find the resonance frequencies of sound waves in a rectangular box having dimensions of $a \times b \times c$.

(*Answer:* $\omega_{l,m,n} = c_s \pi [(l/a)^2 + (m/b)^2 + (n/c)^2]$ with l, m, n nonzero integers.)

9. Make an order of magnitude estimate for the period of the earth resonance. The earth has a radius of 6400 km. Assume the sound velocity of 5×10^3 m/sec.

(*Answer:* About 1 h.)

10. A steel rod is joined to a copper rod at a smooth, flat surface. Steel has a mass density of 7800 kg/m^3 and a Young's modulus of 2.0×10^{11} N/m^2, and copper has a mass density of 8900 kg/m^3 and a Young's modulus of 1.1×10^{11} N/m^2. A sinusoidal sound wavetrain in the steel rod is incident on the boundary.

   (a) Calculate the fraction of reflected and transmitted wave energies relative to the incident wave energy.

   (*Answer:* 1.3%, 98.7%.)

   (b) Calculate the amplitudes of reflected and transmitted waves relative to the amplitude of the incident wave.

   (*Answer:* 0.12, 1.12.)

   (c) What are the mechanical impedance of the steel and copper rods for sound waves?

   (*Answer:* 3.95×10^7, 3.13×10^7 kg/m$^2 \cdot$sec.)

11. Glass and aluminum have approximately the same sound speed, 5.0×10^3 m/sec. From this, can we conclude that sound waves incident on a glass–aluminum boundary suffer no reflection? Explain.

CHAPTER 7

Spherical and Cylindrical Waves;
Waves in Nonuniform Media,
and Multidimensional Waves

7.1. Introduction

We have frequently used the following representative waveform

$$\xi(x, t) = \xi_0 \sin(kx - \omega t) \tag{7.1}$$

to describe all kinds of waves (waves on a string, sound waves in gases and solids, etc.) where ξ_0 is the amplitude of the sinusoidal displacement wavetrain. The amplitude ξ_0 is a constant, which is to be determined by wave sources, such as a speaker for sound waves in air. (Recall that the power associated with mechanical waves is proportional to ξ_0^2.) For a wavetrain described by Eq. (7.1), we observe the same amplitude ξ_0 everywhere along the wavetrain. Such waves are called *one-dimensional* or *plane waves*.

It is a common experience, however, that we hear louder sounds as we get closer to a radio receiver. The radio receiver, on the other hand, can receive radio waves better at places nearer to the broadcasting station. A stone thrown into water creates water waves whose amplitude becomes smaller and smaller as the waves propagate radially outward. All of these examples indicate that wave amplitudes become smaller as waves in extended media propagate away from localized wave sources. In this chapter we study these geometrical effects on wave amplitudes. Also, we briefly study how waves behave in a nonuniform medium in which the wave velocity slowly varies as a function of coordinates. For example, we wish to understand why water waves increase their amplitude as they approach the shore.

7.2. Conservation of Energy Flow, Spherical Waves

As frequently stated, waves carry energy. The amount of energy passing through a unit area (1 m²) in a unit time (1 sec) was defined as the power density or intensity, J/sec·m², or W/m². For example, a plane sound wave described by Eq. (7.1) in either solids or gases has an average power density (or intensity)

$$I = \tfrac{1}{2}\rho_v c_w \omega^2 \xi_0^2 \quad (\text{W/m}^2) \tag{7.2}$$

112

where ρ_v (kg/m^3) is the volume mass density of the medium, c_w the sound velocity, and $\omega = 2\pi\nu$ (rad/sec) the angular frequency of the wave. For plane waves this power density is a constant, since ξ_0 does not depend on the spatial coordinate, x, measured from the wave source. In Fig. 7.1 we show the power

$$P = IA \quad \text{(W)} \tag{7.3}$$

going through the area A (m^2) normal to the wave direction. We assume that the waves are confined within the area A. The power P is provided by a wave source, and should be a constant. Therefore, we may conclude

> *Intensity (I) times area (A) is constant and equal to the power provided by the wave source. The area A is normal to the wave propagation, and covers the region in which waves are present.*

This is reminiscent of water flow through a pipe (Fig. 7.2). The flow rate (L/sec) through any cross section must be the same because of mass conservation. For the energy flow with waves, the principle of energy conservation applies.

Consider now a point wave source radiating waves (and energy) radially outward. We consider two areas, A_1 and A_2, located at radii r_1 and r_2, respectively. Since the area is proportional to radius squared, we have

$$\frac{A_1}{A_2} = \frac{r_1^2}{r_2^2}. \tag{7.4}$$

The wave intensity at each radial position is therefore inversely proportional to the radius squared.

$$\frac{I_1}{I_2} = \frac{r_2^2}{r_1^2} \tag{7.5}$$

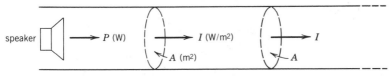

Fig. 7.1.   Sound waves in a uniform pipe have the same intensity everywhere. The power P is given by AI.

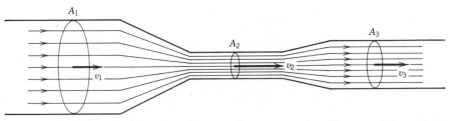

Fig. 7.2.   Water flow through a pipe with nonuniform cross section. Flow rate Av is constant.

since $I_1 A_1 = I_2 A_2 =$ const. Obviously, the wave intensity is larger as we go closer to the wave source, as intuitively expected.

Waves characterized by Eq. (7.5) are called *spherical waves*. Sound waves created by a loudspeaker and radio and TV waves emitted from antennas are typical examples. It is noted here that spherical waves are not necessarily radiated isotropically, or uniformly, in every direction.

Since the intensity is proportional to the wave amplitude squared, ξ_0^2, [Eq. (7.2)], we find that the amplitude of spherical waves is inversely proportional to the distance r from the wave source,

$$\xi_0(r) \sim \frac{1}{r} \tag{7.6}$$

Any vector quantity (velocity, force, field, etc.) associated with spherical waves in uniform media must have this $1/r$ dependence, as required by the energy conservation principle.

Example 1. A loudspeaker is emitting sound waves in air with a power 25 W in every direction (Fig. 7.3). Find (a) the wave intensity at a distance 25 m from the speaker, (b) the amplitude of air molecule displacement waves at the same position. Assume $T = 20°C$, $\rho_v = 1.3 \text{ kg/m}^3$ (air density), and frequency is 500 Hz.

(a) Since the speaker is radiating sound waves in every direction, the area a distance r (m) away from the speaker is $4\pi r^2$ (m²). Then the intensity is

$$I = \frac{P}{A} = \frac{25 \text{ W}}{4\pi \times (25)^2 \text{ m}^2} = 3.2 \times 10^{-3} \text{ W/m}^2$$

$$= 10 \log_{10} \left(\frac{3.2 \times 10^{-3}}{10^{-12}} \right) = 95 \text{ dB}.$$

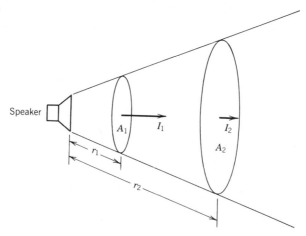

Fig. 7.3. Spherical sound waves. The speaker can be regarded as a point sound source at a sufficiently large distance.

(b) In the expression for the intensity

$$I = \tfrac{1}{2}\rho_v c_w \omega^2 \xi_0^2 \quad (\text{W/m}^2),$$

ρ_v (mass density) $= 1.3 \text{ kg/m}^3$, $c_w = 343 \text{ m/sec} (20°C)$, $\omega = 2\pi \times 500 \text{ rad/sec}$, and $I = 3.2 \times 10^{-3} \text{ W/m}^2$.

Substituting these, we find

$$\xi_0 = \frac{1}{\omega}\sqrt{\frac{2I}{\rho_v c_w}} = 1.2 \times 10^{-6} \text{ m}.$$

As we will see in Problem 6, the wave equation for spherical waves is given by

$$\frac{\partial^2}{\partial t^2}(r\xi) = c_w^2 \frac{\partial^2}{\partial r^2}(r\xi) \tag{7.7}$$

or

$$\frac{\partial^2 \xi}{\partial t^2} = c_w^2 \left(\frac{\partial^2 \xi}{\partial r^2} + \frac{2}{r}\frac{\partial \xi}{\partial r}\right), \tag{7.8}$$

which are fundamentally different from our previous one-dimensional wave equation. The general harmonic solution to these equations is

$$\xi(r, t) = \frac{A}{r}\sin(kr - \omega t), \quad \frac{\omega}{k} = c_w, \tag{7.9}$$

as can be proved by direct substitution. Here c_w is the wave velocity and A is a constant.

7.3. Cylindrical Waves

We do not have good examples of cylindrical waves, which are something between plane waves and spherical waves. To create cylindrical waves, we need a long wave source. In Fig. 7.4, the cylindrical configuration is shown about the

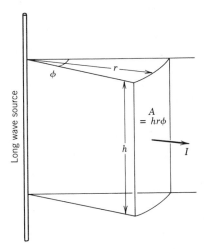

Fig. 7.4. Cylindrical waves created by a long wave source.

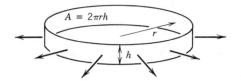

Fig. 7.5. Surface water waves may be regarded as one example of cylindrical waves.

long linear wave source. Since the area A is proportional to the radius r, $AI =$ const requires

$$I \propto \frac{1}{r},$$
(7.10)

or the wave intensity is inversely proportional to the distance (r) from the wave source. Consequently, the amplitude of the wave should be proportional to the square root of the distance,

$$\xi_0 \propto \frac{1}{\sqrt{r}}.$$
(7.11)

Surface waves on water are probably the best example we can think of although the waves are not really cylindrical. They can exist only on the surface of water without deeply extending toward the bottom. However, since the energy associated with the surface waves can only propagate on water surface radially outward away from a wave source, we can still consider a cylinder with a small height h in Fig. 7.4, in which the surface waves are confined (Fig. 7.5).

Cylindrical waves are described by the following differential equation

$$\frac{\partial^2 \xi}{\partial t^2} = c_w^2 \left(\frac{\partial^2 \xi}{\partial r^2} + \frac{1}{r} \frac{\partial \xi}{\partial r} \right).$$
(7.12)

Unfortunately, this does not allow a solution like

$$\frac{A}{\sqrt{r}} \sin (kx - \omega t)$$
(7.13)

although this approximately gives a correct answer at sufficiently large distance r from a wave source. Knowledge of Bessel functions is required to analyze cylindrical waves fully.

7.4. Nonuniform Wave Medium

We have studied in Chapter 4 the transverse waves on a string under a tension T. The tension is of course the same everywhere on the string and the velocity of the transverse waves

$$c_w = \sqrt{\frac{T}{\rho_l}}$$

is also the same everywhere on the strong. The string subject to a tension is a *uniform medium* for the transverse waves. Consider now a string hanging from a ceiling by its own weight (see Fig. 7.6). In practice, we have to use a string relatively heavy so that the string hangs straight. Let the linear mass density of the string be ρ_l (kg/m). At a point distance x (m) away from the lower end, the tension T is given by

$$T = g\rho_l x, \tag{7.14}$$

which now depends on the coordinate x! The velocity for the transverse waves then also depends on x and is given by*

$$c_w(x) = \sqrt{\frac{g\rho_l x}{\rho_l}} = \sqrt{gx}. \tag{7.15}$$

The string vertically hung is a *nonuniform medium* for transverse waves.

In Fig. 7.6 the propagation of a pulse created at the top end is qualitatively

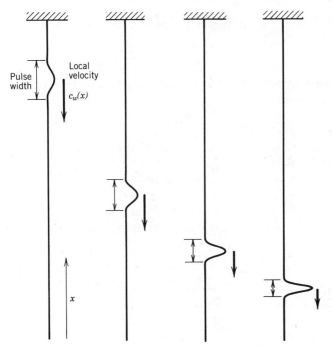

Fig. 7.6.   Transverse waves on a string hanging from a ceiling. The tension T is proportional to x, the distance measured from the lower end. The pulse sent out downward slows down as it propagates. Also the pulse width becomes narrower.

---

*A wave equation is derived in Problem 4 in this chapter.

shown. Observe the following: As the pulse approaches the lower end,

1. The pulse propagates more slowly.
2. The pulse width becomes narrower.
3. The pulse height (amplitude) becomes larger.

The pulse is squeezed as it approaches the lower end and the amplitude becomes very large when it reaches the lower end. This is understandable since the portion of the pulse behind the peak always tends to catch up with the peak, and thus squeezing takes place.

Example 2.   Surface Waves Near a Beach.   The velocity of surface waves on water is given by

$$c_w = \sqrt{gh}$$

where h is the depth of water and g is the gravitational acceleration. Discuss how a pulse behaves as it approaches the beach. Assume a plane bottom. (See Fig. 7.7.)

If the bottom is planar, the depth is proportional to the distance x from the beach and given by

$$h = x \tan \theta$$

where θ is the angle between the water surface and the bottom. Then the velocity of the surface waves is

$$c_w = \sqrt{g \tan \theta \cdot x}$$

which is mathematically identical to Eq. (7.15). Therefore exactly the same things happen as in the case of the vertical string: The pulse increases its amplitude and becomes narrower as it approaches the beach.

Surfers use this pulse-steepening phenomenon. The final fate of the pulse when it hits the beach is catastrophic, an event with which we are familiar.

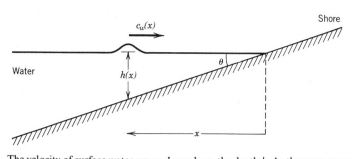

Fig. 7.7.   The velocity of surface water waves depends on the depth h. As the wave approaches the shore, it slows down.

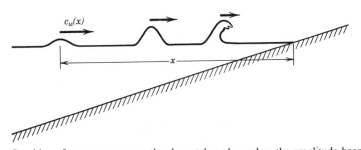

Fig. 7.8.   Breaking of water waves near the shore takes place when the amplitude becomes too large.

The amplitude becomes so large that the waves break down, as shown in Fig. 7.8.

Surfers enjoy big waves, but for those living near seashore, tidal waves (tsunami) created by earthquakes are terrifying. Although small when created by an earthquake far away, waves can gain enormous amplitude as they approach seashores. Tidal waves as high as 30 m have been recorded.

7.5. Multidimensional Waves

Waves we have been analyzing all have only one spatial variable (x for plane waves, r for spherical and cylindrical waves) and such waves are called one-dimensional waves. However, it often (very often, in fact) becomes necessary to introduce more than one spatial variable in analyzing wave phenomena. An example of "two-dimensional" waves is the standing waves on the membranes of a drum. In contrast to the one-dimensional counterpart, namely a string with both ends clamped, the frequencies of standing waves on a membrane under tension are not in general integer multiples of the fundamental frequency.

Since analyses of a circular membrane require the knowledge of Bessel functions, which are beyond the scope of this book, we model it by a rectangular membrane having an area $a \times b$, with its four edges rigidly clamped. We also assume that the membrane is under a surface tension T_s (N/m) everywhere.

The wave equation for the transverse displacement $\xi(x, y, t)$ is now given by

$$\frac{\partial^2 \xi}{\partial t^2} = c_w^2 \left(\frac{\partial^2 \xi}{\partial x^2} + \frac{\partial^2 \xi}{\partial y^2} \right), \tag{7.16}$$

where $c_w^2 = T_s/\rho_s$, with ρ_s now the surface mass density (kg/m^2) of the membrane. (Derivation of this wave equation is left as an exercise, Problem 10.) We wish to find standing-wave solutions to this wave equation.

The standing-wave solution to a clamped string was in the form

$$\xi(x, t) = \xi_0 \sin \left(\frac{n\pi}{L} x \right) \cos (\omega_n t) \tag{7.17}$$

where $n = 1, 2, 3, \ldots$ (integer), L is the string length, and $\omega_n = n\omega_1$ with $\omega_1 = \pi c_w/L$

the fundamental frequency. (See Section 6.2.) The appearance of sin $[(n\pi/L)x]$ is understandable since the string clamped at $x=0$ and L (both ends) requires that $\xi=0$ at $x=0$ and L at any time. In the case of the rectangular membrane whose edges are all clamped, we must have

$$\xi=0 \quad \text{at } x=0 \text{ and } a$$

$$\xi=0 \quad \text{at } y=0 \text{ and } b$$

at any time. Then we may try the following function for the displacement,

$$\xi(x, y, t)=\xi_0 \sin\left(\frac{m\pi}{a}x\right) \sin\left(\frac{n\pi}{b}y\right) \cos(\omega t) \tag{7.18}$$

and see what oscillation frequency ω will satisfy the wave equation, Eq. (7.16). Noting

$$\frac{\partial^2 \xi}{\partial x^2}=-\left(\frac{m\pi}{a}\right)^2 \xi 2$$

$$\frac{\partial^2 \xi}{\partial y^2}=-\left(\frac{n\pi}{b}\right)^2 \xi^2,$$

$$\frac{\partial^2 \xi}{\partial t^2}=-\omega^2 \xi^2,$$

and substituting these into Eq. (7.16), we find (derivation is left as an exercise)

$$\omega^2=c_w^2\left[\left(\frac{m\pi}{a}\right)^2+\left(\frac{n\pi}{b}\right)^2\right] \tag{7.19}$$

where m, n are nonzero integers. This already indicates that the higher harmonic frequencies of standing waves in a membrane are not in general integer multiples of the fundamental $(m=1, n=1)$ frequency

$$\omega(m=1, n=1)=c_w\pi\sqrt{\frac{1}{a^2}+\frac{1}{b^2}}.$$

If $a>b$, the next resonance frequency is given by

$$\omega(m=2, n=1)=c_w\pi\sqrt{\frac{4}{a^2}+\frac{1}{b^2}},$$

which cannot be an integer multiple of the fundamental frequency. The third harmonic frequency is given by

$$\omega(m=1, n=2)=c_w\pi\sqrt{\frac{1}{a^2}+\frac{4}{b^2}}, \quad \left(\sqrt{\frac{8}{3}}b>a>b\right)$$

and so on. Although some harmonic frequencies $(m=n=2, m=n=3,\ldots)$ are integer multiples of the fundamental frequency, most are not even related to the fundamental frequency through rational numbers. This explains why sounds coming out of a metal plate hit with a hammer are rather uncomfortable to our ears.

In the case of a circular membrane (like the one in a drum), the ratio between any harmonic frequency and the fundamental is an irrational number. The fundamental frequency of a clamped circular membrane of a radius a is

$$\omega_1 \simeq 2.405 \frac{c_w}{a}.$$

The second harmonic frequency is $\omega_2 = 3.832\ (c_w/a)$, the third $\omega_3 = 5.136\ (c_w/a)$, and so on. Here $c_w = \sqrt{T_s/\rho_s}$ is the velocity of the transverse waves on the membrane.

Problems

1. A speaker is radiating spherical sound waves with a power 5 W. The radiation is limited within a cone with a 20° angle as shown. Within the cone, the radiation can be assumed to be uniform. (See Fig. 7.9)

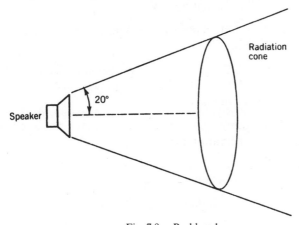

Fig. 7.9. Problem 1.

  (a) What is the power density 10 m away from the speaker?
  (b) At what distance does the intensity become 10^{-6} W/m²?
  (*Answer:* 0.13 W/m², 3.6 km.)

2. Assuming that water waves obey the law of cylindrical waves, find the amplitude of water waves 50 m away from a source. The waves have an amplitude of 15 cm when 10 m away from the source.

3. A radio station is emitting spherical waves at 50 kW.
  (a) Find the intensity 1 mi away from the station assuming isotrop radiation.

(b) Electromagnetic waves are characterized by electric and magnetic fields as we will see later. What is the ratio between the electric field at 1 mi and that at 10 mi?

(c) A radio has the lowest receivable intensity of 3 μW/m^2. How far can one bring the radio to listen to the station?

(*Answer:* 1.6×10^{-3} W/m^2, 10:1, 23 miles.)

4. (a) Following the procedure used for the transverse waves on a string (Chapter 4, Section 4.6), show that the differential equation the transverse waves on a vertical free string should satisfy is given by

$$\frac{\partial^2 \xi}{\partial t^2} = gx \frac{\partial^2 \xi}{\partial x^2} + g \frac{\partial \xi}{\partial x}$$

where ξ is the displacement and x is measured from the lower end of the string.

(b) Under what conditions can we use Eq. (7.10) for the velocity of the transverse waves on the string?

5. A whip used by animal tamers is nonuniform. Explain why in terms of the amplitude of a pulse created on a whip.

6. Show that $\xi(r, t) = (A/r) \sin (kr - \omega t)$ satisfies the spherical wave equation, either Eq. (7.7) or Eq. (7.8).

7. Show that $\xi(r, t) = (A/\sqrt{r}) \sin (kr - \omega t)$ approximately satisfies the cylindrical wave equation, Eq. (7.12). What condition(s) must be imposed for the solution to be sufficiently accurate?

8. Discuss the wave-steepening mechanism of water waves on a beach in terms of energy conservation principle.

9. We saw that water waves slow down as they approach a beach. How can a surfer get accelerated by water waves then?

10. Derive Eq. (7.16), the two-dimensional wave equation for a membrane.

11. A rectangular membrane has edges 20 cm and 30 cm, a surface tension of 5 N/m, and a surface mass density of 40 g/m^2.

(a) What is the velocity of transverse waves on the membrane?

(b) Calculate the lowest-order standing-wave frequency.

(c) What are the higher-order resonance frequencies?

12. (a) Show that the wave equation for three-dimensional sound waves is given by

$$\frac{\partial^2 \xi}{\partial t^2} = c^2 \nabla^2 \xi,$$

where

$$\nabla^2 = \frac{\partial^2}{\partial x^2} + \frac{\partial^2}{\partial y^2} + \frac{\partial^2}{\partial z^2} \text{ (Laplacian).}$$

(b) Discuss the resonance frequencies of sound waves in a rectangular box having a volume $a \times b \times c \text{ (m}^3)$.

CHAPTER 8

Doppler Effect of Sound Waves
and Shock Waves

8.1. Introduction

The tone or frequency of a police car siren appears to change from higher to lower frequency when the car is passing by, even though the frequency of the siren is always constant. The policeman driving the car always hears the same frequency one would hear when the car is not running. Such an apparent change in the frequency caused by the motion of wave sources (the siren in this case) relative to an observer is called the Doppler effect. (C. J. Doppler, 1803–1853, an Austrian physicist, actually discovered the effect in light waves.) For sound waves in air, we have three agents: the sound source (the siren in the preceding example) emitting a wavetrain with a certain frequency, the observer, and the wave medium, air. All can be moving relative to others.

Some objects, such as supersonic planes, can travel faster than sounds. So-called sonic booms created by these objects are alternatively known as shock waves. Shock waves contain large amounts of energy concentrated over a narrow spatial range and could even cause mechanical damage.

In this chapter we study these phenomena caused by the motion of sound sources relative to air and an observer.

8.2. Stationary Sound Source and Moving Observer

Let us first consider a stationary sound source emitting a sinusoidal wavetrain with a frequency v_0. We assume that air is stationary, too; that is, we have no winds. An observer is moving away from the sound source with a velocity u_0 relative to the wave medium, air (Fig. 8.1). We denote the sound velocity by

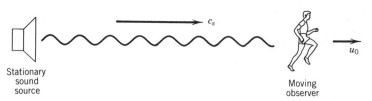

Stationary
sound
source

Moving
observer

Fig. 8.1. Stationary sound source and moving observer. The net sound velocity for the observer is $c_s - u_0$.

124

c_s. Since the observer is moving in the same direction as the sound wave, the apparent sound velocity relative to the observer is

$$c_s - u_0. \tag{8.1}$$

We assume that the observer is moving with a subsonic velocity, $u_0 < c_s$, so that Eq. (8.1) is a positive quantity. If the wave source is stationary, it is emitting sound waves in every direction with the same wavelength λ, determined from (Fig. 8.2)

$$\lambda = \frac{c_s}{\nu_0}. \tag{8.2}$$

The observer, moving or not moving, should always observe the *same wavelength*. However, if he is moving, the apparent sound velocity becomes $c_s - u_0$. Therefore *the frequency* ν' he hears can be found from

$$\lambda = \frac{c_s}{\nu_0} = \frac{c_s - u_0}{\nu'}$$

or solving for ν',

$$\nu' = \frac{c_s - u_0}{c_s} \nu_0. \tag{8.3}$$

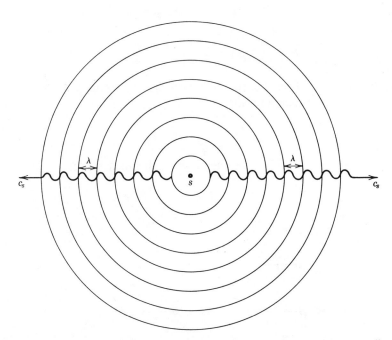

Fig. 8.2.   Wave patterns created by a stationary sound source S are symmetric in every direction or isotropic.

As expected, the apparent frequency heard by the observer is lower than the true frequency if the observer is moving away from the source. The case of an observer approaching the source can be given simply by changing the sign of u_0 ($u_0 < 0$), and the apparent frequency is higher. You should watch the sign convention chosen for the velocity u_0. $u_0 > 0$ if the observer is moving away from the source. If the observer is traveling with the sound speed, $u_0 = c_s$, the frequency becomes zero, since the observer is riding on the waves and does not detect anything varying with time.

Let us derive Eq. (8.3) from an alternative point of view (Fig. 8.3). Assume that the observer is a distance l (m) apart from the second source at a certain instant, say $t = 0$ sec. One second later the observer has moved a distance $u_0 \times 1$ sec $= u_0$ (m) and the distance between the source and the observer is $l + u_0$ (m) at this instant. The time t_0 required for the sound wave emitted at $t = 0$ to reach the observer can be found from

$$c_s t_0 = l + u_0 t_0$$

or

$$t_0 = \frac{l}{c_s - u_0} \text{ (sec)}. \tag{8.4}$$

Similarly, the time t_1 at which the wave emitted at $t = 1$ sec reaches the observer is found from

$$c_s(t_1 - 1) = l + u_0 + u_0(t_1 - 1)$$

or

$$t_1 = \frac{l + c_s}{c_s - u}. \tag{8.5}$$

During the period of 1 sec, the source has emitted v_0 waves, but the observer hears v_0 waves in the time duration of

$$t_1 - t_0 = \frac{c_s}{c_s - u_0} \text{ (sec)},$$

which is longer than 1 sec. In 1 sec the observer then hears

$$\frac{c_s}{c_s - u_0} v_0$$

waves. This is identical to Eq. (8.3).

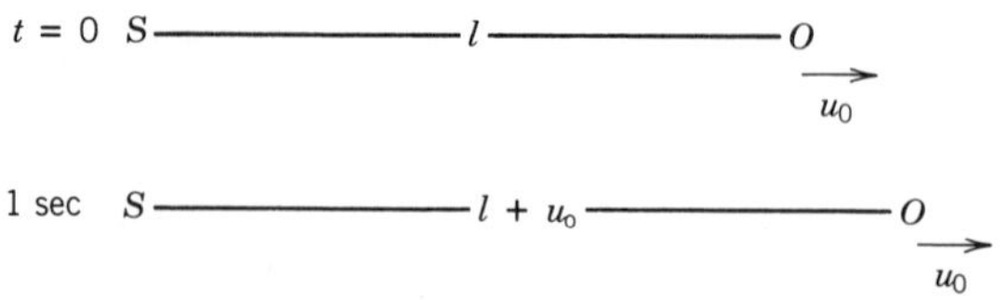

Fig. 8.3.   At the instant $t = 0$, the observer is a distance l from the source, and 1 sec later, the observer is $l + u_0$ away.

Example 1.   A car moving at a velocity 50 mi/hr is passing by a stationary sound source of a frequency 500 Hz (Fig. 8.4). The closest distance between them is 20 m. Find the apparent frequency heard by the driver as a function of the distance x. Assume $c_s = 340$ m/sec.

In this case, the velocity of the car is not directed toward the sound source, and we have to find the component of the velocity vector directed toward the source (Fig. 8.5). It is given by

$$u_0 \cos \theta = u_0 \frac{x}{\sqrt{x^2 + 20^2}} = 22 \frac{x}{\sqrt{x^2 + 20^2}} \text{ m/sec},$$

where we have converted the velocity of 50 mi/hr into 22 m/sec. Then the apparent (Doppler-shifted) frequency is

$$v'(x) = v_0 \frac{c_s + u_0 \cos \theta}{c_s}$$

$$= 500 \left(1 + 0.065 \frac{x}{\sqrt{x^2 + 20^2}} \right) \text{ Hz}.$$

You should plot this as a function of x for $-100 \leqslant x \leqslant 100$ m. At $x = 0$, the car is moving purely perpendicular to the wave and at the instant when the car passes this point, the driver hears the true frequency, 500 Hz.

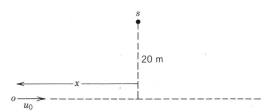

Fig. 8.4.   Example 1. Observer moving along a line not interesecting the source.

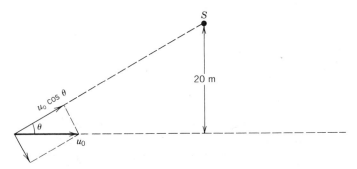

Fig. 8.5.   The velocity component toward (or away from) the source is responsible for Doppler shift.

8.3. Moving Sound Source and Stationary Observer

Next we consider an opposite case in which the sound source is moving toward a stationary observer with a velocity u_s. You may wonder why this case should be different from the previous case discussed in Section 8.2, since it seems what matters is only the relative velocity between the source and the observer and an observer approaching to a stationary sound source should hear the same frequency as a stationary observer hearing the source approaching him. This argument is wrong, however, for sound waves *although it is correct for electromagnetic waves.*

Let the sound source be a distance l (m) away from the observer at a certain instant say, $t=0$ sec. Since the observer is not moving, it takes $t_0 = l/c_s$ sec for the sound wave emitted at $t=0$ to reach the observer. One second later the source is $l - u_s \times 1 = l - u_s$ (m) away from the observer, and the time at which the sound wave emitted at $t=1$ sec reaches the observer is

$$t_1 = 1 + \frac{l - u_s}{c_s} \text{ (sec).}$$

During the period of 1 sec, the sound source has emitted v_0 waves, which is heard by the observer in

$$t_1 - t_0 = 1 - \frac{u_s}{c_s} \text{ (sec).}$$

Therefore the apparent (Doppler-shifted) frequency to be heard by the observer is

$$v' = v_0 \frac{1}{t_1 - t_0} = \frac{c_s}{c_s - u_s} v_0. \tag{8.6}$$

Compare this with Eq. (8.3). The velocity of the sound source u_s appears in the denominator in contrast to Eq. (8.3).

In the case of a stationary sound source and a moving observer, the sound velocity appears to change because of the relative motion, but the wavelength λ remains the same. Here the sound velocity remains the same, but the wavelength appears to change. In Fig. 8.6 circles indicate the locations of waves emitted by the source when it was at A, B, C, ..., and so on, equally separated. We can clearly see that waves are squeezed in the region between the sound source and the observer, whereas in the region behind the source, the wavelength is elongated.

Example 2. A sound source of a frequency 1000 Hz is approaching an observer who also has a sound source of the same frequency. When the observer hears 5 beats a second, what is the velocity of the moving sound source? Assume the sound velocity is 340 m/sec.

We should recall that the beat phenomenon is caused by two waves of slightly different frequencies, and the difference appears as the beats [Eq. (2.49)].

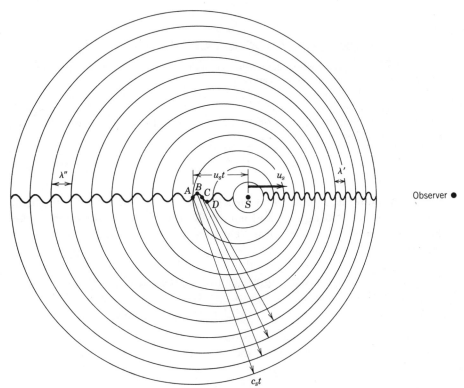

Fig. 8.6.   Moving source and stationary observer. The wavelength changes are due to the motion of the source.

Let the velocity of the moving source be u_s. From Eq. (8.6), the Doppler-shifted frequency v' is

$$v' = \frac{c_s}{c_s - u_s} \, v_0,$$

where $v_0 = 1000$ Hz and $c_s = 340$ m/sec. Then the frequency difference is

$$\Delta v = v' - v_0 = \frac{u_s}{c_s - u_s} \, v_0.$$

Solving for u_s, we find

$$u_s = \frac{\Delta v}{v_0 + \Delta v} \, c_s = 1.7 \text{ m/sec.}$$

8.4. General Expression for Doppler-Shifted Frequency

We can generalize the two cases: (a) a stationary sound source and moving observer (Eq. (8.3) and (b) a moving sound source and stationary observer,

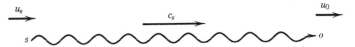

Fig. 8.7.   Both source and observer are moving relative to medium (air). u_s and u_0 are positive if they are in the same direction as the sound velocity.

Eq. (8.6). If both (source and observer) are moving (Fig. 8.7), the apparent frequency to be observed is given by

$$v' = \frac{c_s - u_0}{c_s - u_s} v_0. \qquad (8.7)$$

Again you should be careful about the sign convention adopted here. We have assigned positive velocities (u_s and u_0) for those in the same direction as the sound velocity c_s. If any of these velocities are actually against the sound velocity, a negative value must be substituted.

Example 3.   A police car is approaching a solid wall with a velocity 30 m/sec. If it is sounding a siren of a frequency 1500 Hz, what frequency of reflected sound waves does the policeman hear? Assume $c_s = 340$ m/sec.

The wall can be regarded as an observer who hears the Doppler-shifted frequency of the approaching siren given by Eq. (8.6),

$$v' = \frac{c_s}{c_s - u_s} v_0,$$

where $u_s = 30$ m/sec and $v_0 = 1500$ Hz. When sound waves are reflected, *the wall now acts as a new sound source with the preceding frequency v'.* The policeman is an observer who is *approaching* the sound source with the velocity u_s. Then Eq. (8.3) applies, with the sign of u_s flipped, and the frequency observed by the policeman is

$$v'' = \frac{c_s + u_s}{c_s} v'.$$

Substituting v', we find

$$v'' = \frac{c_s + u_s}{c_s - u_s} v_0 = \frac{340 + 30}{340 - 30} \times 1500 \text{ Hz}$$

$$= 1790 \text{ Hz}.$$

Notice that the frequency of the reflected wave is doubly Doppler-shifted in the preceding example.

Example 4.   Show that if a wind of a velocity u_w exists in the direction of the sound velocity c_s, the Doppler-shifted frequency Eq. (8.7) should be rewritten as

$$v' = \frac{c_s + u_w - u_0}{c_s + u_w - u_s} v_0. \tag{8.8}$$

We recall the fact that the sound velocity c_s is relative to the sound wave medium, namely, air. Therefore if the medium is moving or a wind exists, the sound velocity relative to the ground becomes $c_s + u_w$, where u_w is the wind velocity chosen to be positive if the wind is directed in the same direction as the sound wave. Equation (8.7) should then be modified as

$$v'' = \frac{c_s + u_w - u_0}{c_s + u_w - u_s} v_0$$

where u_0 and u_s are the velocities relative to the ground.

The Doppler effect in electromagnetic waves is given by

$$v' = \frac{c - u}{c} v_0, \tag{8.9}$$

where $c = 3.0 \times 10^8$ m/sec is the speed of light in free space and u is the relative velocity between a wave source with frequency v_0 and an observer, which is assumed to be much smaller than the velocity of light, $|u| \ll c$. For electromagnetic waves, only the relative velocity matters, and it is immaterial which is moving. This surprising fact stems from the relativity theory developed by Einstein, who discovered that electromagnetic waves in free space have entirely different properties from waves in material media, such as sound waves in air. According to the relativity, light velocity in vacuum is constant ($c = 3.0 \times 10^8$ m/sec) irrespective of the velocity of the light source.

8.5. Shock Waves

If the velocity of the sound source approaches the velocity of sound waves in air, Eq. (8.6) predicts that the Doppler-shifted frequency v' becomes large and finally diverges when the source velocity is further increased up to the sound velocity. What happens if the velocity of the sound source exceeds the velocity of sound? When an object is traveling through air with a velocity u larger than the speed of sound c_s, the object is said to have a supersonic velocity. The ratio u/c_w is called the Mach number, after Ernst Mach (1838–1916, Austrian physicist and philosopher). A supersonic object has a Mach number larger than 1.0.

Let us see if we can draw wave patterns similar to Fig. 8.6 when the source velocity u_s is larger than the sound velocity c_s (Fig. 8.8). Suppose that the source is at point A at $t = 0$. After time t (sec), the waves emitted at A are on a sphere with a radius $c_s t$ (m). Since $u_s > c_s$, the distance traveled by the source $AS = u_s t$ is larger than the distance traveled by the sound waves. The waves emitted at successive points, B, C, and so on, are on the line $A'S$, where the circles are

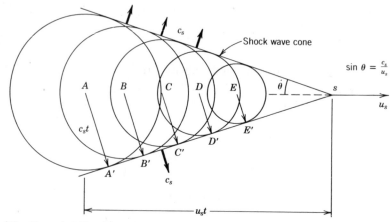

Fig. 8.8.   Example 5. Shock waves created by a sound source moving faster than the sound velocity.

most crowded. We can see that sound waves are all confined in a cone, and outside this cone no sound waves are present and thus detected. The sound velocity c_s is normal to the cone surface.

The energy of the sound waves is mostly concentrated on the cone surface, where the wave pattern circles are most crowded. When the cone surface hits an observer, he detects the sudden arrival of a large amplitude pulse, which is known as a shock wave or sonic boom.

Example 5.   A supersonic plane is traveling horizontally 1 km above the ground. An observer on the ground experiences a shock wave approximately 5 sec after the plane passed directly overhead. Estimate the Mach number of the plane. Assume $c_s = 340$ m/sec.

In Fig. 8.8 the angle θ between the axis AS and the cone surface $A'S$ is given by

$$\sin \theta = \frac{c_s}{u_s}, \quad \text{or} \quad \theta = \sin^{-1}\left(\frac{c_s}{u_s}\right).$$

However, $\sin \theta$ is also equal to $h/[\sqrt{(5u_s)^2 + h^2}]$ (Fig. 8.9). Then

$$\frac{c_s}{u_s} = \frac{h}{\sqrt{(5u_s)^2 + h^2}}.$$

Solving for u_s, we find

$$u_s = \frac{c_s h}{\sqrt{h^2 - 5c_s^2}} \quad \text{or} \quad \frac{u_s}{c_s} = \frac{h}{\sqrt{h^2 - 5c_s^2}}$$

Substituting $h = 10^3$ m, $c_s = 340$ m/sec, we find

$$\text{Mach number} = \frac{u_s}{c_s} = 1.54.$$

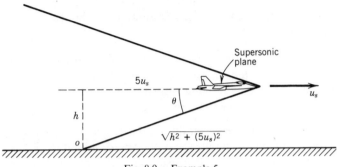

Fig. 8.9. Example 5.

Supersonic objects are necessarily associated with shock waves, or they always create shock waves, which enormously increase the drag force acting on supersonic objects, since shock waves drain energy from them. Shock waves cannot be avoided no matter how well plane bodies are designed.

Problems

Assume that the sound velocity in air is 340 m/sec in the problems.

1. A police car sounding a siren of a frequency of 800 Hz is approaching a car running in the same direction. The velocity of the police car is 90 mi/hr and that of the car 55 mi/hr. Assume that the air is still.
   (a) What frequency does the car driver hear before the police car passes the car?
   (b) After?
   (*Answer:* 842 Hz, 767 Hz.)

2. Repeat Problem 1 assuming that a steady wind of 30 mi/hr against the cars exists.
   (*Answer:* 844 Hz, 768 Hz.)

3. A whistle of a frequency 600 Hz is placed on the edge of an LP record (radius 15 cm, $33\frac{1}{3}$ rpm). What is the lowest and the highest frequencies heard by an observer who is sufficiently far from the record?
   (*Answer:* 600.9 Hz, 599.1 Hz.)

   Note: The whistle frequency sinusoidally varies between these frequencies being frequency modulated.

4. A boy sounding a whistle with a frequency of 800 Hz is approaching a wall that reflects sounds well. When he hears three beats a second, what is his velocity?
   (*Answer:* 0.64 m/sec.)

5. A bullet is fired with a velocity of 500 m/sec. Find the angle between the shock wave surface and the line of motion of the bullet.
(*Answer:*   43°.)

6. Whenever an object travels faster than the wave characterized by a medium, a shock wave is created. Surface waves on water are slow and you can easily create a shock wave by moving a finger dipped into water fast enough in a horizontal direction. Try this experiment.

7. If a relative velocity u between the source of electromagnetic waves and an observer is sufficiently smaller than the speed of light (which is an electromagnetic wave) $c = 3.0 \times 10^8$ m/sec, the Doppler-shifted frequency is given by

$$v' = \frac{c+u}{c} v_0,$$

where $u > 0$ when the source and the observer are approaching each other and $u < 0$ when receding. When microwaves of a frequency 4.0×10^9 Hz are reflected from a jet fighter, the frequency is changed by 10 kHz. What is the speed of the plane? (This principle is used in highway patrol radar.)
(*Answer:*   375 m/sec.)

8. The exact formula for the Doppler-shifted frequency of electromagnetic waves is

$$v' = \frac{1+\beta}{\sqrt{1-\beta^2}} v_0$$

where $\beta = u/c$ and u is the relative velocity between the light source and the observer ($u > 0$ for approaching, $u < 0$ for receding). When starlight, known to have a wavelength of 5500 Å (1 Å = 10^{-8} cm), observed in the laboratory, appears to have a wavelength of 6500 Å:

(a) What is the speed of the star relative to the earth?

(b) Is the star receding or approaching? (This is known as "redshift" of spectrum lines.)

(*Answer:* $u/c = 0.16$, receding.)

CHAPTER 9

Electromagnetic Waves

9.1. Introduction

We know that electromagnetic waves from AM radio waves to visible and invisible light can propagate through vacuum. That is, vacuum can be a medium for electromagnetic waves. This seemingly obvious fact was long disbelieved and the concept of "ether" prevailed for a long period of time. In mechanical waves such as sound waves in gases and solids, and transverse waves on a string, we have no difficulty in "visualizing" wave motion. In sound waves, for example, molecules move about their equilibrium positions, and we have seen that the motion of the molecules determines the kinetic energy and the displacement of molecules from equilibrium positions determines the potential energy, associated with wave motion. Said conversely, in any media capable of storing kinetic and potential energies, mechanical waves can be produced and propagated.

Close analogy can be found in vacuum. Take a capacitor first. A capacitor can store electrical energy in its volume. Although most capacitors are filled with dielectric materials, this is not essential. Dielectrics can be replaced by air or vacuum; that is, vacuum can store electrical energy. Next, take an inductor, which is capable of storing magnetic energy. Again, the magnetic energy is stored in the volume occupied by the inductor, which can be air or vacuum. Thus we draw an important conclusion. Vacuum is capable of storing electrical and magnetic energies, which correspond to two energies, potential and kinetic, in the case of mechanical waves. In any media capable of storing electrical and magnetic energies, electromagnetic waves can be produced and propagated.

In an ideal vacuum, electromagnetic waves are completely dispersionless; that is, the phase velocity equals the group velocity for any frequencies,

$$\frac{\omega}{k} = \frac{d\omega}{dk} = c = 3.0 \times 10^8 \text{ m/sec.}$$

Can you imagine what would happen to FM radio if this were not the case? If the electromagnetic waves were dispersive, two waves of different frequencies would be received at different times, and the signals would be all mixed up!

Electromagnetic waves can be dispersive, however, in media other than vacuum. In fact, the velocity of visible light in glass is slower than that in vacuum.

135

More important, the velocity of light in glass, depends, although slightly, on wave frequency; that is, the electromagnetic waves in water are not dispersionless. Another example is the propagation of shortwave radio around the earth's surface, being reflected by both the earth's surface and the ionospheric plasma surrounding the earth. As we have seen in mechanical waves, waves can be reflected whenever they enter another medium in which the propagation velocity (more precisely, the impedance) is different. As we will see later, the ionospheric plasma acts as a "soft" boundary. Visible light has no difficulty in penetrating into the ionospheric plasma since the plasma becomes essentially nondispersive at such high frequencies.

In this chapter we derive the wave equation for electromagnetic waves using the basic knowledge we have. In fact, all we need is the knowledge of Kirchhoff's voltage and current theorems. Then we generalize the primitive method using macroscopic Maxwell's equations (namely, Faraday's induction law and Maxwell's displacement current). Finally, you will be briefly introduced to the differential form (or microscopic form) of the Maxwell's equations, which govern all electromagnetic phenomena.

9.2. Wave Equation for an *LC* Transmission Line

A ladder network composed of series inductances and parallel capacitances is called an LC transmission line or delay line (Fig. 9.1). This is an excellent analogue of the distributed mass–spring system (Fig. 9.2) we studied (Chapter 4) and in fact can describe electromagnetic waves under many practical situations.

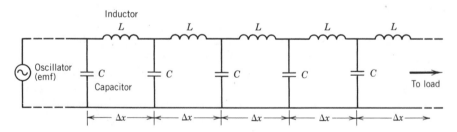

Fig. 9.1.   *LC* transmission line.

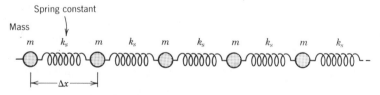

Fig. 9.2.   Mechanical transmission line.

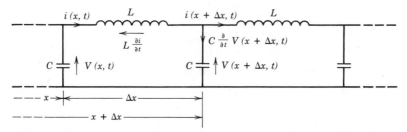

Fig. 9.3.   Voltages and currents at one section of an *LC* transmission line. All quantities, *V*'s and *i*'s, depend on time also.

We pick up one section of the LC line located at x, and assign currents and voltages as shown in Fig. 9.3. Kirchhoff's voltage theorem requires

$$V(x)=L\frac{\partial i(x)}{\partial t}+V(x+\Delta x),\tag{9.1}$$

and the current theorem requires

$$i(x)=C\frac{\partial V(x)}{\partial t}+i(x+\Delta x).\tag{9.2}$$

The first term in the RHS of Eq. (9.2) should have been written as

$$C\frac{\partial}{\partial t}V(x+\Delta x),$$

but if Δx is small, which we assume, we may Taylor-expand $V(x+\Delta x)$ as

$$V(x+\Delta x)\simeq V(x)+\Delta x\frac{\partial V}{\partial x},\tag{9.3}$$

and $(\partial/\partial t)V(x+\Delta x)$ can be well approximated by $(\partial/\partial t)V(x)$. The current at $x+\Delta x$ can also be Taylor-expanded as

$$i(x+\Delta x)\simeq i(x)+\Delta x\frac{\partial i}{\partial x}.\tag{9.4}$$

Substituting Eqs. (9.3) and (9.4) into Eqs. (9.1) and (9.2), we obtain

$$-\Delta x\frac{\partial V}{\partial x}=L\frac{\partial i}{\partial t}\tag{9.5}$$

$$-\Delta x\frac{\partial i}{\partial x}=C\frac{\partial V}{\partial t},\tag{9.6}$$

which may be considered as two simultaneous differential equations for $V(x, t)$ and $i(x, t)$. Next, we differentiate Eq. (9.5) with respect to the spatial coordinate x, and Eq. (9.6) with respect to time t to obtain

$$-\Delta x\frac{\partial^2 V}{\partial x^2}=L\frac{\partial^2 i}{\partial x\,\partial t}\tag{9.7}$$

$$-\Delta x \frac{\partial^2 i}{\partial t\, \partial x} = C \frac{\partial^2 V}{\partial t^2}. \tag{9.8}$$

Since $\partial^2 i/\partial x\, \partial t = \partial^2 i/\partial t\, \partial x$, the current i can be eliminated and we finally obtain a partial differential equation for the voltage $V(x, t)$,

$$\frac{\partial^2 V}{\partial t^2} = \frac{(\Delta x)^2}{LC} \frac{\partial^2 V}{\partial x^2}. \tag{9.9}$$

Exactly the same equation for the current i can be found

$$\frac{\partial^2 i}{\partial t^2} = \frac{(\Delta x)^2}{LC} \frac{\partial^2 i}{\partial x^2}, \tag{9.10}$$

by differentiating Eq. (9.5) with respect to t and Eq. (9.6) with respect to x. (You should work this out by yourself.)

Equations (9.9) and (9.10) are of the form of a wave equation. The propagation velocity is immediately found as

$$c_w = \frac{\Delta x}{\sqrt{LC}}, \tag{9.11}$$

or

$$c_w = \frac{1}{\sqrt{L/\Delta x \cdot C/\Delta x}}. \tag{9.12}$$

But $L/\Delta x$ and $C/\Delta x$ are the inductance and capacitance per unit length of the transmission line. Thus we have found that:

The propagation velocity of voltage and current on a transmission line is determined by the inductance and capacitance per unit length, henry/m, and farad/m.

This conclusion is in fact quite general, and once we know the inductance and capacitance per unit length of any transmission lines, the velocity can readily be found.

Let us go back a little bit in the derivation of the wave equation. We assumed Δx to be small, but did not specify how small it should be. When we wrote Eq. (9.2), we used $C\,[\partial V(x)/\partial t]$ instead of $C\,[\partial V(x + \Delta x)/\partial t]$. This can be done if

$$\left| \Delta x \frac{\partial V}{\partial x} \right| \ll |V|,$$

as is clear from the Taylor expansion for $V(x + \Delta x)$. For a sinusoidal waveform of the voltage, $V(x, t) = V_0 \sin(kx - \omega t)$, with $\omega/k = c_w$, we find

$$\frac{\partial V}{\partial x} = kV_0 \cos(kx - \omega t)$$

$$= \frac{2\pi}{\lambda} V_0 \cos(kx - \omega t).$$

Therefore $|\Delta x \, (\partial V/\partial x)| \ll |V|$ requires that

$$\Delta x \, \frac{2\pi}{\lambda} \ll 1,$$

or roughly speaking, Δx must be smaller than the wavelength λ. Thus when we have a discrete *LC* transmission line as shown in Fig. 9.1, the propagation of voltage and current signals can be described by a dispersionless, linear wave equation only if the preceding condition ($\Delta x \ll \lambda/2\pi$) is satisfied. This is the major limitation of this model, although it is possible to find an exact dispersion relation for the discrete LC transmission line (Example 2).

This limitation seems very severe, but we do not have to worry about it at all in practical transmission lines, which in most cases are *continuous*. We saw that the propagation velocity is determined by the inductance and capacitance *per unit length*; that is, Δx can be taken as small as we wish. The inductance and capacitance per unit length remain as finite quantities no matter how small Δx is chosen. The most familiar transmission line is the one composed of just two conductor wires (Fig. 9.4). The inductance and capacitance per unit length can easily be calculated for such a system (see Example 4).

Another important transmission line is the coaxial cable, which we will study in detail later. Usually, the coaxial cable is filled with dielectric material in order to provide mechanical strength. The dielectric in turn increases the capacitance per unit length. Consequently, the propagation velocity of electromagnetic signals in coaxial cables is smaller than that in those filled with air, or in vacuum. As we will see, coaxial cables filled with air would have the propagation velocity 3.0×10^8 m/sec, which is the velocity of light in vacuum (or air).

Let us now reexamine the dimensions of the vacuum permittivity $\varepsilon_0 = 8.85 \times 10^{-12}$ $C^2/(m^2 \cdot N)$ and permeability $\mu_0 = 4\pi \times 10^{-7}$ N/A^2. The dimensions of ε_0, $C^2/m^2 \cdot N$, can be rewritten as farad/meter, since farad has the dimensions of $C^2/N \cdot m$. Thus the physical meaning of ε_0 is the capacitance per unit length in vacuum. Similarly, μ_0 can be understood as the inductance per unit length in vacuum,

$$\varepsilon_0 = \frac{C}{\Delta x} \text{ (farad/meter)}$$

$$\mu_0 = \frac{L}{\Delta x} \text{ (henry/meter).}$$

Thus the velocity of electromagnetic waves in vacuum can immediately be found as

$$c = \frac{1}{\sqrt{\varepsilon_0 \mu_0}} = 3.0 \times 10^8 \text{ m/sec.}$$

(This is the velocity of electromagnetic waves in unbounded vacuum or air. The wave velocity of electromagnetic waves in bounded media, such as air-filled waveguides, is not given by the preceding expression and waves become

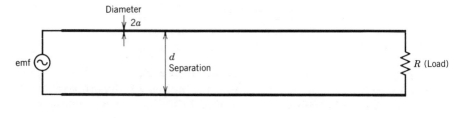

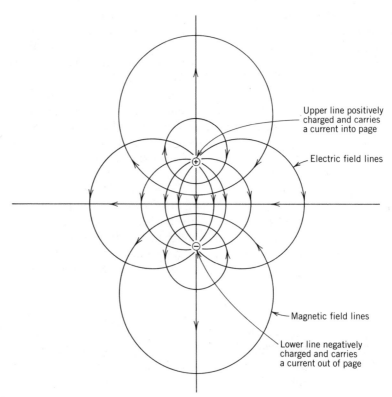

Fig. 9.4. Parallel-wire transmission line for low-frequency electromagnetic waves. Lower figure indicates electric and magnetic field lines.

dispersive.) In dielectric materials, such as glass and water, the permittivity ε is larger, and the velocity of electromagnetic waves in dielectric materials is correspondingly smaller.

Thus we have seen that the simple LC transmission line can model several important media for electromagnetic waves. In the introduction, we emphasized the fact that any media capable of storing both electric and magnetic energies can let electromagnetic waves propagate through them. Since the bulk expression for both energies are $\frac{1}{2}CV^2$ and $\frac{1}{2}Li^2$ with C and L the capacitance and inductance, respectively, we can alternatively state that any medium having both capacitance and inductance is capable of accommodating electromagnetic waves.

Consider a highly conductive metal such as copper. We know that an electrostatic field cannot penetrate into copper. In other words, metals cannot store electrical energy. Thus electromagnetic waves cannot exist in metals with high electrical conductivity, and waves incident on metals are completely reflected. Conductivity in conductors plays much more dominant roles than permittivity for electromagnetic waves. In Section 9.7 we will see that electromagnetic waves in conductors obey a differential equation entirely different from the usual wave equation such as Eq. (9.10). A static field should be distinguished from a dc field, which is dynamic and associated with a current flow in conductors. A dc field can penetrate into a conductor, as we will see in the section on skin effect.

Example 1.   Find the velocity of electromagnetic waves in a coaxial cable filled with Teflon, which has $\varepsilon = 2.0\varepsilon_0$ and $\mu = \mu_0$ (Fig. 9.5).

From

$$c = \frac{1}{\sqrt{\varepsilon\mu_0}}$$

we find

$$c = \frac{1}{\sqrt{2 \times 8.85 \times 10^{-12} \text{ F/m} \times 4\pi \times 10^{-7} \text{ H/m}}}$$

$$= 2.1 \times 10^8 \text{ m/sec.}$$

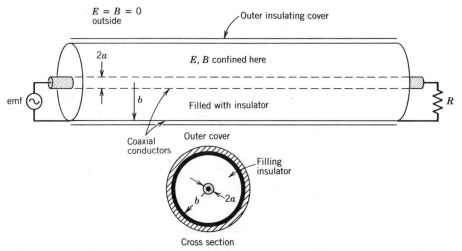

Fig. 9.5.   Coaxial cable for high-frequency electromagnetic waves. There are no fields outside the cable.

Example 2.   (a) For the *LC* transmission line shown in Fig. 9.1, derive a difference equation similar to Eq. (4.7) for a mass–spring transmission line. (b) Show that the exact dispersion relation is given by

$$\omega^2 = 4\omega_0^2 \sin^2\left(\frac{\Delta x k}{2}\right)$$

for a harmonic wave $V(x, t) = V_0 \sin(kx - \omega t)$. Here $\omega_0^2 = 1/LC$.

(a) Let us consider two adjacent units as shown in Fig. 9.6. Applying Kirchhoff's voltage theorem repeatedly, we obtain

$$V(x - \Delta x) = L\frac{\partial}{\partial t} i(x - \Delta x) + V(x) \tag{A}$$

$$V(x) = L\frac{\partial}{\partial t} i(x) + V(x + \Delta x). \tag{B}$$

Subtracting and rearranging give

$$L\frac{\partial}{\partial t}[i(x) - i(x - \Delta x)] = 2V(x) - V(x + \Delta x) - V(x - \Delta x). \tag{C}$$

However, Kirchhoff's current theorem yields

$$i(x - \Delta x) = C\frac{\partial}{\partial t} V(x) + i(x)$$

or

$$i(x) - i(x - \Delta x) = -C\frac{\partial}{\partial t} V(x). \tag{D}$$

Substituting (D) into (C), we obtain

$$LC\frac{\partial^2}{\partial t^2} V(x) = V(x + \Delta x) + V(x - \Delta x) - 2V(x). \tag{E}$$

This is the required difference equation for the voltage. Note that this is mathematically identical to Eq. (4.7), for the mechanical transmission lines.

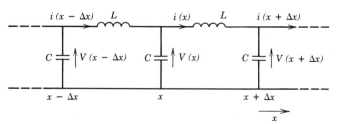

Fig. 9.6.   Example 2.

(b) Let $V(x, t) = V_0 \sin (kx - \omega t)$. Noting

$$\frac{\partial^2}{\partial t^2} V(x, t) = -\omega^2 V_0 \sin (kx - \omega t)$$

$$V(x + \Delta x) - V(x) = 2V_0 \sin \left[\frac{\Delta x k}{2} \right] \cos \left[\left(kx + \frac{\Delta x}{2} k \right) - \omega t \right]$$

$$V(x - \Delta x) - V(x) = -2V_0 \sin \left[\frac{\Delta x k}{2} \right] \cos \left[\left(kx - \frac{\Delta x}{2} k \right) - \omega t \right]$$

and thus

$$V(x + \Delta x) + V(x - \Delta x) - 2V(x) = -4 \sin^2 \left(\frac{\Delta x k}{2} \right) V_0 \sin (kx - \omega t),$$

we find

$$\omega^2 = \frac{1}{LC} 4 \sin^2 \left(\frac{\Delta x k}{2} \right)$$

$$= 4\omega_0^2 \sin^2 \left(\frac{\Delta x k}{2} \right)$$

Note that the long wavelength limit corresponds to $\Delta x k \ll 1$. Approximating $\sin (\Delta x k/2)$ by simply $\Delta x k/2$ yields our previous dispersion relation,

$$\omega^2 = \frac{(\Delta x)^2}{LC} k^2.$$

9.3. Coaxial Cable

A cable composed of two coaxial cylindrical conductors is called a coaxial cable and is frequently used for transmitting electromagnetic signals from one device to another. Coaxial cables can confine electromagnetic waves within themselves and under ideal conditions the electromagnetic waves do not leak out, in contrast to the open, two-parallel-wire transmission line. Parallel-wire transmission lines are typically used for high-power, low-frequency (60-Hz) electromagnetic waves. At such a low frequency, the radiation loss is negligible. As the frequency becomes higher, however, the parallel-wire transmission lines become very ineffective because of the radiation loss; that is, the wave energy can easily leak out from the transmission line. Therefore at high frequencies, the transmission lines must be of closed type, such as coaxial cable and microwave waveguide. (It is a common experience that sound waves in a pipe can propagate a further distance than in free space.)

Suppose that we have a long coaxial cable connected to a dc emf at $t = 0$ at one end (Fig. 9.7). The cable cannot be filled with charge instantly, since the electromagnetic waves should travel with a large, but finite speed. We assume

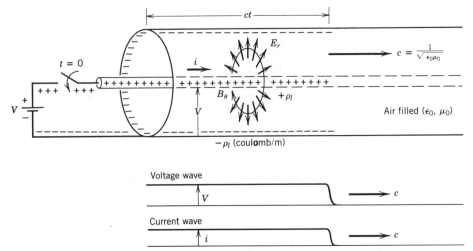

Fig. 9.7.　When a dc power supply is suddenly connected to a coaxial cable, both voltage and current waves start propagating along the cable.

that the space between the inner and outer conductor is filled with air (or vacuum), which has permittivity ε_0 and permeability μ_0. Our purpose here is to find the propagation velocity c of the charge pulse, pretending we do not know the wave equation which we found in the previous section. (Actually, we are cheating ourselves since when we assume that a square charge pulse can exist in the cable, we have to assume that the electromagnetic waves in the coaxial cable are dispersionless; that is, the propagation velocity is independent of the wave frequency. This can only be assured by the wave equation that we are going to find out!)

Let ρ_l be the linear charge density in the region to the left of the pulse front. (This ρ_l should not be confused with the linear mass density we used for mechanical waves.) Since we have cylindrical symmetry, the radial electric field in the space between the inner and outer conductor can easily be found from Gauss' law as (Fig. 9.8)

$$E_r = \frac{\rho_l}{2\pi\varepsilon_0} \frac{1}{r}. \tag{9.13}$$

Next we notice that the current i should be related to the line charge density ρ_l through

$$i = \rho_l c,$$

where c is the velocity at which the charge front is moving. (The wave propagation velocity c should not be confused with the velocity of charge carriers, namely, electrons in the conductors. What propagates at the velocity c is the perturbation in the charge density, and this has nothing to do with the electron velocity, which is extremely small. A similar situation has already been encountered in sound waves in which the wave velocity c_s and the velocity wave

E profile

B profile

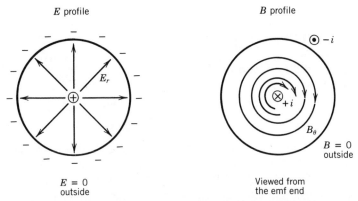

$E = 0$
outside

Viewed from
the emf end

Fig. 9.8.   Electric and magnetic field profile in the coaxial cable.

$\partial\xi/\partial t$ were entirely different physical quantities.) Then the azimuthal magnetic field can be found from Ampere's law as (Fig. 9.8)

$$B_\theta = \frac{\mu_0 i}{2\pi r} = \frac{\mu_0 \rho_l}{2\pi r} c. \tag{9.14}$$

From Eqs. (9.13) and (9.14), we find

$$\frac{E_r}{B_\theta} = \frac{1}{\varepsilon_0 \mu_0 c}. \tag{9.15}$$

Thus if we find one more relationship between the electric field E_r and magnetic field B_θ, we may find the propagation velocity c.

This can be done if we apply Faraday's law to a thin rectangle located at the pulse front at a certain instant (Fig. 9.9). (We neglect unimportant edge effects at the pulse front.) As the pulse propagates to the right, the magnetic flux enclosed by the rectangle increases. Thus an emf is induced along the rectangle

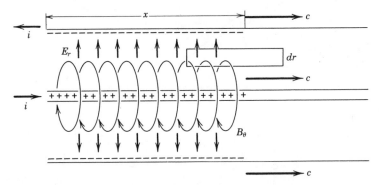

Fig. 9.9.   Faraday's law is applied to the rectangle. The magnetic flux through the rectangle is increasing with time.

edges

$$\oint \mathbf{E} \cdot d\mathbf{l} = -\frac{d}{dt}(Bx\, dr).$$ (9.16)

The integral with a circle on it is called a closed line integral. $\oint \mathbf{E} \cdot d\mathbf{l}$ can be interpreted as

$$\oint E \cos dl,$$ (9.17)

where θ is the angle between $\mathbf{E}$ and $d\mathbf{l}$, and the integration is over a closed loop of arbitrary shape (Fig. 9.10). In the present case the only contribution to the integral comes from the edge AB, since the electric field is perpendicular to the edges BC and DA ($\cos 90° = 0$), and we do not have the field along the edge CD (Fig. 9.11). Then

$$\oint \mathbf{E} \cdot d\mathbf{l} = -E_r\, dr.$$

(Note that along AB, $\theta = 180°$.) The RHS of Eq. (9.16) becomes

$$-\frac{d}{dt}(B_\theta\, drx) = -B_\theta\, dr\frac{dx}{dt} = -B_\theta\, drc.$$

Then

$$E_r = B_\theta c.$$ (9.18)

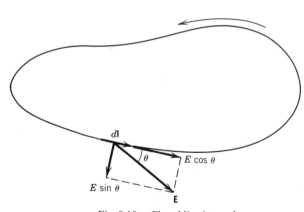

Fig. 9.10.   Closed line integral.

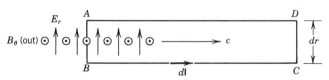

Fig. 9.11.   The only contribution to the line integral comes from the edge AB, where $\mathbf{E} \parallel d\mathbf{l}$.

From Eqs. (9.15) and (9.18), we find

$$c^2 = \frac{1}{\varepsilon_0 \mu_0} = 9.0 \times 10^{16} \quad (\text{m/sec})^2$$

or

$$c = 3.0 \times 10^8 \text{ m/sec},$$

which is the desired velocity of propagation of an electromagnetic pulse in a coaxial cable filled with air.

It is obvious that if the cable is filled with a dielectric material other than air, we have to replace ε_0 by $\varepsilon = \kappa \varepsilon_0$, where κ is the dielectric constant, and the velocity becomes correspondingly smaller. *Caution:* The dielectric constant κ is usually a function of the frequency of electromagnetic waves. For example, water has $\kappa \approx 80$ for static fields ($v = 0$), but for visible light ($v \approx 5 \times 10^{14}$ Hz), $\kappa = 1.8$.

Equation (9.15) can alternatively be derived from Ampere–Maxwell's law, which states that a time-varying electric flux should induce a magnetic field. In other words, a time-varying electric field is equivalent to a current (displacement current). This is exactly how a current can flow through a capacitor. Let us consider again a thin rectangle that is perpendicular to the one we considered before. The electric flux enclosed by the rectangle $ABCD$ induces a magnetic field along $ABCD$ (Fig. 9.12).

$$\oint \mathbf{B} \cdot d\mathbf{l} = \varepsilon_0 \mu_0 \frac{d}{dt} (E_r x \, dl)$$

$$= \varepsilon_0 \mu_0 E_r \, dl \frac{dx}{dt}$$

$$= \varepsilon_0 \mu_0 E_r \, dl \, c. \tag{9.19}$$

Now the only contribution to the integral comes from the edge AB, and we have

$$\oint \mathbf{B} \cdot d\mathbf{l} = B_\theta \, dl. \tag{9.20}$$

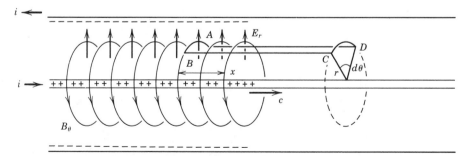

Fig. 9.12. Electric flux through $ABCD$ is increasing with time, which induces a magnetic field through the displacement current.

We then find

$$B_\theta = \varepsilon_0 \mu_0 E_r c,$$

which is identical to Eq. (9.15).

Thus we have been able to derive the velocity of an electromagnetic pulse in a coaxial cable without using the wave equation. Once we find the velocity, we can calculate the current i for a given value of the emf (volt) or find the impedance of the coaxial cable. For this we have to assume a finite radius of the inner conductor a (m) and that of the outer conductor b (m). The electric field at radius r is again found from Gauss' law (Fig. 9.13),

$$2\pi r E = \frac{\rho_l}{\varepsilon_0}, \quad \text{for } a < r < b.$$

The voltage difference between the inner and outer conductor is then given by

$$V = \int_a^b E_r \, dr = \frac{\rho_l}{2\pi\varepsilon_0} \int_a^b \frac{1}{r} \, dr = \frac{\rho_l}{2\pi\varepsilon_0} \ln\left(\frac{b}{a}\right).$$

But $\rho_l = i/c$. Then

$$Z = \frac{V}{i} = \frac{1}{2\pi\varepsilon_0 c} \ln\left(\frac{b}{a}\right) \quad (\Omega).$$

Using $c = 1/\sqrt{\varepsilon_0 \mu_0}$, we find

$$Z = \sqrt{\frac{\mu_0}{\varepsilon_0}} \frac{1}{2\pi} \ln\left(\frac{b}{a}\right) \quad (\Omega). \tag{9.21}$$

Since $\ln(b/a)$ is dimensionless, we see that $\sqrt{\mu_0/\varepsilon_0}$ has the dimension of resistance, Ω. You should check this directly from the dimensions for μ_0 and ε_0.

The characteristic impedance we just found can alternatively be written as

$$Z = \sqrt{\frac{\text{inductance per unit length}}{\text{capacitance per unit length}}}. \tag{9.22}$$

To see this let us find the inductance and capacitance per unit length of the coaxial cable. For every 1 m of the cable, the cable has ρ_l coulombs of charge.

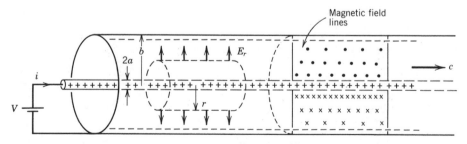

Fig. 9.13. Gauss's law applied to a cylinder with radius r to find the electric field. Magnetic field lines are shown for comparison.

The voltage difference between the conductors is still V. Then

$$C = \frac{\rho_l}{V} = 2\pi\varepsilon_0 \frac{1}{\ln(b/a)} \quad \text{(F/m)}. \tag{9.23}$$

The inductance of the cable can be found if we calculate the total magnetic flux existing between the two conductors, $a < r < b$. Since

$$B = \frac{\mu_0 i}{2\pi r},$$

the magnetic flux per unit length of the cable is

$$\Phi_B = \int_a^b B \, dr = \frac{\mu_0 i}{2\pi} \ln\left(\frac{b}{a}\right) \quad \text{(weber/meter)}.$$

The inductance per unit length is then

$$L = \frac{\Phi_B}{i} = \frac{\mu_0}{2\pi} \ln\left(\frac{b}{a}\right) \quad \text{(H/m)}. \tag{9.24}$$

We have not included the magnetic flux in the inner conductor in the inductance calculation; in other words we have neglected the internal inductance. This is allowed if the current flows only on the surface of the conductor (skin effect). Substituting these values for C and L into Eq. (9.22), we recover

$$Z = \sqrt{\frac{\mu_0}{\varepsilon_0}} \frac{1}{2\pi} \ln\left(\frac{b}{a}\right) \quad (\Omega).$$

You may wonder why we cannot simply use ε_0 and μ_0 as the capacitance and inductance per unit length, particularly since we are considering electromagnetic waves in *air* filling the space between the two conductors. $Z_0 = \sqrt{\mu_0/\varepsilon_0}$ indeed has the meaning of the characteristic impedance of vacuum (or air), but we cannot use this for the coaxial cable. The reason is that the electromagnetic wave in the coaxial cable is *not* a plane wave. We saw that both the electric and magnetic fields depend on the radial position as well as, of course, x, the direction of propagation. For plane waves the impedance indeed becomes simply $\sqrt{\mu_0/\varepsilon_0}$, but whenever the waves are confined geometrically, as in the case of coaxial cable, the impedance must be accordingly modified.

The reason we call Z the characteristic impedance rather than the resistance is that the coaxial cable does not dissipate energy. The cable itself is a *reactive medium* composed of capacitance and inductance. It only provides a medium for electromagnetic wave propagation. If we terminate one end by a resistor having the same value as the characteristic impedance, the energy is most effectively transferred from the cable to the load resistor, without causing any reflection of electromagnetic waves at the terminating end. At this stage you should be reminded of the concept of impedance matching for most efficient energy transfer from a battery having internal resistance R_i. If the load resistance matches the internal resistance R_i, the energy dissipation in the load resistance becomes a maximum. The physics behind this matching is now obvious. If the load resistance matches the characteristic impedance, no wave (and thus energy) reflection occurs, and energy is most efficiently transferred.

Let us now take a look at the energy stored in the coaxial cable. Since we know both the electric and magnetic field as functions of the radius r, we can immediately find the local energy densities:

$$\text{Electric E.D.} = \tfrac{1}{2}\varepsilon_0 E_r^2 = \frac{\rho_l^2}{8\pi^2\varepsilon_0 r^2} \quad (\text{J/m}^3) \tag{9.25}$$

$$\text{Magnetic E.D.} = \frac{1}{2\mu_0} B_\theta^2 = \frac{\mu_0 \rho_l c^2}{8\pi^2 r^2} \quad (\text{J/m}^3) \tag{9.26}$$

But $c^2 = 1/\varepsilon_0\mu_0$. Thus we find that the electric energy density is equal to the magnetic energy density just as in the case of mechanical waves in which the kinetic energy density equals the potential energy density. We conclude that for nondispersive electromagnetic waves, the same amount of energy is stored in both the electric and magnetic fields.

The total energy density is

$$W = 2 \times \frac{\rho_l^2}{8\pi^2\varepsilon_0 r^2}$$

$$= \frac{\rho_l^2}{4\pi^2\varepsilon_0 r^2} \quad (\text{J m}^3). \tag{9.27}$$

Then the energy per unit length of the coaxial cable is

$$\int_a^b W \cdot 2\pi r \, dr = \frac{\rho_l^2}{2\pi\varepsilon_0} \ln\left(\frac{b}{a}\right) \quad (\text{J/m}),$$

and the power is

$$P = c \frac{\rho_l^2}{2\pi\varepsilon_0} \ln\left(\frac{b}{a}\right) = \frac{V^2}{Z} = i^2 Z \quad (\text{W}), \tag{9.28}$$

which is equal to the power supplied by the emf, Vi, as can easily be shown.

Example 3. Determine the ratio between the outer and inner conductor radii b/a of a Teflon- $(\varepsilon = 2\varepsilon_0)$ filled, 50-Ω coaxial cable.

The impedance is given by

$$Z = \sqrt{\frac{\mu_0}{\varepsilon}} \frac{1}{2\pi} \ln\left(\frac{b}{a}\right) \quad (\Omega).$$

Then

$$\ln\left(\frac{b}{a}\right) = 2\pi \sqrt{\frac{\varepsilon}{\mu_0}} Z$$

$$= 2\pi \sqrt{\frac{2\varepsilon_0}{\mu_0}} Z.$$

Substituting $\sqrt{\mu_0/\varepsilon_0} = 377\ \Omega$ and $Z = 50\ \Omega$, we obtain

$$\ln\left(\frac{b}{a}\right) = 2\pi\sqrt{2} \times \frac{50}{377}$$

$$= 1.178$$

or

$$\frac{b}{a} = 3.25.$$

As long as this ratio is used, a Teflon-filled cable has a 50-Ω impedance irrespective of its actual size, large or small.

Example 4. Calculate the characteristic impedance of a parallel-wire transmission line. Assume that the conductors have a common radius a and are separated by a distance d much larger than the radius a.

To find the capacitance per unit length of the transmission line, we let the wires carry $+\rho_l$ (C/m) and $-\rho_l$ (C/m) line charges, respectively. In the plane containing the wires, the electric field is given by (Fig. 9.14)

$$E_r = \frac{\rho_l}{2\pi\varepsilon_0}\left(\frac{1}{r} + \frac{1}{d-r}\right),$$

where r is the distance from the positively charged wire. Then the potential difference between the wires is

$$V = \int_a^{d-a} E_r\,dr = \frac{\rho_l}{2\pi\varepsilon_0}\int_a^{d-a}\left(\frac{1}{r} + \frac{1}{d-r}\right)dr$$

$$= \frac{\rho_l}{\pi\varepsilon_0}\ln\left(\frac{d-a}{a}\right) \simeq \frac{\rho_l}{\pi\varepsilon_0}\ln\left(\frac{d}{a}\right)$$

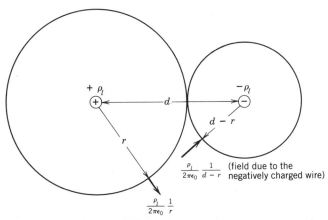

Fig. 9.14. Example 4.

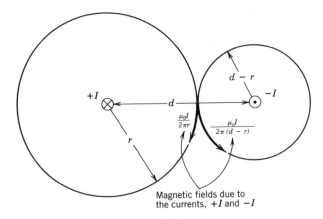

+I

−I

d − r

d

$\frac{\mu_0 I}{2\pi r}$

$\frac{\mu_0 I}{2\pi (d - r)}$

r

Magnetic fields due to
the currents, +I and −I

Fig. 9.15. Example 4.

and the capacitance per unit length of the transmission line becomes

$$\frac{C}{l} = \frac{\rho_l}{V} = \frac{\pi \varepsilon_0}{\ln (d/a)} \quad \text{(F/m)}.$$

The inductance per unit length can be found in a similar manner. For this purpose, we let the wires carry currents $+I$ (A) and $-I$ (A), respectively. The magnetic field at a distance r from the positive current is (Fig. 9.15)

$$B = \frac{\mu_0 I}{2\pi} \left(\frac{1}{r} + \frac{1}{d-r} \right).$$

Then the magnetic flux linked to unit length of the transmission line is

$$\frac{\Phi_B}{l} = \frac{\mu_0 I}{2\pi} \int_a^{d-a} \left(\frac{1}{r} + \frac{1}{d-r} \right) dr$$

$$\simeq \frac{\mu_0 I}{\pi} \ln \left(\frac{d}{a} \right) \quad \text{(W/m)}.$$

The inductance per unit length is thus

$$\frac{L}{l} = \frac{\Phi_B}{lI} = \frac{\mu_0}{\pi} \ln \left(\frac{d}{a} \right) \quad \text{(H/m)}.$$

Finally, the characteristic impedance obtains as

$$Z = \sqrt{\frac{L/l}{C/l}} = \frac{1}{\pi} \sqrt{\frac{\mu_0}{\varepsilon_0}} \ln \left(\frac{d}{a} \right) \quad (\Omega).$$

9.4. Poynting Vector

In the previous section we found the energy density in the coaxial cable is

$$W = \frac{\rho_l^2}{4\pi^2 \varepsilon_0 r^2} \quad \text{(J/m}^3\text{)}. \tag{9.29}$$

Then the power density (energy flow per unit time and per unit area) is

$$cW = \frac{\rho_l^2 c}{4\pi^2 \varepsilon_0 r^2} \quad (W/m^2)$$

$$= \frac{\rho_l}{2\pi\varepsilon_0 r} \times \frac{\mu_0 \rho_l v}{2\pi r} \frac{1}{\mu_0}$$

$$= E_r \times \frac{B_\theta}{\mu_0}.$$

$$= E_r \times H_\theta. \tag{9.30}$$

The quantity $\mathbf{E} \times \mathbf{H} \, (= \mathbf{E} \times \mathbf{B}/\mu_0)$ is called the Poynting vector and is denoted by $\mathbf{S}$. Its physical meaning is simply the power density (W/m^2). Furthermore, since $\mathbf{S}$ is a vector, it also tells us in which direction the electromagnetic energy is flowing. In the coaxial cable the electric field is radially outward and the magnetic field is in the azimuthal direction as shown, and the direction of $\mathbf{E} \times \mathbf{B}$ is indeed toward the load consistent with the direction of the energy flow (Fig. 9.16).

From the arguments given above, it is now apparent that energy is *not* transferred *in* the conductors. Rather, the energy is transferred through the space between the conductors. The role of the conductors is only to confine the electromagnetic energy in the space between the conductors, thus preventing leakage. This is not surprising if you remember that vacuum (or air) is an excellent medium for electromagnetic waves. In contrast, conductors are an extremely poor medium for electromagnetic waves, as we have briefly seen before.

Let us reexamine the energy flow in some simple cases in terms of the Poynting vector, $\mathbf{S} = \mathbf{E} \times \mathbf{B}/\mu_0$. Suppose a dc current is flowing in a resistor rod with resistance R (Fig. 9.17). The potential drop across the resistor is $V = iR$, and the electric field in the resistor is $E = V/l$. The magnetic field is azimuthal, and at the surface of the rod (use Ampere's law),

$$B = \frac{\mu_0 i}{2\pi a}.$$

Then the Poynting vector $\mathbf{S}$ is radially inward, and its magnitude is

$$S = \frac{EB}{\mu_0} = \frac{V}{l} \frac{i}{2\pi a} \quad (W/m^2).$$

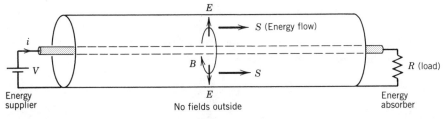

Fig. 9.16. Electric field, magnetic field, and Poynting vector in a coaxial cable. The Poynting vector is directed from the emf to the load everywhere in the cable.

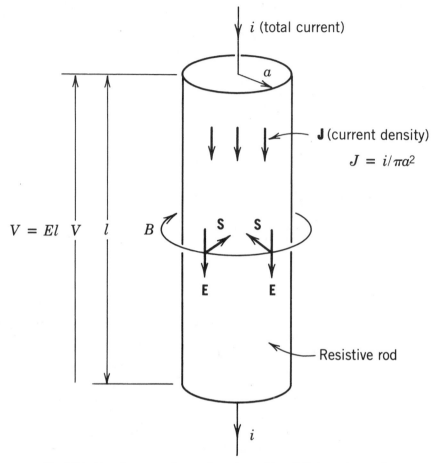

Fig. 9.17. Poynting vector S around a resistor. S is radially inward everywhere.

The total power flow through the surface is

$$P = S \times 2\pi al = Vi \quad \text{(W)},$$

which is consistent with the power being dissipated in the resistor rod. In this case again, the energy is not carried in the conductor. Rather it is carried by the electromagnetic fields (dc) through the space surrounding the conductor and the resistor.

Next consider a parallel plate capacitor C that is being charged (Fig. 9.18). We know the following relationships:

$$i(t) = \frac{dq}{dt},$$

$$q = CV.$$

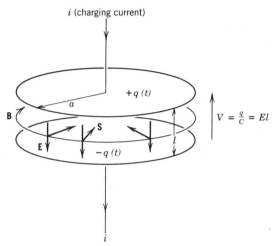

Fig. 9.18. Poynting vector in a capacitor being charged.

We also know that no conduction current is flowing through the capacitor, but a displacement current is, which is given by

$$i = \frac{dq}{dt} = C\frac{dV}{dt}.$$

Then the rate of energy storage in the capacitor is

$$P = iV = C\frac{dV}{dt}\,V = \frac{d}{dt}(\tfrac{1}{2}CV^2)\quad(\text{W}).$$

Ampere's law still holds for displacement currents. Then the magnetic field at the edge of the capacitor is

$$B = \frac{\mu_0 i}{2\pi a} = \frac{\mu_0}{2\pi a}\,C\frac{dV}{dt}.$$

The electric field is simply $E = V/l$. Then the Poynting vector at the edge is

$$S = \frac{EB}{\mu_0} = \frac{V}{l}\frac{1}{2\pi a}\,C\frac{dV}{dt}\quad(\text{W/m}^2),$$

and we again have

$$2\pi a l S = \frac{d}{dt}(\tfrac{1}{2}CV^2).$$

The preceding argument is correct only if the rate of charging is sufficiently slow. If the capacitor is rapidly charged, $E = V/l$ does not hold anymore. However, Poynting theorem still applies.

From these examples it is now clear that electromagnetic energy transfer is achieved in the form of the Poynting vector. An electric field or magnetic field alone cannot cause the flow of electromagnetic energy. We must have both.

Furthermore, since the Poynting vector is the vector product between **E** and **B**, the Poynting vector vanishes if **E** is purely parallel to **B**, even though both are nonzero. In most electromagnetic phenomena we study, **E** is usually perpendicular to **B**, and you do not have to worry about this complication.

9.5. Plane Electromagnetic Waves in Free Space

We know that radio or TV signals can be transmitted through air (or vacuum). We also know that solar energy is transmitted from the sun to the earth through vacuum in the form of radiation of various frequencies or wavelengths. Electromagnetic waves have a tremendously vast frequency range. Surprisingly enough, they all propagate with the same speed, $c = 3.0 \times 10^8$ m/sec, the velocity of light in vacuum (or air, with negligible error).

In Section 9.2 we almost got the wave equation for electromagnetic waves in free space. If we take $L/\Delta x = \mu_0$, and $C/\Delta x = \varepsilon_0$, we may indeed conclude that electromagnetic waves in free space should have the velocity

$$c = \frac{\Delta x}{\sqrt{LC}} = \frac{1}{\sqrt{\varepsilon_0 \mu_0}} = 3.0 \times 10^8 \text{ m/sec.}$$

However, it is far from obvious that we can analyze the electromagnetic waves in free space, which is a continuous medium for the waves, in terms of discrete inductances and capacitances. Also, the analogy does not tell us what the electric and magnetic field should be, which we need when we calculate the Poynting vector associated with the electromagnetic waves. Here we directly derive the wave equation for electromagnetic waves in free space using the fundamental laws in electromagnetism, namely, Faraday's induction law, and the Ampere–Maxwell displacement current law.

For simplicity, we assume a one-dimensional or plane wave that is propagating in the x direction, which is also the direction of energy flow. From the arguments on the Poynting vector, if we assume an electric field in the y direction, E_y, we must have the magnetic field in the z direction, as shown in Fig. 9.19. Let us assume a thin rectangle in the x-y plane having a length of 1 m and width Δx (Fig. 9.20). We apply Faraday's law to the rectangle. Since the magnetic flux enclosed by the rectangle is $\Phi_B = B_z(x)\Delta x \times 1$, we have

$$\oint \mathbf{E} \cdot d\mathbf{l} = E(x + \Delta x) \times 1 - E(x) \times 1$$

$$\simeq \Delta x \frac{\partial E}{\partial x}.$$

This must be equal to $-\partial \Phi_B / \partial t = -\Delta x (\partial B / \partial t)$. Then

$$-\frac{\partial E}{\partial x} = \frac{\partial B}{\partial t}. \tag{9.31}$$

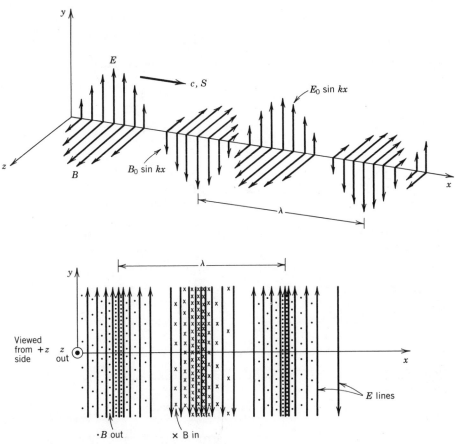

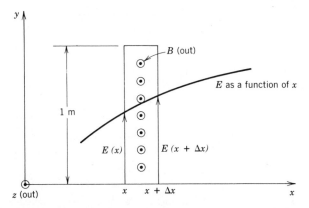

Fig. 9.19. Plane electromagnetic wave propagating in the x direction ("snapshot"). In the upper part of the figure the vector length corresponds to the local field intensity. In the lower part the number of field lines per unit length corresponds to the local field intensity.

Fig. 9.20. The magnetic flux through the rectangle is varying with time.

157

Next consider a thin rectangle in the x-z plane through which an electric flux

$$\Phi_E = E\,\Delta x \times 1$$

penetrates (Fig. 9.21). The Ampere–Maxwell law requires that

$$\oint \mathbf{B}\cdot d\mathbf{l} = B(x)\times 1 - B(x+\Delta x)\times 1$$

$$= \varepsilon_0\mu_0\frac{\partial\Phi_E}{\partial t} = \varepsilon_0\mu_0\,\Delta x\,\frac{\partial E}{\partial t}\,,$$

or

$$-\frac{\partial B}{\partial x} = \varepsilon_0\mu_0\frac{\partial E}{\partial t}\,. \tag{9.32}$$

Differentiating Eq. (9.31) and Eq. (9.32) with respect to x and t respectively and eliminating B, we find

$$\frac{\partial^2 E}{\partial t^2} = \frac{1}{\varepsilon_0\mu_0}\frac{\partial^2 E}{\partial x^2}\,. \tag{9.33}$$

Similarly, we can find

$$\frac{\partial^2 B}{\partial t^2} = \frac{1}{\varepsilon_0\mu_0}\frac{\partial^2 B}{\partial x^2}\,. \tag{9.34}$$

Equations (9.33) and (9.34) are our desired wave equations for the electric and magnetic fields in free space.

The velocity of propagation is immediately found as

$$c = \frac{1}{\sqrt{\varepsilon_0\mu_0}} = 3.0\times 10^8 \text{ m/sec.}$$

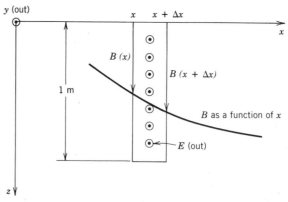

Fig. 9.21. The electric flux through the rectangle is time varying, which induces a magnetic field along the edges (Ampere–Maxwell law for the displacement current).

If the medium has an electric permittivity ε and magnetic permeability μ, the velocity should accordingly be modified to

$$c = \frac{1}{\sqrt{\varepsilon\mu}}.$$

Let us see whether for the electromagnetic waves the condition that the electric energy be equal to the magnetic energy is satisfied. For this, it is sufficient to see if the energy densities are the same. The electric energy density is

$$\tfrac{1}{2}\varepsilon_0 E^2 \quad (\text{J/m}^3)$$

and the magnetic energy density is

$$\frac{1}{2\mu_0} B^2 \quad (\text{J/m}^3).$$

Assuming a harmonic wave of the form

$$E(x, t) = E_0 \sin (kx - \omega t)$$

$$B(x, t) = B_0 \sin (kx - \omega t),$$

we find from Eq. (9.31) that

$$-kE_0 \cos (kx - \omega t) = -B_0 \cos (kx - \omega t),$$

or

$$kE_0 = \omega B_0.$$

Since $\omega/k = c$, we have

$$E(x, t) = cB(x, t)$$

which was found earlier in Section 9.3. Then

$$\tfrac{1}{2}\varepsilon_0 E^2 = \tfrac{1}{2}\varepsilon_0 c^2 B^2$$

$$= \frac{1}{2\mu_0} B^2.$$

Indeed both energy densities are equal. In the preceding derivation we have assumed a harmonic wave. However, this is not essential and the equipartition of energy holds for any traveling electromagnetic waves of arbitrary form.

Let us next calculate the Poynting vector for the case of harmonic waves. Since E is in the y direction and B in the z direction

$$\mathbf{S} = \mathbf{E} \times \frac{\mathbf{B}}{\mu_0}$$

is directed along the x axis, which is the direction of the wave, and thus the energy flow. The magnitude of $\mathbf{S}$ is

$$\frac{EB}{\mu_0}.$$

Substituting $E = E_0 \sin (kx - \omega t)$, and $B = B_0 \sin (kx - \omega t)$, we find

$$S = \frac{E_0 B_0}{\mu_0} \sin^2 (kx - \omega t)$$

$$= c\varepsilon_0 E_0^2 \sin^2 (kx - \omega t) \quad (\text{W/m}^2).$$

The root mean square value of S is

$$S_{rms} = \tfrac{1}{2} c\varepsilon_0 E_0^2 = c\varepsilon_0 E_{rms}^2 = \sqrt{\frac{\varepsilon_0}{\mu_0}} \, E_{rms}^2, \qquad (9.35)$$

which is composed of $\tfrac{1}{2} c\varepsilon_0 E_{rms}^2$ and $\tfrac{1}{2} c(B_{rms}^2/\mu_0)$ as required from the equipartition of energy between electric and magnetic fields.

The characteristic impedance for electromagnetic waves in vacuum can still be defined by

$$Z = \sqrt{\frac{\text{inductance/m}}{\text{capacitance/m}}}$$

$$= \sqrt{\frac{\mu_0}{\varepsilon_0}} = 377 \, \Omega. \qquad (9.36)$$

This quantity is important for antenna engineering. As we will see in Chapter 10, the impedance of antennas used for electromagnetic radiation (called the *radiation resistance*) is proportional to this quantity.

Since the energy density associated with the electromagnetic wave is

$$\varepsilon_0 E_{rms}^2 = \tfrac{1}{2}\varepsilon_0 E_{rms}^2 + \frac{1}{2\mu_0} B_{rms}^2 \quad (\text{J/m}^3),$$

we conjecture that the electromagnetic waves have a pressure $(\text{J/m}^3 = \text{N/m}^2)$ given by the same expression. The pressure associated with electromagnetic waves is called the *radiation pressure*. [In gas dynamics the pressure exerted by a gas on a wall *that can absorb all the particles impinging on it* is given by $\tfrac{1}{2} n k_B T$ where $n \, (\text{m}^{-3})$ is the molecule density, k_B is the Boltzmann constant, and $T \, (\text{K})$ is the absolute temperature. For a wall that can elastically reflect particles, the pressure exerted is $2 \times \tfrac{1}{2} n k_B T = n k_B T$. The radiation pressure we found above corresponds to the one-way momentum transfer, or the pressure exerted on an absorbing wall. For a perfect reflector the pressure exerted is twice as large, in complete analogy to the gas pressure.] Usually, this radiation pressure is negligibly small. However, the radiation pressure is actually the measure of the rate of momentum transfer, and it should be realized that electromagnetic waves carry momentum as well as energy. Furthermore, a circularly polarized electromagnetic wave can carry an angular momentum.

Example 5. A monochromatic light source radiates at an rms power of 30 W isotropically. (a) What is the rms electric field at a distance of 5 m from the source? (b) What is the magnetic field (rms) at the same distance? (c) Find the radiation force exerted on a mirror placed normal to radiation at the same distance. Assume that the mirror has an area of $10 \times 10 \, \text{cm}^2$.

(a) The rms Poynting vector is

$$S_{rms} = \frac{power}{4\pi r^2} = \frac{30\ W}{4\pi(5)^2 m^2} = 0.0955\ W/m^2.$$

This should be equal to E_{rms}^2/Z, where $Z = \sqrt{\mu_0/\varepsilon_0} = 377\ \Omega$. Then

$$E_{rms} = \sqrt{S_{rms}Z} = \sqrt{0.0955\ W/m^2 \times 377\ \Omega}$$

$$= 6.0\ V/m.$$

(b) The rms magnetic field can be found from

$$B_{rms} = E_{rms}/c = \frac{6.0\ V/m}{3 \times 10^8\ m/sec} = 2.0 \times 10^{-8}\ tesla.$$

(c) The radiation pressure exerted on a reflector is

$$2 \times \varepsilon_0 E_{rms}^2 = (N/m^2).$$

Then the force is

$$F = 2\varepsilon_0 E_{rms}^2 A$$

$$= 2 \times 8.85 \times 10^{-12}\ F/m \times (6.0\ V/m)^2 \times 0.01\ m^2$$

$$= 6.4 \times 10^{-12}\ N \quad \text{(negligibly small)}.$$

Example 6. The Poynting vector $S = E \times B/\mu_0$ is interpreted as the intensity of electromagnetic waves, W/m^2, which is equivalent to power density or energy flux density. How do you interpret the following quantities? (a) S/c, (b) $r \times S/c$, (c) $r \times S/c^2$, where r is the distance from the radiation source.

(a) Since $S = c\varepsilon_0 E_{rms}^2$, S/c is the energy density. However, this is alternatively interpreted as the radiation pressure (note $J/m^3 = N/m^2$), and furthermore as the rate of momentum transfer per unit area, that is, the momentum flux density.

(b) $r \times$ (momentum vector) is the angular momentum. Therefore $r \times S/c$ may be interpreted as the angular momentum flux density. For pure plane waves this quantity is identically zero since r and S are parallel.

(c) The quantity S/c^2 may be interpreted as the momentum density. Therefore $r \times S/c^2$ becomes angular momentum density.

9.6. Reflection of Electromagnetic Waves

The concept of characteristic impedance introduced in previous sections can greatly simplify our understanding of the reflection phenomena of electromagnetic waves. It allows us to quantize the concept of "hard" and "soft" boundaries that we introduced for mechanical waves.

Suppose a transmission line having a characteristic impedance $Z\ (\Omega)$ and a load resistance $R\ (\Omega)$ is connected to an emf (dc) at $t = 0$ (Fig. 9.22). As we have

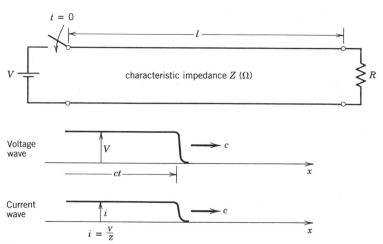

Fig. 9.22. A transmission line having a characteristic impedance Z and a load resistance R is suddenly connected to an emf V. Both voltage and current pulses start propagating. The emf supplies energy that is stored in the transmission line and also dissipated in the load.

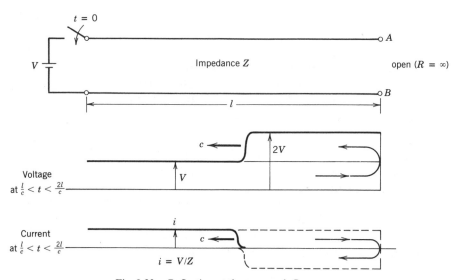

Fig. 9.23. Reflection at the open end, $R = \infty$.

seen before, both voltage and current pulses start propagating toward the load with a constant velocity c, which of course is determined by the medium. If the line is l m long, it takes the pulse l/c sec to reach the load. The question here is what would happen when and after the pulse front reaches the load? In Fig. 9.22 we consider two extreme cases $R = \infty$ (open) and $R = 0$ (short). If $R = Z$, then there should be no reflection at the load, and in this case the power transfer is most efficiently accomplished (impedance matching).

Open End $(R = \infty)$.

In the case of an open end, no current can flow from the point A to B since the resistance is infinitely large. That is, the current at $x = l$ must be zero at all times. For this to be possible we must have a negative current of equal amplitude propagating toward the emf, after the pulse front reaches the open end (Fig. 9.23).

The voltage, on the other hand, behaves in a completely different manner. At the open end, no voltage drop occurs, and the reflected voltage pulse must have the same polarity as before. Thus, after reaching the open end, a voltage of 2 V appears at the end.

This can be argued in terms of electric and magnetic fields associated with electromagnetic waves. Since the reflected pulse must have a Poynting vector directed *toward* the emf, the electric field (corresponding to the voltage) must have the same polarity as the incident pulse if the magnetic field (corresponding to the current) is reversed,

$$\mathbf{S}_i = \mathbf{E} \times \frac{\mathbf{B}}{\mu_0} \quad \text{directed to the right}$$

$$\mathbf{S}_r = \mathbf{E} \times \frac{(-\mathbf{B})}{\mu_0} \quad \text{directed to the left.}$$

Thus we may conclude that at the open end the reflected voltage wave has the same polarity as the incident wave, while the reflected current wave has an opposite polarity with respect to the incident current wave. (This statement actually holds as long as the end is terminated by a resistance R larger than the characteristic impedance Z.) Of course, all the incident energy is reflected back at the boundary, and we have no waves beyond $x = l$.

Closed or Shorted End $(R = 0)$.

In the case of a closed or shorted end (Fig. 9.24) the polarity of the voltage is reversed, but the polarity of the current remains unchanged, since the voltage at the shorted end must be zero at any time, and the Poynting vector of reflected wave must be directed toward the emf end.

Something seems wrong with this argument. If we short-circuit the terminals of a battery, a tremendously large current should flow, but Fig. 9.24 indicates that the current only doubles. How do we explain this?

To answer this we have to find out what happens when the reflected current pulse reaches the emf, after time $2l/c$. If the internal resistance of the emf is small, which is usually the case, the reflected current pulse again sees a closed end at the emf. Thus, after $t = 2l/c$, we have a current amplitude $3V/Z$ propagating toward the closed end. This multiple reflection process continues; the current amplitude builds up and eventually becomes dangerously large enough to destroy either the emf or the transmission line itself. This is the physics behind

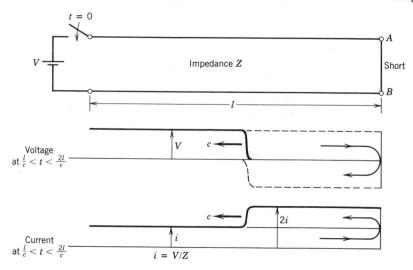

Fig. 9.24. Reflection at the closed end, $R=0$.

short-circuiting, which in any case should be avoided in the laboratory. Of course, the time required for the current buildup depends on the length of the transmission line, but you can easily see that the time scale is extremely short because of the fast propagation velocity, which is close to or equal to the speed of light.

In both cases ($R=\infty$ and 0) discussed above, the reflection is complete or 100%. No energy can go beyond the open or closed end except for small radiation that can be observed as leakage. For a finite load resistance other than the characteristic impedance, incomplete reflection occurs; that is, some energy is dissipated in the load, and the rest is reflected. For example, if $R=2Z$, the voltage reflection is $V/3$, and the current reflection is $-V/3Z$. In general, the voltage reflection coefficient is given by

$$\Gamma_V = \frac{R-Z}{R+Z} \tag{9.37}$$

and the current reflection coefficient is given by

$$\Gamma_I = \frac{Z-R}{R+Z} = -\Gamma_V. \tag{9.38}$$

Example 7.　Derive Eq. (9.37), the voltage reflection coefficient, by using the energy method we employed for the reflection coefficient of mechanical waves, Eq. (6.26).

Let the incident voltage be V_i and the reflected voltage be V_r. The power associated with the incident voltage is given by V_i^2/Z, and that associated with the reflected wave is V_r^2/Z. The resistor R dissipates energy at the rate $(V_i+V_r)^2/R$, where V_i+V_r is the voltage to appear at the terminated end. Then the power conservation principle requires that

$$\frac{V_i^2}{Z} = \frac{V_r^2}{Z} + \frac{(V_i + V_r)^2}{R}.$$

Solving for V_r, we obtain

$$V_r = \frac{R-Z}{R+Z} V_i,$$

which defines the voltage reflection coefficient

$$\Gamma_V = \frac{R-Z}{R+Z}.$$

Example 8. In Fig. 9.25 discuss how the voltage wave develops after the switch is closed.

When the switch is closed, a voltage pulse having an amplitude $10 \times 50/(50+50) = 5$ V starts propagating down the transmission line. (Note that the 50-Ω transmission line essentially acts as a 50-Ω resistor for the *initial* pulse, and the voltage dividing principle applies.) It takes the initial pulse $30m/c = 10^{-7}$ sec to reach the load end, where the voltage reflection coefficient is

$$\Gamma_V = \frac{20-50}{20+50} = -0.43.$$

Then, after 10^{-7} sec, a negative pulse of 5 V $\times (-0.43) = -2.1$ V propagates toward the emf end. When this negative pulse reaches the emf end, where impedance matching exists, it is completely absorbed by the 50-Ω resistor, and no more reflection occurs. Then a steady state is achieved at a transmission line voltage of $5 - 2.1 = 2.9$ V. This voltage is consistent with what we expect from the dc theory, which yields the resistor voltage of

$$10 \times \frac{20}{50+20} \text{ V} = 2.9 \text{ V}.$$

However, as we have seen, it takes a finite (although short) time to establish this dc, or steady state. The evolution of the voltage is shown in Fig. 9.26. (The evolution of the current wave is left as an exercise.)

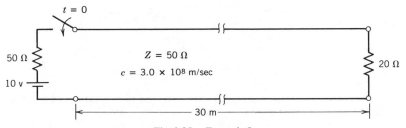

Fig. 9.25.   Example 8.

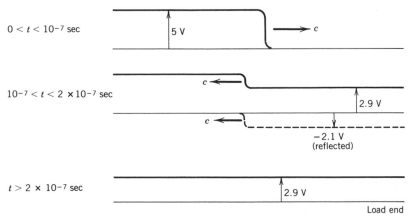

$0 < t < 10^{-7}$ sec

5 V

c

$10^{-7} < t < 2 \times 10^{-7}$ sec

c

2.9 V

c

−2.1 V
(reflected)

$t > 2 \times 10^{-7}$ sec

2.9 V

Load end

Fig. 9.26.   Example 8.

Since the characteristic impedance for plane electromagnetic waves is given by

$$ Z = \sqrt{\frac{\mu}{\varepsilon}}, $$

the waves are reflected whenever they enter a medium with different permittivity and/or permeability (Fig. 9.27). In most media we may assume that the permeability is unchanged, being equal to μ_0. However, ε, the permittivity, can easily vary. The velocity of visible light in water and glass is smaller than that in vacuum. (In water, $c/1.33$, and in glass $c/1.5$.) This is due to the larger values of permittivity in water and glass than ε_0. Thus the characteristic impedance of water and glass with respect to visible light is smaller than that in vacuum, 377 Ω. It is then obvious that the electric field associated with the reflected light has opposite polarity with respect to the incident light, since the reflection coefficient for the voltage (and thus the electric field) is negative. In other words, water (or glass) acts as a hard medium for the electric field associated with light.

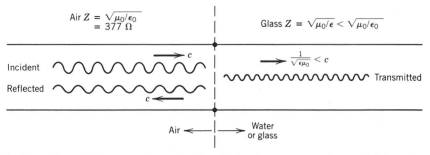

Air $Z = \sqrt{\mu_0/\epsilon_0}$
$= 377\ \Omega$

Glass $Z = \sqrt{\mu_0/\epsilon} < \sqrt{\mu_0/\epsilon_0}$

c

$\frac{1}{\sqrt{\epsilon\mu_0}} < c$

Incident

Reflected

Transmitted

c

Air

Water
or glass

Fig. 9.27.   Transmission line analogue of light reflection from water (or glass), which has a lower characteristic impedance than that in air.

This polarity change at a hard boundary is quite analogous to the case of mechanical waves, particularly the transverse waves on a string under tension. The polarity change of the electric field at hard boundaries plays important roles in Chapter 11 on interference.

9.7. Electromagnetic Waves in Matter

As we have seen, electromagnetic waves in vacuum are dispersionless and the propagation velocity c is independent of the wave frequency. This beautiful nondispersive nature breaks down for electromagnetic waves in matter. We have already seen that the velocity of electromagnetic waves in matter is different from $c = 3.0 \times 10^8$ m/sec. Usually, it becomes smaller than c, as in the case of the light velocity in glass, for example. We also learned that the change in the velocity causes the change in the characteristic impedance, which in turn causes the reflection of electromagnetic waves whenever they try to penetrate into a different medium.

Let us consider shortwave radio, which can be received at a place to which the waves cannot travel directly (along a straight line), or a point "out of sight." Shortwave radio actually uses the reflection of electromagnetic waves by the ionospheric plasma surrounding the earth, and by the surface of the earth itself, which is a good conductor (Fig. 9.28). Plasmas are ionized gases in which equal amounts of negative (electrons) and positive (ions) charges coexist, maintaining gross charge neutrality. The ionospheric plasma is produced by the radiation (X-ray, ultraviolet) from the sun and trapped by the earth's magnetic field. The plasma frequently becomes visible as aurora, depending on the solar activity.

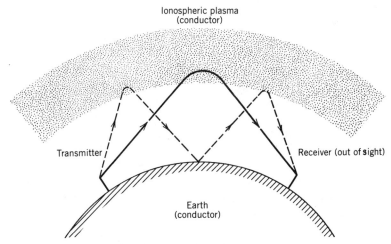

Fig. 9.28.   Reflection of radio waves from the ionospheric plasma. The earth is a good conductor too for radio waves and multiple reflection can occur. The plasma and the earth form a waveguide for suitable frequencies.

In plasmas, charged particles are highly mobile and can easily be accelerated by the electric field associated with the electromagnetic waves. The motion of charged particles (mainly electrons because their mass is much smaller than that of ions) create a conduction current, which as we know induces a magnetic field in addition to the magnetic field associated with the waves. The magnetic field thus induced in turn induces an electric field according to Faraday's law. As we will see, this electric field induced by the electron motion always tends to oppose the electric field of the incident wave. In other words, the incident wave would encounter difficulties whenever they try to penetrate into a plasma. Under certain conditions the wave just gives up and is reflected from the plasma.

Exactly the same argument can be applied to the reflection from a metal surface. Metals are characterized by the presence of large numbers of free electrons, or conduction electrons. For example, copper has about 10^{23} free electrons in 1 cm^3. Again these free electrons can well respond to an electric field or react to the externally applied field, thus preventing the field from penetrating.

As far as the wave equation for electromagnetic waves is concerned, all we have to do is just add another current in the Maxwell–Ampere law, which had only the displacement current density

$$\varepsilon_0 \frac{\partial E}{\partial t} \quad (A/m^2)$$

in vacuum. In the presence of a conduction current density, J (A/m^2), the total current density thus becomes

$$\varepsilon_0 \frac{\partial E}{\partial t} + J,$$

and the Maxwell–Ampere equation now becomes

$$-\frac{\partial B}{\partial x} = \mu_0 \left(\varepsilon_0 \frac{\partial E}{\partial t} + J \right). \tag{9.39}$$

Then if we can somehow express the conduction current J in terms of either E or B, our problem will be solved.

To do this let us consider how the electrons are accelerated by the electric field E. Since the electric field E exerts a force $-eE$ ($e = 1.6 \times 10^{-19}$ C) on an electron, the equation of motion for the electron can be written as

$$m \frac{\partial u}{\partial t} = -eE \quad (u = \text{electron velocity}). \tag{9.40}$$

We used the partial derivative $\partial/\partial t$ since we expect that the electrons would follow the wave motion of the electric field, which is a function of both the spatial coordinate x and time t. If we multiply Eq. (9.40) by the electron density n_0 (m^{-3}) and the charge of the electron $-e$, we obtain

$$m \frac{\partial J}{\partial t} = n_0 e^2 E, \tag{9.41}$$

since the conduction current density J is given by

$$J = -n_0 eu \quad (A/m^2).$$

On the other hand, the time derivative of Eq. (9.39) is

$$-\frac{\partial^2 B}{\partial t\, \partial x} = \mu_0 \left(\varepsilon_0 \frac{\partial^2 E}{\partial t^2} + \frac{\partial J}{\partial t} \right). \tag{9.42}$$

Substituting $\partial J/\partial t$ from Eq. (9.41) into Eq. (9.42), we find

$$-\frac{\partial^2 B}{\partial t\, \partial x} = \mu_0 \varepsilon_0 \left(\frac{\partial^2 E}{\partial t^2} + \frac{n_0 e^2}{m\varepsilon_0} E \right). \tag{9.43}$$

The Faraday's induction equation is unchanged,

$$-\frac{\partial E}{\partial x} = \frac{\partial B}{\partial t}, \tag{9.44}$$

or taking the spatial derivative,

$$-\frac{\partial^2 E}{\partial x^2} = \frac{\partial^2 B}{\partial x\, \partial t}. \tag{9.45}$$

Eliminating $\partial^2 B/\partial x\, \partial t$ between Eqs. (9.43) and (9.45), we finally obtain

$$\frac{\partial^2 E}{\partial x^2} = \mu_0 \varepsilon_0 \left(\frac{\partial^2 E}{\partial t^2} + \frac{n_0 e^2}{m\varepsilon_0} E \right). \tag{9.46}$$

This is our desired differential equation for electromagnetic waves in a plasma. Of course, it reduces to the wave equation we derived before if we have no plasma electrons, $n_0 = 0$. Also note that in the equation of motion, we have assumed free acceleration of electrons, or neglected collisions between electrons and other particles in the plasma. We still call Eq. (9.46) a wave equation although it has one additional term due to the conduction current.

To see that the term due to the plasma electrons indeed tends to reduce the total electric field, let us assume a harmonic wave,

$$E(x, t) = E_0 \sin (kx - \omega t).$$

Obviously,

$$\frac{\partial^2 E}{\partial t^2} = -\omega^2 E_0 \sin (kx - \omega t).$$

But the last term in Eq. (9.46) always has an opposite sign with respect to the term $\partial^2 E/\partial t^2$, since the quantity $n_0 e^2/m\varepsilon_0$ is positive definite. Thus the reaction effect of the plasma electrons is clearly seen. Plasma electrons tend to reduce the effective amplitude of the electric field, and wave propagation from vacuum into a plasma must encounter some difficulty.

The quantity

$$\frac{n_0 e^2}{m\varepsilon_0}$$

has the dimension of sec^{-2}, as can easily be checked. (Do this.) The square root of this is called the plasma (angular) frequency and is written as

$$\omega_p = \left(\frac{n_0 e^2}{m\varepsilon_0}\right)^{1/2} \quad \text{(rad/sec)}. \tag{9.47}$$

The corresponding plasma frequency in hertz is

$$\nu_p = \frac{1}{2\pi}\left(\frac{n_0 e^2}{m\varepsilon_0}\right)^{1/2} \quad \text{(Hz)} \tag{9.48}$$

$$= 8.97\sqrt{n_0} \quad \text{(Hz)} \tag{9.49}$$

where the electron density n_0 is in m^{-3}. The ionospheric plasma has the electron density in the range of $10^{11} \sim 10^{13}$ m^{-3}. The corresponding plasma frequency range is $2.8 \sim 28$ MHz. In metals the electron density is much higher and of the order of 10^{29} m^{-3}. The plasma frequency of metals is then around 3×10^{15} Hz, which is higher than the frequency of visible light, 5×10^{14} Hz.

Let us now find the dispersion relation, or the relationship between ω and k of electromagnetic waves in a plasma. For a harmonic wave of the form

$$E(x, t) = E_0 \sin (kx - \omega t),$$

Eq. (9.46) yields (Problem 18)

$$\omega^2 = k^2 c^2 + \omega_p^2. \tag{9.50}$$

A rough sketch of ω as a function of k is shown in Fig. 9.29. Note that we now have a minimum frequency of electromagnetic waves that can exist in a plasma.

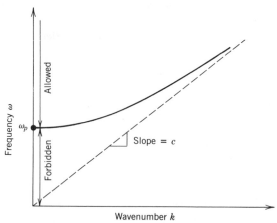

Fig. 9.29. Dispersion relation for electromagnetic waves in a plasma.

The cutoff frequency is given by the plasma frequency. We have no solutions for ω below ω_p; that is, waves with a frequency below ω_p cannot exist in a plasma. Now we see why a plasma can reflect certain electromagnetic waves. Waves of a frequency well above the plasma frequency have no difficulties in penetrating into a plasma, but those with a frequency below the plasma frequency must be completely reflected at the boundary. Visible light has a frequency around 5×10^{14} Hz, which is much higher than the plasma frequency of the ionospheric plasma. Thus light should have no difficulties in penetrating through the ionospheric plasma. On the other hand, shortwave radio communication uses the frequency band between 3 and 30 MHz, which is subject to reflection by the ionosphere (Fig. 9.30).

The dispersion relation we obtained has a peculiar property. The phase velocity ω/k becomes

$$\frac{\omega}{k} = \frac{\sqrt{k^2 c^2 + \omega_p^2}}{k} = c \sqrt{1 + \frac{\omega_p^2}{c^2 k^2}}, \tag{9.51}$$

which is faster than the speed of light, c. Is this not in grave contradiction to what Einstein found? According to him nothing can travel faster than light. What travels with the electromagnetic waves? Energy does. But we know that energy is carried at a group velocity $d\omega/dk$, rather than the phase velocity. The group velocity for $\omega = \sqrt{k^2 c^2 + \omega_p^2}$ is

$$\frac{d\omega}{dk} = \frac{kc^2}{\sqrt{k^2 c^2 + \omega_p^2}} = \frac{kc^2}{\omega}, \tag{9.52}$$

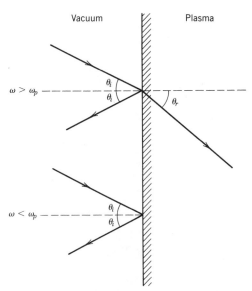

Fig. 9.30.   Wave reflection from a plasma. Note that the refraction angle θ_r is *larger* than the incident angle θ_i for a plasma, in contrast to the refraction of visible light at the surface of water or glass. This means that we cannot make a converging lens using a plasma.

which can never exceed the velocity of light, c. Thus the dispersion relation does not contradict Einstein's relativity theory.

Example 9. The dispersion relation Eq. (9.50) can be used to measure the average plasma electron density. Let a high-frequency (microwave) electromagnetic wave whose frequency is higher than the plasma frequency go through a plasma slab having an electron density n and a width d (Fig. 9.31). In the absence of the plasma, the phase change due to the propagation over the distance d is $k_0 d = (\omega/c)d$, where $k_0 = \omega/c$.

(a) What is the phase change in the presence of the plasma?

(b) Show that when $\omega^2 \gg \omega_p$, the difference between $(\omega/c)d$ and the phase change found in (a) become proportional to the electron density.

(a) From Eq. (9.50), we obtain

$$k = \frac{1}{c}\sqrt{\omega^2 - \omega_p^2}$$

Then

$$kd = \frac{d}{c}\sqrt{\omega^2 - \omega_p^2} = \frac{d\omega}{c}\sqrt{1 - (\omega_p/\omega)^2}.$$

(b) If $\omega^2 \gg \omega_p^2$, we may expand $\sqrt{1 - (\omega_p/\omega)^2}$ as

$$\sqrt{1 - \left(\frac{\omega_p}{\omega}\right)^2} \simeq 1 - \frac{1}{2}\left(\frac{\omega_p}{\omega}\right)^2 \quad \text{(binomial expansion)}.$$

Then

$$kd \simeq \frac{d\omega}{c}\left[1 - \frac{1}{2}\left(\frac{\omega_p}{\omega}\right)^2\right].$$

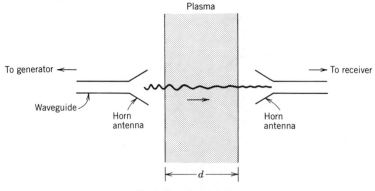

Fig. 9.31. Example 9.

The difference between $k_0 d$ and kd becomes

$$k_0 d - kd = \frac{1}{2} \frac{d}{c} \frac{\omega_p^2}{\omega}$$

in which d, c, and ω are fixed parameters. Then the phase difference is proportional to ω_p^2, or the electron density. (This phase difference can be measured electronically, thus enabling us to determine the plasma electron density. This diagnostic technique is widely used in plasma research.)

We saw that electromagnetic fields cannot penetrate into metals. Then how can we use, say, copper wires as electrical conductors at all? For the free electrons in metals to move, we must have an electric field in metals. The dispersion relation we derived tells us no electromagnetic fields with frequencies below the plasma frequency of metals can exist in metals. For copper, $f_p \simeq 3 \times 10^{15}$ Hz, but we frequently use copper wires for transmitting 60-Hz (or 50-Hz) electromagnetic waves (commercial electrical power) from power plants to wherever it is used. When we state that electromagnetic waves cannot penetrate into metals, we should have in mind perfectly conducting metals, or ideal conductors with zero resistivity. Although such conductors have been realized as superconductors, their commercial availability is still remote, and we will have to use conventional conductors for decades to come. Copper at room temperature has a small but finite resistivity, as we all know. Its value is 1.7×10^{-8} $\Omega \cdot$m at 20°C and is one of the smallest among conventional metals. If the resistivity is finite, a dc electric field can fully penetrate into a metal. The electric field E_{dc} is related to the current density through

$$E_{dc} = \eta J_{dc}. \qquad (9.53)$$

(You should check that this equation is dimensionally correct.) This is the microscopic form of Ohm's law. If we multiply Eq. (9.53) by the length l of a rod, we find (Fig. 9.32)

$$E_{dc} l = V_{dc} = \eta l J_{dc}$$

$$= \frac{\eta l}{A} A J_{dc}$$

$$= R I_{dc},$$

which is the macroscopic form of the well-known Ohm's law.

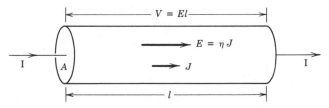

Fig. 9.32.   Current density J and electric field E in a resistive rod.

Now we will see what happens to electromagnetic waves in a metal if it has a finite resistivity. Equation (9.41) was obviously for the case of infinite conductivity or zero resistivity. But we now know that in the case of a dc field, or when $\partial J/\partial t = 0$ ($J \neq 0$!) Eq. (9.53) must hold. Then, Eq. (9.41) should be modified as

$$m\frac{\partial J}{\partial t} = n_0 e^2 [E - \eta J]. \tag{9.54}$$

This equation is also derivable directly from the equation of motion in which we assume a finite collision frequency for the electrons,

$$m\frac{\partial u}{\partial t} = -eE - mv_c u,$$

where v_c ($\sec^{-1}$) is a measure of how frequently electrons collide with either ions or neutrals to lose momentum. Multiplying this by $-n_0 e$, we obtain

$$m\frac{\partial J}{\partial t} = n_0 e^2 E - mv_c J$$

$$= n_0 e^2 \left[E - \frac{mv_c}{n_0 e^2} J \right]$$

which, when compared with Eq. (9.54), indicates that the resistivity η is given by

$$\eta = \frac{mv_c}{n_0 e^2} \quad (\Omega \cdot m).$$

Therefore the resistivity is a direct consequence of electron collisions. For copper, $\eta = 1.7 \times 10^{-8}\ \Omega \cdot m$, and $n_0 \simeq 10^{29}\ m^{-3}$. Substituting $m = 9.1 \times 10^{-31}$ kg and $e = 1.6 \times 10^{-19}$ coulombs, we estimate the collision frequency of electrons in copper as

$$v_c = \frac{\eta n_0 e^2}{m} = \frac{1.7 \times 10^{-8} \times 10^{29} \times (1.6 \times 10^{-19})^2}{9.1 \times 10^{-31}}\ \sec^{-1}$$

$$= 4.8 \times 10^{13}\ \sec^{-1}$$

This collision frequency in metals usually increases with temperature, and this explains why metal resistivity increases with temperature.

In the limit of no time variation ($\partial/\partial t = 0$, or $\omega = 0$) in Eq. (9.54), we indeed recover $E = \eta J$ as required by Eq. (9.53). Thus our equations for three field quantities, J, E, and B, are Eq. (9.54) and

$$-\frac{\partial E}{\partial x} = \frac{\partial B}{\partial t} \tag{9.44}$$

$$-\frac{\partial B}{\partial x} = \mu_0 \left(\varepsilon_0 \frac{\partial E}{\partial t} + J \right). \tag{9.39}$$

As before, to derive a wave equation for one field quantity, say J, we have to eliminate E and B from these three equations.

The magnetic field B can easily be eliminated as before:

$$\frac{\partial^2 E}{\partial x^2} = \mu_0 \left(\varepsilon_0 \frac{\partial^2 E}{\partial t^2} + \frac{\partial J}{\partial t} \right). \tag{9.55}$$

From Eq. (9.54), we have

$$E = \frac{m}{n_0 e^2} \frac{\partial J}{\partial t} + \eta J. \tag{9.56}$$

Substituting Eq. (9.56) into (9.55), we obtain

$$\frac{\partial^2}{\partial x^2} \left(\frac{m}{n_0 e^2} \frac{\partial J}{\partial t} + \eta J \right) = \varepsilon_0 \mu_0 \frac{\partial^2}{\partial t^2} \left(\frac{m}{n_0 e^2} \frac{\partial J}{\partial t} + \eta J \right) + \mu_0 \frac{\partial J}{\partial t}. \tag{9.57}$$

This looks terrible but don't worry. The term $m/n_0 e^2 (\partial J/\partial t)$ can safely be neglected compared with the term ηJ under most practical conditions. Assuming a harmonic current form, $J(x, t) = J_0 \sin(kx - \omega t)$, we see that the amplitude of $m/ne^2 (\partial J/\partial t)$ is $(m\omega/n_0 e^2)J_0$ which is to be compared with ηJ_0. For copper, we have $n_0 \simeq 1 \times 10^{29}$ m^{-3} and $\eta = 1.7 \times 10^{-8}$ $\Omega \cdot$m. Even in the frequency range of microwaves, $\omega = 10^{10} - 10^{12}$ rad/sec, $m\omega/n_0 e^2$ is

$$\frac{9.1 \times 10^{-31} \times 10^{12}}{1 \times 10^{29} \times (1.6 \times 10^{-19})^2} = 3.6 \times 10^{-10} \ \Omega \cdot \text{m},$$

which is much less than the resistivity of copper. Thus the term $(m/n_0 e^2) \times \partial J/\partial t$ can be neglected compared with ηJ, in Eq. (9.57), and we have

$$\frac{\partial^2 J}{\partial x^2} = \varepsilon_0 \mu_0 \frac{\partial^2 J}{\partial t^2} + \frac{\mu_0}{\eta} \frac{\partial J}{\partial t}. \tag{9.58}$$

Next compare the two terms in the RHS. The amplitude of the first term is $\varepsilon_0 \mu_0 \omega^2 J_0$ and that of the second term is $\mu_0 \omega J_0/\eta$. We see that the first term can safely be neglected even in the microwave frequency range, $\omega \simeq 10^{12}$ sec^{-1}. Thus the original differential equation, Eq. (9.57), has now been greatly simplified as

$$\frac{\partial^2 J}{\partial x^2} = \frac{\mu_0}{\eta} \frac{\partial J}{\partial t}. \tag{9.59}$$

Of course, the other fields, E and B, should be described by the same differential equation,

$$\frac{\partial^2 E}{\partial x^2} = \frac{\mu_0}{\eta} \frac{\partial E}{\partial t} \quad \text{and} \quad \frac{\partial^2 B}{\partial x^2} = \frac{\mu_0}{\eta} \frac{\partial B}{\partial t}. \tag{9.60}$$

A differential equation of the preceding form is called a *diffusion equation*. It describes, for example, how rapidly an ink drop placed in water spreads (or diffuses) and is one of the most important equations we have in physics. We shall study this equation later in Chapter 15.

The solution of Eqs. (9.59) and (9.60) can be found for a harmonic "wave" without too much difficulty. The term *wave* is used here in a somewhat broader sense than before. The equations contain the first-order time derivatives only, and do not have the second-order time derivatives present in the usual wave equations

$$\frac{\partial^2 E}{\partial t^2} = c_w^2 \frac{\partial^2 E}{\partial x^2}.$$

Let us assume a solution of the form

$$E(x, t) = A(x) \sin (kx - \omega t), \tag{9.61}$$

for the differential equation for the electric field,

$$\frac{\partial^2 E}{\partial x^2} = \frac{\mu_0}{\eta} \frac{\partial E}{\partial t}. \tag{9.62}$$

Since

$$\frac{\partial E}{\partial x} = \frac{dA}{dx} \sin (kx - \omega t) + kA \cos (kx - \omega t)$$

$$\frac{\partial^2 E}{\partial x^2} = \frac{d^2 A}{dx^2} \sin (kx - \omega t) + 2k \frac{dA}{dx} \cos (kx - \omega t) - k^2 A \sin (kx - \omega t)$$

and

$$\frac{\partial E}{\partial t} = -\omega A \cos (kx - \omega t),$$

Eq. (9.62) becomes

$$\left(\frac{d^2 A}{dx^2} - k^2 A\right) \sin (kx - \omega t) + \left(2k \frac{dA}{dx} + \frac{\omega \mu_0}{\eta} A\right) \cos (kx - \omega t) = 0 \tag{9.63}$$

Since Eq. (9.63) must hold for any values of x and t, we must have

$$\frac{d^2 A}{dx^2} - k^2 A = 0 \tag{9.64}$$

and

$$2k \frac{dA}{dx} + \frac{\omega \mu_0}{\eta} A = 0. \tag{9.65}$$

These are mere ordinary differential equations. We assume $A(x) = A_0 e^{-\gamma x}$. Then

$$\frac{dA}{dx} = -\gamma A_0 e^{-\gamma x}$$

and

$$\frac{d^2 A}{dx^2} = \gamma^2 A_0 e^{-\gamma x}.$$

Substituting these into Eqs. (9.64) and (9.75), we find

$$\gamma^2 = k^2 \tag{9.66}$$

and

$$-2\gamma k + \frac{\omega\mu_0}{\eta} = 0. \tag{9.67}$$

Then

$$\gamma = k = \sqrt{\frac{\omega\mu_0}{2\eta}},$$

and the electric field becomes

$$E(x, t) = A_0 e^{-kx} \sin(kx - \omega t), \tag{9.68}$$

where

$$k = \sqrt{\frac{\omega\mu_0}{2\eta}}. \tag{9.69}$$

A sketch of the preceding solution at a certain time (snapshot) is shown in Fig. 9.33. We see that the electric field in the metal spatially damps in an exponential manner. The envelope becomes $1/e$ $(e = 2.7182) = 0.37 = 37\%$ of the value at the metal surface $(x = 0)$ at a distance

$$x_0 = \sqrt{\frac{2\eta}{\omega\mu_0}}. \tag{9.70}$$

This quantity, x_0, is called the *skin* depth and indicates how deep an electromagnetic wave can penetrate into a metal. In the limit of $\omega \to 0$ (dc), x_0 becomes infinitely large, being consistent with our experience. A dc field does not have any difficulty in penetrating into a metal, although it takes a finite time (skin time) for a dc field to penetrate. As ω increases, the skin depth becomes finite, being inversely proportional to the square root of the frequency. Even at 60 Hz, the skin depth of copper is only 8.5 mm. Thus it is meaningless to make the diameter of high-voltage transmission lines much more than the skin depth, since the current is practically limited within a skin depth. As the frequency becomes higher, the skin depth becomes smaller (Fig. 9.34). At 1 MHz (AM radio waves), the skin depth is 0.07 mm; that is, the current can flow practically only at the surface of copper.

You must wonder why a higher resistivity results as a less spatial damping, as indicated by Eq. (9.70). This is quite opposite to the case of an *LCR* circuit, in which a higher resistance (R) makes the oscillation damp faster (Fig. 9.35). Let us see if we can construct an equation similar to Eq. (9.62) using the preceding circuit as a unit element of a transmission line. As before, Kirchhoff's voltage and current theorems yield (Fig. 9.36)

$$V(x) = V(x + \Delta x) + L\frac{\partial i}{\partial t} + Ri$$

$$i(x) = i(x + \Delta x) + C\frac{\partial V}{\partial t}$$

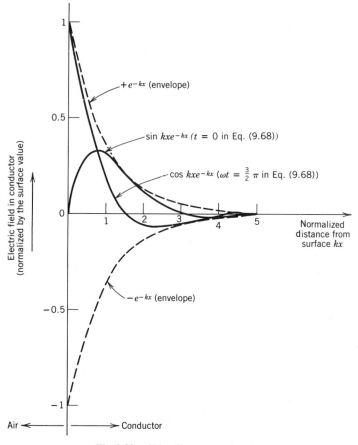

Fig. 9.33. Skin effect at metal surface.

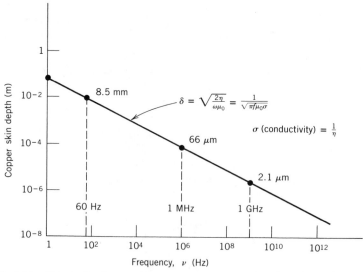

Fig. 9.34. Skin depth of copper at room temperature as a function of wave frequency.

178

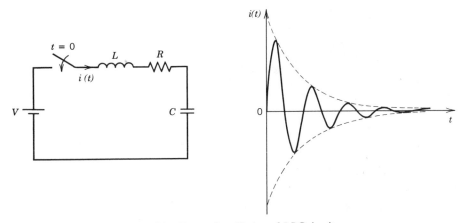

Fig. 9.35.   Damped oscillation of *LRC* circuit.

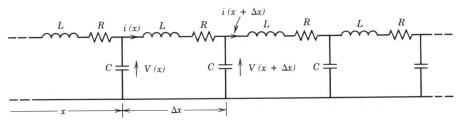

Fig. 9.36.   An incorrect model transmission line for simulating skin effect.

or

$$-\Delta x \frac{\partial V}{\partial x} = L \frac{\partial i}{\partial t} + Ri \qquad (9.71)$$

$$-\Delta x \frac{\partial i}{\partial x} = C \frac{\partial V}{\partial t} . \qquad (9.72)$$

From these, we find

$$\frac{\partial^2 V}{\partial x^2} = \frac{C}{(\Delta x)^2} \left(L \frac{\partial^2 V}{\partial t^2} + R \frac{\partial V}{\partial t} \right), \qquad (9.73)$$

which is clearly of the form of Eq. (9.20), in which we neglected the term

$$\varepsilon_0 \mu_0 \frac{\partial^2 J}{\partial t^2}$$

compared with $\mu_0/\eta(\partial J/\partial t)$. In other words, we neglected the term associated with the capacitance ε_0 (F/m). In Eq. (9.73), we cannot do this, since if we neglect the capacitance ($C \to 0$), we lose all the RHS, ending up with a trivial equation

$$\frac{\partial^2 V}{\partial x^2} = 0.$$

Thus our model is obviously wrong.

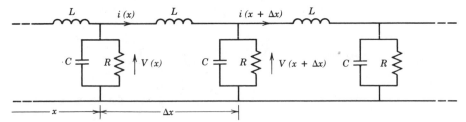

Fig. 9.37.   Correct model: Resistance in parallel with capacitance.

A correct model is the one that has a resistance in parallel with the capacitance. For the transmission line (see Fig. 9.37), we have

$$-\Delta x \frac{\partial V}{\partial x} = L \frac{\partial i}{\partial t} \tag{9.74}$$

$$-\Delta x \frac{\partial i}{\partial x} = \frac{V}{R} + C \frac{\partial V}{\partial t}. \tag{9.75}$$

These yield the following differential equation for the current i (derive this):

$$\frac{\partial^2 i}{\partial x^2} = \frac{L}{(\Delta x)^2} \left(C \frac{\partial^2 i}{\partial t^2} + \frac{1}{R} \frac{\partial i}{\partial t} \right). \tag{9.76}$$

This looks better, since we now have the capacitance term separated from the resistance term. In fact, if we define $L/\Delta x$ (inductance per unit length), $C/\Delta x$ (capacitance per unit length), and $1/R \, \Delta x$ (conductance per unit length), and replace $L/\Delta x$ by μ_0, $C/\Delta x$ by ε_0, and $1/R \, \Delta x$ by $1/\eta$, Eq. (9.38) exactly corresponds to Eq. (9.20). In the transmission line shown (Fig. 9.37), it is obvious that as the resistance increases, the line dissipates less energy (since the resistance is in parallel!) and the energy can be transferred further spatially, indicating less spatial damping!

Problems

1. An LC transmission line has the following parameters:

$$\frac{L}{\Delta x} = 1.0 \times 10^{-4} \text{ H/m}, \quad \frac{C}{\Delta x} = 20 \times 10^{-12} \text{ F/m}.$$

Find the velocity of electromagnetic waves on the transmission line. What is the impedance?

2. It is required that the velocity of electromagnetic waves in a coaxial cable be one-half of the velocity of light in vacuum. What dielectric material should be used? Assume $\mu = \mu_0$.

3. Consider the inductance of an LC transmission line having a leakage (or stray) capacitance C_2 (see Fig. 9.38).

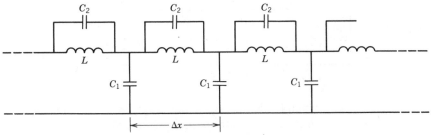

Fig. 9.38. Problem 3.

(a) Show that the differential equation for the voltage is given by

$$\frac{\partial^2 V}{\partial t^2} = \frac{(\Delta x)^2}{LC_1}\frac{\partial^2 V}{\partial x^2} + \frac{C_2}{C_1}(\Delta x)^2 \frac{\partial^4 V}{\partial t^2 \partial x^2}$$

(b) Assuming a harmonic voltage $V(x, t) = V_0 \sin(kx - \omega t)$, show that the dispersion relation is

$$\omega^2 = \frac{1}{1 + (C_2/C_1)(\Delta x)^2 k^2}\frac{(\Delta x)^2}{LC_1} k^2.$$

Note: This indicates that if there is a stray capacitance across the inductance in each section, the waves are no longer dispersionless. The phase velocity ω/k is now different from the group velocity $d\omega/dk$, as can easily be checked. (Do this.) The dispersion relation can model interesting waves we encounter in several branches of physics. For example, shallow water waves can be well approximated by this dispersion relation. Another example is the sound wave in an ionized gas (plasma), which is called the ion acoustic wave. (We will study this in Chapter 15).

4. A coaxial cable has an inner radius of 2×10^{-4} m and an outer radius of 3×10^{-3} m and is filled with a dielectric material of $\kappa = 2.0$. Find

(a) The velocity of electromagnetic waves in the cable.

(b) The characteristic impedance of the cable.

(Answer: 2.1×10^8 m/sec, 115 Ω.)

5. A coaxial cable is desired to have 50-Ω characteristic impedance. If the radius of the outer conductor is 2 mm, and the propagation velocity is to be $0.7c$, what dielectric material should be used? What is the inner radius? (Answer: $\varepsilon = 2.04\varepsilon_0$, 0.6 mm.)

6. From the dimensions for E and B, show that the Poynting flux S indeed has the dimension of watts per square meter. (Introduce a vector $H = B/\mu_0$ having a dimension of amperes per meter.)

7. Parallel-wire transmission lines are most commonly used for low-frequency (including dc) power transfer. In the schematic diagram of Fig. 9.39, sketch roughly the electric and magnetic field profiles, and show that the Poynting vector is directed from the emf to the load everywhere.

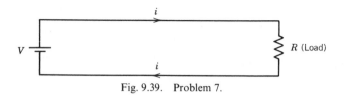

Fig. 9.39.   Problem 7.

8. A long densely wound solenoid is storing magnetic energy. Discuss the mechanism of energy flow in terms of the Poynting vector.

9. In Eq. (9.31), the electric field is in the y direction and the magnetic field is in the z direction. Show that Eq. (9.31) is equivalent to

$$\begin{vmatrix} \mathbf{i} & \mathbf{j} & \mathbf{k} \\ \partial/\partial x & \partial/\partial y & \partial/\partial z \\ 0 & E & 0 \end{vmatrix} = -\frac{\partial B}{\partial t}\mathbf{k}$$

if we assume $\partial E/\partial z = 0$. Similarly, Eq. (9.32) may be written as

$$\begin{vmatrix} \mathbf{i} & \mathbf{j} & \mathbf{k} \\ \partial/\partial x & \partial/\partial y & \partial/\partial z \\ 0 & 0 & B \end{vmatrix} = \varepsilon_0\mu_0\frac{\partial E}{\partial t}\mathbf{j}$$

with $\partial B/\partial y = 0$. In general,

$$\begin{vmatrix} \mathbf{i} & \mathbf{j} & \mathbf{k} \\ \partial/\partial x & \partial/\partial y & \partial/\partial z \\ A_x & A_y & A_z \end{vmatrix} = \left(\frac{\partial A_z}{\partial y} - \frac{\partial A_y}{\partial z}\right)\mathbf{i} + \left(\frac{\partial A_x}{\partial z} - \frac{\partial A_z}{\partial x}\right)\mathbf{j} + \left(\frac{\partial A_y}{\partial x} - \frac{\partial A_x}{\partial y}\right)\mathbf{k},$$

and this vector differentiation is written as

$$\mathbf{\nabla} \times \mathbf{A} \text{ (or curl } \mathbf{A}).$$

Using $\mathbf{\nabla}$, we may generalize Eq. (9.31) as

$$\mathbf{\nabla} \times \mathbf{E} = -\frac{\partial \mathbf{B}}{\partial t}$$

and Eq. (9.32) as

$$\mathbf{\nabla} \times \mathbf{B} = \varepsilon_0\mu_0\frac{\partial \mathbf{E}}{\partial t}.$$

These are called Maxwell's equations in free space in which no conduction currents can exist. In the presence of a conduction current, the second equation should be generalized as

$$\mathbf{\nabla} \times \mathbf{B} = \mu_0\left(\varepsilon_0\frac{\partial \mathbf{E}}{\partial t} + \mathbf{j}\right),$$

where $\mathbf{j}$ is the conduction current density (A/m²).

10. A giant laser pulse has a power density 10^{20} W/m². Calculate the rms value of the electric field associated with the laser pulse.
(*Answer:* $E_{rms} = 1.94 \times 10^{11}$ V/m.)

11. A radio station is emitting 50 kW radio waves spherically.* Find the rms value of the electric field 1 mi from the station.
(*Answer:* $E_{rms} = 0.76$ V/m.)

12. A 1-g target completely absorbs the energy of laser pulse (500 MW, 10-nsec duration). Find the momentum to be gained by the target and the velocity.
(*Answer:* $p = 1.67 \times 10^{-8}$ kg·m/sec, $v = 1.67 \times 10^{-5}$ m/sec.)

13. Discuss how the current pulse develops after the switch is closed for $R = 25 \, \Omega$, 50 Ω, 100 Ω (Fig. 9.40). Note that there is no reflection at the emf end. Does the final current reduce to what you expect?

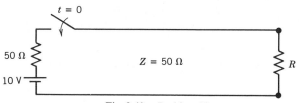

Fig. 9.40. Problem 13.

14. Repeat Problem 13 for the case of Fig. 9.41. The current should eventually approach 1 A. Does it?

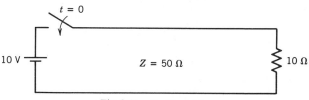

Fig. 9.41. Problem 14.

15. Derive Eq. (9.38).

16. A coaxial cable has the same characteristic impedance as free space, 377 Ω. Can we conclude that electromagnetic waves reaching an open end of the cable should suffer no reflection since the impedances are the same? (The answer is no. Why?)

17. Derive Eq. (9.50).

---

*As we will see in Chapter 10, a radio station cannot radiate spherically or in every direction.

18. The dispersion relation of electromagnetic waves in a plasma [Eq. (9.50)] does not allow a solution for ω below the plasma frequency ω_p. Suppose an electromagnetic wave of a frequency much less than ω_p is incident on a plasma. Neglecting the term $\partial^2 E/\partial t^2$ compared with $\omega_p^2 E$ in Eq. (9.8), show that the solution for E can be written as

$$E = E_0 e^{-\gamma x} \sin \omega t,$$

where $\gamma = \omega_p/c$. The quantity c/ω_{pe} is called the skin depth of a plasma with no resistivity. Evaluate this quantity for the ionospheric plasma assuming $n_0 = 10^{12}\ \mathrm{m}^{-3}$.

19. Discuss the tunnel effect of electromagnetic waves through a plasma (Fig. 9.42). If the plasma is semi-infinite, or if the thickness d is much larger than the skin depth c/ω_p found in Problem 18, the incident wave is 100% reflected. However, if the thickness d is shorter than the skin depth, some energy can go through the plasma, and the reflection is no longer 100%. Can you see why?

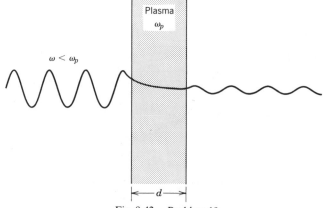

Fig. 9.42. Problem 19.

20. Show that the energy density of a plane electromagnetic wave in a plasma with amplitude of electric field E_0 is given by

$$\tfrac{1}{4}\varepsilon_0 E_0^2 \left(1 + \frac{\omega_p^2}{\omega^2} + \frac{\omega^2 - \omega_p^2}{\omega^2} \right) = \tfrac{1}{2}\varepsilon_0 E_0^2 \text{ rms.} \qquad (A)$$

Equation (A) indicates the energy partition among the electric field energy, kinetic energy of electrons, and magnetic field energy.

21. Any waves tend to be bent toward a region of lower phase velocity. Usually, the electron density in the ionospheric plasma increases with the altitude. Consider an oblique incident wave. See why a wave with a frequency higher than the plasma frequency can be reflected back toward the earth by the ionosphere (Fig. 9.43).

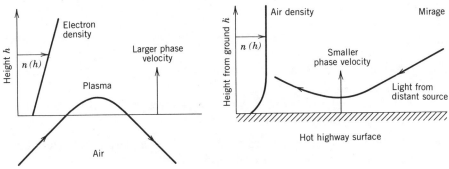

Fig. 9.43. Problem 21.

Note: This bending phenomenon is known as wave refraction. In fact, shortwave radio with frequencies higher than the plasma frequency of the ionosphere can be effectively reflected. The cutoff phenomenon at the plasma frequency can be well defined only for normal incidence. Of course, the refraction depends on the wave frequency, and for waves with frequencies much higher than ω_p, the refraction becomes ineffective. Visible light can go through the ionosphere at any angle of incidence.

A mirage is caused by the same mechanism. (On a hot summer day the highway surface often becomes a perfect mirror, as you must have experienced.) The speed of light in air is slightly less than in vacuum, or the speed of light in air decreases as the air density increases. Any gas becomes less dense as it is heated. Thus the air density at the surface of a hot highway is less than that above where the air temperature is lower. Then the speed of light decreases with the height and light is refracted toward the region of slower velocity.

22. Derive Eqs. (9.73) and (9.76).

23. There is a simple reason for the LCR line in Fig. 9.36 to be incorrect and that in Fig. 9.37 to be correct. Explain why in terms of the transverse nature of electromagnetic waves.

24. Show that the electromagnetic waves in a plasma can be modeled by the following transmission line if L, C_0, and C_1 are chosen such that (see

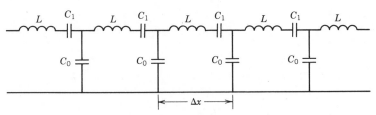

Fig. 9.44. Problem 24.

Fig. 9.44)

$$\frac{(\Delta x)^2}{LC_0} = c^2, \qquad \omega_p^2 = \frac{1}{LC_1}.$$

25. Find the wavelength of electromagnetic waves at which copper becomes transparent for the waves.

    (*Answer:* $\lambda < 3 \times 10^{-10}$ m. X rays have a wavelength shorter than this and can penetrate into copper.)

26. A beam of electromagnetic waves is incident on a plasma lens. Does the lens act as a converging lens or diverging lens?

27. Using the diffusion equation for a resistive medium

$$\frac{\partial^2 E}{\partial x^2} = \frac{\mu_0}{\eta} \frac{\partial E}{\partial t},$$

    estimate the time required for a dc electric field to penetrate fully into a copper slab 2 cm thick. Copper resistivity is $1.7 \times 10^{-8}\, \Omega \cdot$m.

    (*Answer:* For a thickness d, penetration time $\simeq \mu_0 d^2/\eta \simeq 30$ msec. This is the exponentiation time and should be regarded as an order of magnitude estimate.)

28. The resistivity of the earth falls in the range 10^{-2}–$10^2\, \Omega \cdot$m. Assuming $\eta = 1\, \Omega \cdot$m, evaluate the earth skin depth as a function of frequency. (If the skin depth is much shorter than the wavelength in free space, the earth can be regarded as a good conductor and becomes an effective reflector for electromagnetic waves. For example, an antenna erected from the ground with a height h is effectively $2h$ long because of wave reflection, which is responsible for creation of an image of a conductor above the ground.)

CHAPTER 10

Radiation of Electromagnetic Waves

10.1. Introduction

So far we have been discussing how electromagnetic waves behave in some media once they are created, without investigating how they can be created. Radio and TV waves are transmitted by antennas (and received by antennas, too). In this chapter we learn physical mechanisms behind radiation.

10.2. Fields Associated with Stationary Charge and Charge Moving with a Constant Velocity

Suppose we have a point radiation source emitting electromagnetic waves spherically or isotropically in every direction. We saw in the section on Poynting vector that for the point source the Poynting flux is proportional to r^{-2} ($r=$ distance from the source), which of course indicates that the total power is conserved,

$$4\pi r^2 S = \text{power radiated by the source (const)}, \qquad (10.1)$$

unless the medium is absorbing energy. Since the Poynting flux was given by

$$S = c\varepsilon_0 E_{\text{rms}}^2, \qquad (10.2)$$

we find that the amplitude of the electric field must be proportional to r^{-1},

$$E \propto \frac{1}{r} \qquad (10.3)$$

for spherical electromagnetic waves. This $1/r$ dependence is in contrast to the static electric field due to a point charge

$$E_{\text{st}} = \frac{1}{4\pi\varepsilon_0} \frac{q}{r^2}. \qquad (10.4)$$

which has r^{-2} dependence (Coulomb's law). Therefore it is obvious that the radiation electric field cannot be due to static charge, or we can conclude that stationary charge cannot radiate electromagnetic waves. We can draw this conclusion from an alternative point of view. The Poynting vector is the product

187

between the electric field E and the magnetic field B/μ_0. If a charge is not moving, we have no current and thus no magnetic field. The stationary electric field cannot induce the magnetic field either. Then the Poynting vector associated with a stationary charge must be zero, and we have no energy flow or radiation.

What about a charge in motion with a constant velocity? We now have both electric field and magnetic field, and the Poynting vector is expected to take a finite (nonzero) value. Suppose a positive charge q is moving in $+x$ direction with a constant velocity (nonrelativistic)

$$v = \frac{dx}{dt} = \text{const} \ll c \text{ (speed of light).}$$

We consider a circular disk with radius R located at $x = 0$ on the x axis facing perpendicular to the moving charge, as shown in Fig. 10.1. At a certain instant the electric field at the edge of the disk is given by the Coulomb's law,

$$E = \frac{q}{4\pi\varepsilon_0} \frac{1}{R^2 + x^2} \tag{10.5}$$

and its x component is

$$E_x = \frac{q}{4\pi\varepsilon_0} \frac{-x}{(R^2 + x^2)^{3/2}} \tag{10.6}$$

(Notice that when the charge is to the left of the disk, x is negative, and E_x is positive, as is clear from the figure.) Then the total electric flux *through* the disk is

$$\Phi_E = \int_0^R E_x 2\pi R \, dR = \frac{qx}{2\varepsilon_0} \int_0^R \frac{R}{(R^2 + x^2)^{3/2}} \, dR$$

$$= -\frac{qx}{2\varepsilon_0} \left(\frac{1}{|x|} - \frac{1}{\sqrt{R^2 + x^2}} \right),$$

which is time-varying, since x is varying with time. Now we can apply the Maxwell–Ampere law for this time-varying electric flux to find the magnetic

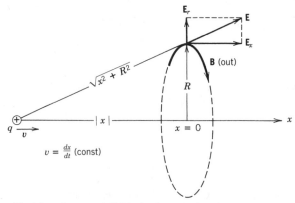

Fig. 10.1.   Electric and magnetic fields due to a moving charge. Velocity v is constant.

field at the edge of the disk.

$$2\pi RB = \varepsilon_0 \mu_0 \frac{d\Phi_E}{dt}$$

$$= \varepsilon_0 \mu_0 \frac{dx}{dt} \frac{d\Phi_E}{dx}$$

$$= \frac{\mu_0 qv}{2} \frac{R^2}{(R^2+x^2)^{3/2}} \quad \text{or} \quad B = \frac{\mu_0 qv}{4\pi} \frac{R}{(R^2+x^2)^{3/2}}. \tag{10.7}$$

(A quicker way to find this magnetic field is to use the Biot–Savart law,

$$d\mathbf{B} = \frac{\mu_0}{4\pi} \frac{i \, d\mathbf{l} \times \mathbf{r}}{r^3}$$

in which we replace $i \, d\mathbf{l}$ by $q\mathbf{v}$, and take $r = \sqrt{R^2+x^2}$. This is a legitimate procedure and you can always replace $i \, d\mathbf{l}$ by $q\mathbf{v}$ in the Biot–Savart law whenever you want to find the magnetic field due to a charge moving at a constant velocity.)

Since both electric and magnetic fields are now found, we can calculate the Poynting vector at any point around the moving charge. Consider a sphere centered at the charge at a certain instant as in Fig. 10.2. The electric field is normal to the sphere everywhere and the magnetic field is tangent to the sphere, being normal to the electric field. Therefore the Poynting vector is tangent to the sphere. In other words, the Poynting vector never penetrates through the surface, and we conclude that energy cannot be radiated by a charge moving with a constant velocity.

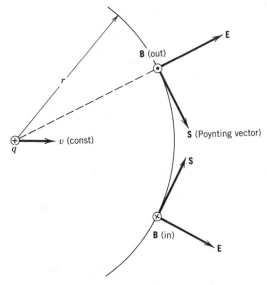

Fig. 10.2. The Poynting vector associated with a charge moving with a constant velocity cannot go through a sphere centered at the charge. The charge cannot radiate.

Example 1.   Using the electric field Eq. (10.5) and the magnetic field Eq. (10.7), calculate the x component of the Poynting vector and then the total energy flow rate through an infinitely large plane placed at x normal to the x axis. Discuss the result.

The x component of the Poynting vector is given by (See Figure 10.1)

$$S_x = E_r \frac{B_0}{\mu_0}$$

$$= \frac{q}{4\pi\varepsilon_0} \frac{R}{(R^2 + x^2)^{3/2}} \frac{qv}{4\pi} \frac{R}{(R^2 + x^2)^{3/2}}$$

$$= \frac{q^2 v}{16\pi^2 \varepsilon_0} \frac{R^2}{(R^2 + x^2)^3} \quad (\text{W/m}^2)$$

Then the total energy flow rate becomes

$$\text{Power} = \int_0^\infty S_x \, 2\pi R \, dR$$

$$= \frac{q^2 v}{8\pi\varepsilon_0} \int_0^R \frac{R^3}{(R^2 + x^2)^3} \, dR.$$

The integration can be carried out by successive use of integration by parts (try this), and the power becomes

$$\frac{q^2 v}{32\pi\varepsilon_0} \frac{1}{x^2} \quad (\text{W}),$$

where $|x|$ is the instant location of the charge measured from the plane. Noting

$$v = \frac{dx}{dt},$$

we may rewrite the power as

$$-\frac{q^2}{32\pi\varepsilon_0} \frac{d}{dt}\left(\frac{1}{x}\right) = -\frac{d}{dt}\left[\frac{q^2}{32\pi\varepsilon_0 x}\right].$$

However, the quantity of $q^2/32\pi\varepsilon_0 x_0$ ($x_0 > 0$) is nothing but the electric energy stored in the space in the region $x > x_0$. (Problem). Therefore the power calculated from the Poynting flux can be interpreted as the flow rate of electrostatic energy stored in space and has nothing to do with radiation energy. (The magnetic energy is of the order of $(v/c)^2 \times$ electric energy, and for nonrelativistic velocity $v \ll c$, it is negligibly small.)

Thus we have seen that (1) a stationary charge cannot radiate electromagnetic waves and (2) a charge with a constant *velocity* (not speed, since a constant speed could mean the case in which a charge is going along a circular orbit, which can radiate energy!) cannot either. What other situation can we have? We have not considered a more general case in which charged particles are subject to acceleration (or deceleration). This is exactly the case in which we can have radiation of electromagnetic waves. In radio antennas, electrons are forced to go back and forth in harmonic manner. Electrons are accelerated back and forth by a signal generator connected to antennas, which radiate electromagnetic waves.

10.3. Radiation Fields Due to an Accelerated (or Decelerated) Charge

Suppose a positive charge q originally at rest at point A is accelerated in the x direction as shown in Fig. 10.3. The acceleration lasts for Δt seconds only until the charge reaches point B, after which the charge moves with a constant

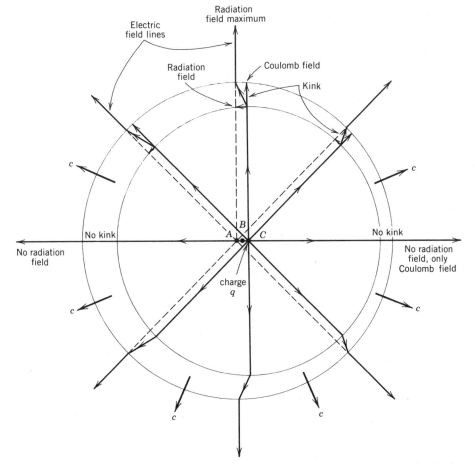

Fig. 10.3. A charge under acceleration does radiate electromagnetic waves. Notice the kinks in the electric field lines.

velocity v. We assume the velocity is much less than the speed of light (nonrelativistic). Let us see how the electric lines of force associated with the charge look. The electric lines of force of a charge stationary or moving with a constant velocity are just radially outward (Coulomb field). Thus the lines of force when the charge was at A, and at a point after B, say C, are all radially outward, although they are not concentric. Since the lines of force must be continuous, these nonconcentric lines must be connected somehow. Therefore the effect of the acceleration appears as kinked electric lines of force as shown. *The kinks, which are disturbance in the electric field lines caused by the acceleration, propagate with the speed of light.* It takes Δt sec for the accelerated charge to move from A to B. The separation between the larger circle and the smaller circle is approximately $c\,\Delta t = $ const; if the charge is accelerated and moves slowly enough, relativistic effects are negligible. In the kinks, we obviously have electric field components perpendicular to the Coulomb fields. These transverse components are responsible for radiation.

Consider a point Q in Fig. 10.4, normal to the velocity of the charge at a certain instant. Let t be the time after the charge passes point B (end of acceleration). At Q we have two electric fields. E_0, the radial component, is the Coulomb field and is given by

$$E_0 = \frac{q}{4\pi\varepsilon_0} \frac{1}{(ct)^2}. \tag{10.8}$$

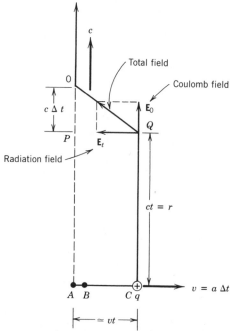

Fig. 10.4.   Coulomb field E_0 and radiation field $\mathbf{E}_t$ at $90°$ relative to the motion of the charge.

The other is the radiation field E_T, which is transverse to the Coulomb field. For the triangle OPQ, the edge $OP = c\,\Delta t$ (const) indicates the duration in which we have radiation. Since

$$\frac{OP}{PQ} = \frac{c\,\Delta t}{vt} = -\frac{E_0}{E_T}, \tag{10.9}$$

we find

$$E_T = -\frac{q}{4\pi\varepsilon_0 c^2}\frac{v}{\Delta t}\frac{1}{r}. \tag{10.10}$$

This is exactly what we wanted. The transverse (or radiation) electric field is proportional to the acceleration $v/\Delta t$ and $1/r$! [The minus sign in Eq. (10.9) is due to the direction of E_T, which is opposite to the direction of acceleration.]

At an arbitrary point, we have fields at arbitrary angle θ;

$$E_T = -\frac{q}{4\pi\varepsilon_0}\frac{v\sin\theta}{c^2\,\Delta t r} \tag{10.11}$$

as is clear from Fig. 10.5. In general, the radiation electric field due to a charge under acceleration $\mathbf{a}$ is given by

$$E_T = -\frac{q}{4\pi\varepsilon_0 c^2}\frac{a\sin\theta}{r} \tag{10.12}$$

where the direction of $\mathbf{E}_T$ is normal to the radial vector $\mathbf{r}$, and θ is the angle between $\mathbf{r}$ and $\mathbf{a}$. Notice that the field depends on the sign of charge q. If q is negative, the direction of the field must be reversed. See Fig. 10.6.

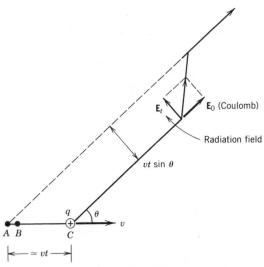

Fig. 10.5.   Fields at arbitrary angle θ.

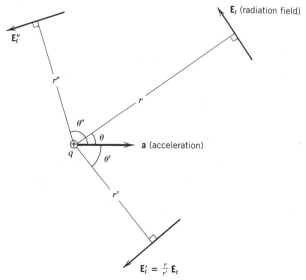

Fig. 10.6.   General relationship among $q(>0)$, $\mathbf{a}(>0$ to the right), and E_t (radiation field).

One point must be considered here with some caution. The acceleration a is in general a function of time. For example, if the charge moves back and forth harmonically, the acceleration must vary harmonically too. Since the radiation reaches the point P after time r/c, what an observer sees at point P is the electric field due to the acceleration r/c sec before! Thus the correct expression for the electric field is

$$E_T(r,\,t)= - \frac{q}{4\pi\varepsilon_0 c^2}\frac{\sin\theta}{r}a_{t-r/c}.$$
(10.13)

The electric field is called the *retarded electric field*. $a_{t-r/c}$ indicates the acceleration r/c seconds before relative to the electric field. It properly takes into account the effect of finite propagation time of electromagnetic radiation.

The magnetic field associated with the radiation electric field can easily be calculated from (see Sections 9.2 and 9.3)

$$E_T = cB_T$$
(10.14)

or

$$B_T(r,\,t)=\frac{q}{4\pi\varepsilon_0 c^2}\frac{\sin\theta}{r}a_{t-r/c},$$
(10.15)

which must be normal to the electric field, and the radial vector $\mathbf{r}$. The radiation electric and magnetic fields $\mathbf{E}_T$, $\mathbf{B}_T$, and Poynting vector $\mathbf{S}$ are due to an accelerated charge (Fig. 10.7). Poynting vector $\mathbf{S}$ is radially outward, and its magnitude is given by

$$S=\frac{q^2}{16\pi^2\varepsilon_0 c^3}\frac{\sin^2\theta}{r^2}a^2_{t-r/c}.$$
(10.16)

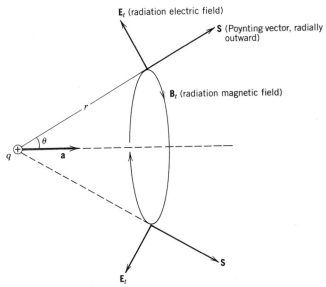

Fig. 10.7.   Radiation electric and magnetic fields, E_t, B_t, and Poynting vector S due to an accelerated charge.

Example 2.   Show that the instantaneous radiation power emitted by a charge q subject to an acceleration a is

$$P = \frac{1}{4\pi\varepsilon_0}\frac{2q^2a^2}{3c^3} \quad \text{(W)}.$$

The Poynting vector is the local power density (W/m²). Then if we integrate the Poynting vector over the entire spherical surface having a radius r, we should obtain the total power. To carry out the surface integration, we note that the area of the thin circular belt having a radius $r \sin \theta$ and a width $r\, d\theta$ (see Fig. 10.8) is

$$dA = 2\pi r \sin \theta\; rd\theta$$
$$= 2\pi r^2 \sin \theta\; d\theta$$

Then the power through this differential area is

$$dP = S\, dA \quad \text{(W)}$$
$$= S\, 2\pi r^2 \sin \theta\; d\theta$$

Substituting the expression for S [Eq. (10.16)] and integrating, we find,

$$P = \int dP = \frac{q^2a^2}{8\pi\varepsilon_0 c^3}\int_0^\pi \sin^3 \theta\; d\theta$$

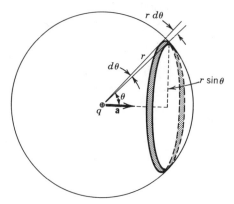

Fig. 10.8.   Surface of integration.

However,

$$\int_0^\pi \sin^3 \theta \, d\theta = \int_{-1}^1 \sin^2 \theta \, d(\cos \theta)$$

$$= \int_{-1}^1 (1 - x^2) \, dx \; (x = \cos \theta)$$

$$= 2 - \tfrac{2}{3} = \tfrac{4}{3}$$

Then the total power becomes

$$P = \frac{q^2 a^2}{8\pi\varepsilon_0 c^3} \times \frac{4}{3}$$

$$= \frac{1}{4\pi\varepsilon_0} \frac{2q^2 a^2}{3c^3} \quad \text{(W)}.$$

Note that the acceleration appears as a^2, and the formula is equally applicable for the case in which the charge is subject to deceleration ($a < 0$).

The Poynting vector $\mathbf{S}$ has a strong angular dependence $\sin^2 \theta$, which becomes maximum at an angle perpendicular to the acceleration. This is called *directivity* of radiation intensity. All antennas have directivity, and it is in fact impossible to let an antenna radiate isotropically, or equally in every angular direction.

10.4. Radiation from an Oscillating Dipole and Dipole Antenna

Let us apply these results to more practical situations. The first example we choose is an oscillating dipole, in which two equal but opposite charges undergo harmonic oscillations. Let the amplitude of oscillations (Fig. 10.9) be x_0 (m),

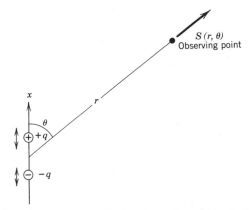

Fig. 10.9.   Oscillating dipole radiates electromagnetic fields.

and assume oscillations for the positive charge

$$x = x_0 \sin \omega t$$

and for the negative charge

$$x = -x_0 \sin \omega t.$$

Then the acceleration for the positive charge is

$$a_+ = \frac{d^2 x}{dt^2} = -x_0 \omega^2 \sin \omega t \qquad (10.17)$$

and

$$a_- = +x_0 \omega^2 \sin \omega t \qquad (10.18)$$

for the negative charge. Then the Poynting flux at a point distance r (m) away from the dipole is

$$S = 4 \times \frac{q^2 x_0^2}{16\pi^2 \varepsilon_0 c^3} \frac{\sin^2 \theta}{r^2} \omega^4 \sin^2 \omega t, \qquad (10.19)$$

and its rms value is*

$$S_{\text{rms}} = \frac{q^2 x_0^2 \omega^4}{8\pi^2 \varepsilon_0 c^3} \frac{\sin^2 \theta}{r^2} \quad (\text{W/m}^2). \qquad (10.20)$$

Notice in Fig. 10.10 that the power is most effectively radiated in the direction perpendicular to the dipole, as in the case of radiation intensity due to a single charge. In fact, any antenna has a strong angular dependence, which is desirable for commercial radio or TV stations.

---

*Note that the power is proportional to ω^4, or λ^{-4}. This explains why the sky is blue. Atmospheric molecules are dipole radiators excited by sunlight.

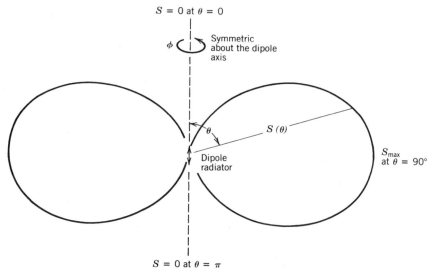

Fig. 10.10.　Angular dependence of the Poynting vector due to an oscillating dipole. S is maximum at $\theta = 90°$ and is symmetric about the dipole axis.

Next, consider a short radio antenna. We assume that the *length of the antenna is much shorter than the wavelength of the radio wave*. This assumption means that the current on the antenna is of the same phase everywhere, and the whole antenna is just oscillating with a certain frequency, ω. Of course, what oscillates are the free electrons in the antennas.

Since we have many electrons, we have to add up the radiation electric fields due to the individual electrons. But the story is much simpler, since we already have the Poynting flux due to an oscillating charge (see Eq. (10.16)). Suppose that the antenna is l (m) long, has a cross section of A (m^2), and has a free electron density of n (m^{-3}). The total number of free electrons is Aln and the charge is $q = eAln$ (coulombs). Let a current $I_0 \sin \omega t$ flow through the antenna. Then the velocity of each electron is

$$v = \frac{l}{q} I_0 \sin \omega t \qquad (10.21)$$

since $I = enAv$. Therefore the acceleration of each electron is

$$a = \frac{dv}{dt} = \frac{l}{q} I_0 \omega \cos \omega t, \qquad (10.22)$$

or

$$qa = l\omega I_0 \cos \omega t. \qquad (10.23)$$

Substituting this into Eqs. (10.13) and (10.16), we find

$$E_T = \frac{l\omega I_0}{4\pi\varepsilon_0 c^2} \frac{\sin \theta}{r} \cos \omega t \qquad (10.24)$$

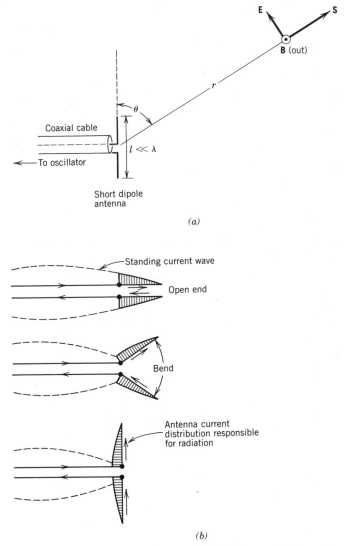

Fig. 10.11. (a) Radiation fields due to a short ($\ll \lambda$) dipole antenna. (b) Formation of a dipole antenna from an open-ended transmission line.

and

$$S = \frac{(l\omega I_0)^2}{16\pi^2\varepsilon_0 c^3} \frac{\sin^2\theta}{r^2} \cos^2\omega t \qquad (10.25)$$

Let us evaluate how much total power the antenna is radiating. For this we have to integrate the Poynting flux over a spherical surface.

The power going through the thin ring with an area $2\pi r \sin\theta \times r\,d\theta$ (see

Fig. 10.12) is

$$dP = S_{ave} \times 2\pi r^2 \sin\theta \, d\theta$$

$$= \frac{(l\omega I_0)^2}{32\pi^2\varepsilon_0 c^3} \frac{\sin^2\theta}{r^2} 2\pi r^2 \sin\theta \, d\theta$$

$$= \frac{(l\omega I_0)^2}{16\pi\varepsilon_0 c^3} \sin^3\theta \, d\theta. \tag{10.26}$$

Then the total power is

$$P = \int_0^\pi \frac{(l\omega I_0)^2}{16\pi\varepsilon_0 c^3} \sin^3\theta \, d\theta$$

$$= \frac{(l\omega I_0)^2}{16\pi\varepsilon_0 c^3} \int_0^\pi \sin^3\theta \, d\theta. \tag{10.27}$$

The integration was performed in Example 2 and its value is 4/3. Therefore the total power radiated by the short antenna is

$$P = \frac{(l\omega I_0)^2}{16\pi\varepsilon_0 c^3} \times \frac{4}{3} = \frac{(l\omega I_0)^2}{12\varepsilon_0 c^3} \quad \text{(W)}. \tag{10.28}$$

This expression enables us to define the so-called radiation resistance through

$$R_{rad} I_{rms}^2 = \tfrac{1}{2} R_{rad} I_0^2 = P, \tag{10.29}$$

or

$$R_{rad} = \frac{(l\omega)^2}{6\pi\varepsilon_0 c^3} \quad (\Omega). \tag{10.30}$$

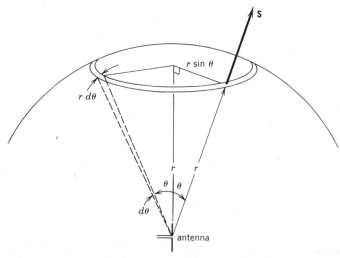

Fig. 10.12.   Integration of Poynting vector S over the entire sphere yields the total power radiated by the antenna.

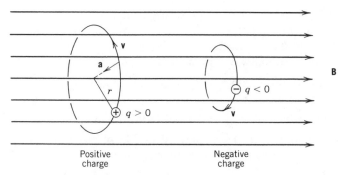

Fig. 10.13. A charged particle in a magnetic field undergoes cyclotron motion and radiates since the particle is under centripetal acceleration.

Since

$$c = \frac{1}{\sqrt{\varepsilon_0 \mu_0}}, \qquad \omega = 2\pi\nu = \frac{2\pi c}{\lambda},$$

we can rewrite the preceding expression as

$$R_{rad} = \frac{2\pi}{3} \sqrt{\frac{\mu_0}{\varepsilon_0}} \left(\frac{l}{\lambda}\right)^2 = 790 \left(\frac{l}{\lambda}\right)^2 \quad (\Omega) \qquad (10.31)$$

where $\sqrt{\mu_0/\varepsilon_0} = 377\,\Omega$ is the characteristic impedance of free space. Remember that the expressions for power and radiation resistance are all subject to the restriction, $l \ll \lambda$ (short antenna). The generator connected to the antenna has to supply the power radiated away in order to maintain a steady state.

In summary, we have seen that electromagnetic waves can be created (or radiated) only if electric charges are accelerated (or decelerated). A charge with a constant velocity (not speed!) cannot radiate electromagnetic waves. As an example of radiation from a charge with a constant *speed*, consider a charged particle in a magnetic field (Fig. 10.13). Charged particles undergo circular motion about a magnetic field line. The centrifugal force mv^2/r is balanced by the Lorentz force qvB, which of course continuously accelerates the particle. The speed v is constant, however. The velocity $\mathbf{v}$ keeps changing its direction because of the centripetal acceleration. The charged particle indeed radiates electromagnetic waves, known as synchrotron radiation.

Problems

1. In the electron gun of an oscilloscope, a 20-kV potential is applied between anode and cathode, which are 5 cm apart. Estimate the maximum radiation electric field at a point 1 m away from the gun assuming there are about 4×10^7 electrons in the gun.

   (*Answer:* 4.5×10^{-2} V/m.)

2. We have seen that for creation of electromagnetic waves, we must have charge acceleration perpendicular to the direction of the Poynting vector. This is due to the *transverse* nature of electromagnetic waves. Discuss how we can create sound waves in air that are *longitudinal*.

3. An AM radio station of 1 MHz frequency uses an antenna 20 m long placed well above the ground.

  (a) What is the radiation resistance of the antenna?

  (b) If the station is to be operated at 50 kW power, what rms current should be supplied to the antenna?

  (*Answer:*   3.5 Ω, 120 A.)

Note: If an antenna l (m) high is erected above the ground, its "effective" length is $2l$ (m). The reason is that the earth is a mirror (reflector) for electromagnetic waves. This "method of images" will be studied in more advanced electromagnetic classes.

4. X-ray (short wavelength electromagnetic waves, $\lambda \simeq 10^{-10}$–10^{-9} m) can be created when energetic electrons hit a surface of hard metal such as tungsten. Explain qualitatively the radiation mechanism. (Also see Chapter 14.)

5. An electron having an energy of 10 keV ($= 10^4 \times 1.6 \times 10^{-19}$ J) undergoes cyclotron motion in a magnetic field of 1 T.

  (a) What is the acceleration on the electron?

  (b) Evaluate the initial rate of electron energy loss caused by synchrotron radiation.

  (*Answer:* $a = 1.0 \times 10^{19}$ m/sec^2, 6.2×10^{-16} W $= 3.9$ keV/sec. This is the initial loss rate. The electron gradually loses energy and the cyclotron radius becomes smaller.)

CHAPTER 11

Interference and Diffraction

11.1. Introduction

A harmonic wave in the typical form we used in previous chapters is given by

$$A \sin (kx - \omega t),$$

which is characterized by the amplitude A and the frequency ω or the wavelength $\lambda = 2\pi/k$. If we have only one wave source, the preceding expression is, in general, sufficient. However, as soon as we have more than one wave source, the total amplitude at a given observation point must be the sum of contributions from each wave source. What is important here is the phase difference between harmonic waves, which we have not considered in detail so far. For example, if two waves of equal amplitude and frequency having no phase difference are added, the amplitude is doubled, and the intensity is quadrupled. But if the two waves are out of phase or have $+\pi$ or $-\pi$ phase difference, the total amplitude becomes zero. Therefore, depending on the phase difference, the total amplitude and thus the wave intensity can vary.

Interference is the most fundamental nature of all wave phenomena. If one physical quantity exhibits interference, that quantity should have wave nature.

In this chapter we study interaction among more than one wave, sometimes infinitely many waves.

11.2. Interference Between Two Harmonic Waves

Suppose we have two point wave sources separated by a distance d, as shown in Fig. 11.1. We assume that the two sources radiate waves at an exactly identical frequency. Such a situation can easily be realized if, for example, two identical radio antennas are connected to a common signal generator.

Consider a point P at which we place a wave detector. The distances between P and the wave sources are x_1 and x_2, respectively. At P we simultaneously detect two different waves, since

1. The field amplitudes, which are proportional to $1/x_1$ and $1/x_2$, respectively, are different.

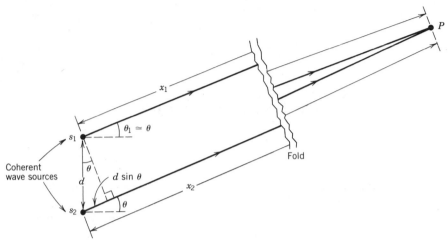

Fig. 11.1.   Waves created by two coherent sources can interfere with each other. Here we assume x_1, $x_2 \gg d$, λ (wavelength), and in this case, the path difference can be approximated by $d \sin \theta$.

2.   The phases $2\pi x_1/\lambda$ and $2\pi x_2/\lambda$, which are proportional to x_1 and x_2 are different.

The amplitude difference may not be important if the point P is far away, so that

$$x_1, x_2 \gg d.$$

The phase difference given by

$$\phi = \frac{2\pi}{\lambda}(x_2 - x_1) \quad (\lambda = \text{wavelength}) \tag{11.1}$$

is, however, important and plays major roles in interference and diffraction. The quantity $x_2 - x_1$ is called the path difference, and in the diagram the path difference is given approximately by

$$d \sin \theta, \tag{11.2}$$

if $x_2, x_1 \gg d$.

   Consider two harmonic waves

$$E_1 = E_0 \sin (kx_1 - \omega t), \qquad k = \frac{2\pi}{\lambda} \tag{11.3}$$

$$E_2 = E_0 \sin (kx_2 - \omega t), \tag{11.4}$$

created by the sources S_1 and S_2, respectively. Using the phase difference given by Eq. (11.1), we may rewrite Eqs. (11.3) and (11.4) as

$$E_1 = E_0 \sin (kx_1 - \omega t) \tag{11.3}$$

$$E_2 = E_0 \sin (kx_1 - \omega t + \phi). \tag{11.5}$$

At the point P, the sum of two fields are observed. Thus the total field becomes

$$E = E_1 + E_2 = 2E_0 \sin\left(kx_1 - \omega t + \frac{\phi}{2}\right) \cos\left(\frac{\phi}{2}\right), \tag{11.6}$$

which is still a propagating wave. (Note $\sin A + \sin B = 2 \sin[(A+B)/2] \cos[(A-B)/2]$.) Its amplitude, however, strongly depends on the phase difference. The effective amplitude is

$$E_m = 2E_0 \left|\cos\left(\frac{\phi}{2}\right)\right|, \tag{11.7}$$

and can vary between 0 and $2E_0$, depending on the phase difference. The maximum amplitude $2E_0$ is realized when $|\cos(\phi/2)| = 1$, or

$$\phi = 0, \pm 2\pi, \pm 4\pi, \ldots$$

$$= 2m\pi \quad (m = \text{integer}) \tag{11.8}$$

and the minimum amplitude 0 when $\cos(\phi/2) = 0$, or

$$\phi = \pm\pi, \pm 3\pi, \ldots = (2m+1)\pi. \tag{11.9}$$

In terms of $d \sin\theta$, we have

$$d \sin\theta = m\lambda \quad \text{for maxima}, \tag{11.10}$$

and

$$d \sin\theta = (m+\tfrac{1}{2})\lambda \quad \text{for minima}. \tag{11.11}$$

These results can be found intuitively if we graphically superpose two waves as shown in Fig. 11.2.

Consider the setup for demonstrating sound wave interference (Fig. 11.3). The path difference in this case is $2d$, and if

$$2d = m\lambda,$$

the sound intensity should be maximum at the receiving end, and if

$$2d = (m+\tfrac{1}{2})\lambda,$$

the sound intensity should be minimum, or zero under ideal conditions.

The recipe for various interference and diffraction phenomena we are going to study is nothing more than Eq. (11.6), that is, superposition of two or more harmonic waves with equal amplitude. If we have many wave sources, we have to add up more waves to find the total field, but the mathematics involved is no more than adding up sinusoidal functions.

The distinction between interference and diffraction is not clear. Both are caused by the interaction among more than one wave. Conventionally, if we have relatively few wave sources, we use the term *interference*, and if we have to add up many (sometimes infinitely large number) waves, *diffraction* is used.

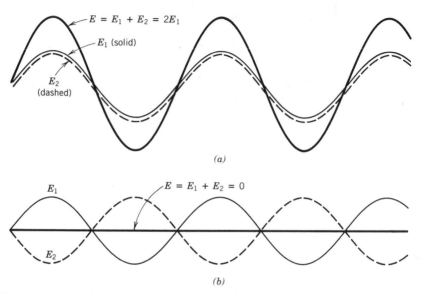

Fig. 11.2.   Graphical superposition of two waves: (*a*) In phase—two waves add up and the amplitude is maximum. (*b*) Out of phase—two waves cancel each other and the amplitude is zero.

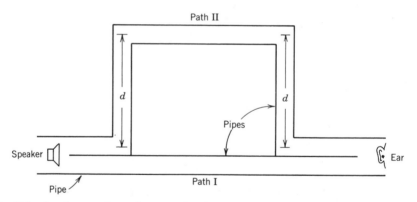

Fig. 11.3.   Interference of sound waves going through two different paths. Depending on d, the received intensity varies.

11.3. Young's Experiment

We know that light is an electromagnetic *wave*. The wavelengths of the visible light spectrum* range from 4000 Å (or 4×10^{-7} m) to 7000 Å (7×10^{-7} m). (1 Å (angstrom) $= 10^{-10}$ m.)

---

*Light having wavelengths shorter than 4000 Å (violet) is called ultraviolet light. Wavelengths longer than 7000 Å (red) are called infrared.

This seemingly obvious fact was not so obvious before Young did the famous experiment in 1801–1803, known as Young's double-slit experiment. The experiment clearly demonstrated the wave nature of light. (We will learn that light also behaves as particles, known as photons. In fact, both wave and particle nature coexist in light, or any moving physical object. For example, energetic electrons also have particle and wave nature. As we will see later, electron microscopes can "see" better than optical microscopes, since "wavelengths of electron waves" are much shorter than those of visible light. Quantum mechanics has been able to unify wave and particle nature.)

The principle of Young's experiment is already shown in Fig. 11.1. Instead of two point sources, as in Fig. 11.1, Young used two narrow parallel slits, illuminated by a monochromatic light source as shown in Fig. 11.4. If the slit opening is narrow enough, an interesting thing happens. The slits act as if they were new light sources. If the light source is placed on the bisector line of the two slits, these slits become light sources of equal phase, since the slits are equal distances away from the light source (no path difference). The slits act as line light sources, rather than point sources. Therefore light emitted from the slits consists of cylindrical, rather than spherical, waves. However, we do not have to worry about this, since all we need is just the path difference. Of course, the slits can be replaced by two pinholes, but this only complicates the analysis.

That narrow openings (such as the slits) act as new light sources can be understood from Huygens' principle, which may be stated as follows: All points on a wavefront act as new point sources. To illustrate this principle, let us consider a trivial case, a plane wave. As shown in Fig. 11.5, the new wavefront

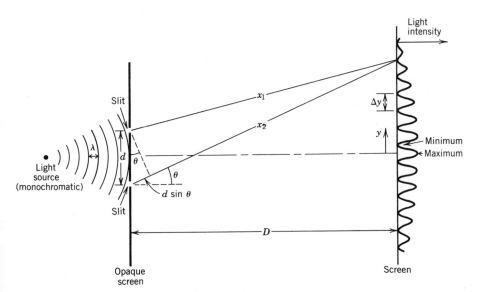

Fig. 11.4a. Arrangement of Young's double-slit experiment. In practice, the screen distance D is much larger than the slit separation d.

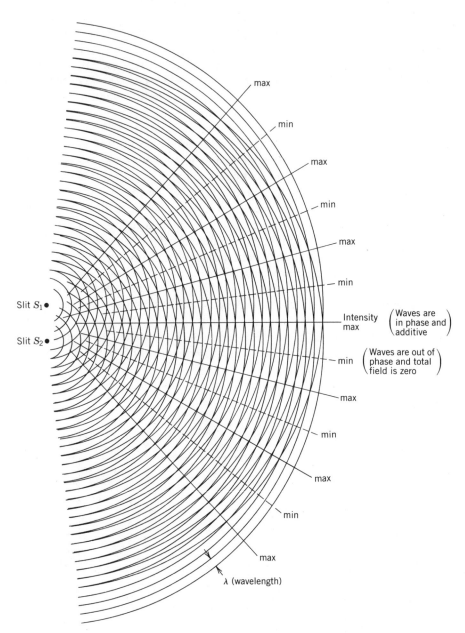

Fig. 11.4b. Qualitative illustration of interference mechanism. Two families of concentric circles indicate radiation from each slit and are drawn with the same interval corresponding to the wavelength λ. Where the two families of circles intersect, the waves from each slit are in phase, and the wave intensity becomes maximum. Intensity minima occurs between neighboring maxima.

208

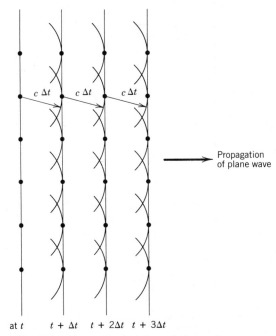

at t $t + \Delta t$ $t + 2\Delta t$ $t + 3\Delta t$

Fig. 11.5.   Huygens' principle applied to a plane wave.

is formed as a tangent plane to all the waves created by the points on the old wavefront. The new wavefront then further creates another wavefront. This process continues and is observed as wave propagation.

Suppose the plane wave encounters an obstacle with a small opening (Fig. 11.6). In the opening, we can have only a few Huygens' points (only one is shown in the figure), and behind the opening, the wave is not a plane wave anymore. If the opening is very small it acts as a point wave source, and behind the opening, we essentially have a spherical wave. The image formed on a screen is then widely spread. Later, we actually calculate this intensity profile for the case of a narrow slit in the section on diffraction.

Now we return to Young's experiment. Since the light intensity is proportional to E^2, where E is the total field, we obtain from Eq. (11.7)

$$I(\theta) = I_0 \cos^2 \left(\frac{\phi}{2} \right), \tag{11.12}$$

where

$$\phi = \frac{2\pi d}{\lambda} \sin \theta.$$

If a screen is placed at a distance $D \, (\gg d)$ parallel to the slits, we find

$$I(y) = I_0 \cos^2 \left(\frac{\pi d}{\lambda} \frac{y}{D} \right), \tag{11.13}$$

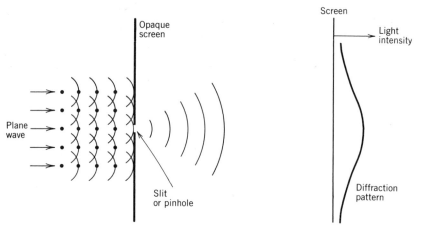

Fig. 11.6.   A plane wave is converted into a spherical wave when going through a pinhole, or into a cylindrical wave through a long slit.

where

$$\sin \theta \simeq \tan \theta = \frac{y}{D} \, (\ll 1)$$

is used, assuming a small angle θ. Then the maxima are located at

$$y = 0, \; \pm \frac{\lambda D}{d}, \; \pm 2 \frac{\lambda D}{d}, \ldots$$

and the minima at

$$y = \pm \frac{1}{2} \frac{\lambda D}{d}, \; \pm \left(1 + \frac{1}{2}\right) \frac{\lambda D}{d}, \; \pm \left(2 + \frac{1}{2}\right) \frac{\lambda D}{d}, \ldots$$

The separation between neighboring maxima (or minima) is

$$\Delta y = \frac{\lambda D}{d}. \tag{11.14}$$

Therefore by measuring Δy, D, and d, the wavelength λ can be determined.

The experiment works better for narrower slit openings. Larger slit openings complicate the intensity pattern formed on the screen because of diffraction effects, which we will study later. If the opening is increased further, the interference pattern disappears and we simply have two slit images, although somewhat blurred. This is due to the fact that for a large opening, the slits do not behave as line sources.

Example 1.   In Young's double-slit arrangement, assume the slit separation d is 0.1 mm and slit-screen distance D is 50 cm. If a separation between neighboring maxima (or minima) of 2.5 mm is observed, what is the wavelength of light illuminating the slits?

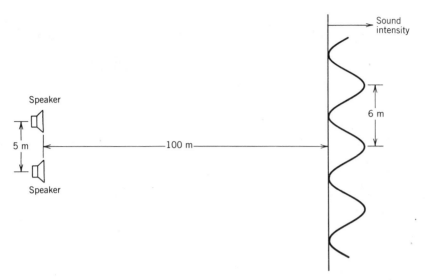

Fig. 11.7. Example 2.

From Eq. (11.14), we find

$$\lambda = \frac{\Delta y d}{D} = \frac{0.25 \text{ cm} \times 0.01/\text{cm}}{50 \text{ cm}}$$

$$= 5.0 \times 10^{-5} \text{ cm} = 5000 \text{ Å}.$$

Example 2. Two loudspeakers connected to a common audio amplifier are 5 m apart (Fig. 11.7). As one walks along a straight path 100 m away from the speakers, at what spatial period does the intensity vary? Assume λ (wavelength) $= 30$ cm.

From Eq. (11.14)

$$\Delta y = \frac{\lambda D}{d} = \frac{0.3 \times 100}{5.0} \text{ m} = 6.0 \text{ m}$$

11.4. Multislit Structure

Let us see what would happen if the number of slits is increased. We assume that all slits are equally spaced and illuminated by a common, monochromatic light source. The case of six slits is shown in Fig. 11.8. The phase difference between two neighboring waves is

$$\delta = \frac{2\pi d}{\lambda} \sin \theta. \qquad (11.15)$$

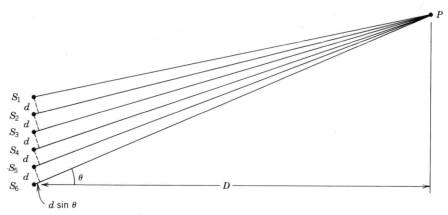

Fig. 11.8.   Six coherent light sources are equally spaced along a straight line.

Thus the total electric field is given by

$$E = E_0 \left[\sin(kx - \omega t) + \sin(kx - \omega t + \delta) \right.$$
$$+ \sin(kx - \omega t + 2\delta) + \sin(kx - \omega t + 3\delta)$$
$$\left. + \sin(kx - \omega t + 4\delta) + \sin(kx - \omega t + 5\delta) \right] \qquad (11.16)$$

We would add these six terms term by term, but there is a more elegant way to do this. In alternating current circuit theory, we learned that two oscillating voltages V_1 and V_2 with a phase difference ϕ can be added vectorially ("phasors"). (See Fig. 11.9.) If we use this technique for the case of double slit, we immediately find that the amplitude of the total electric field E is given by

$$E = 2E_0 \cos\left(\frac{\phi}{2}\right)$$

consistent with Eq. (11.7), which is shown in Fig. 11.10. Notice that the vectors are introduced here just for mathematical convenience and have nothing to do with the electric field vector (true vector) associated with the light. The true electric field vectors are all in the same direction, if the observation point is far away from the light sources.

Now we add up the six fields using ac theory. Since the phase difference (δ) between two neighboring fields is the same everywhere, the six vectors form an

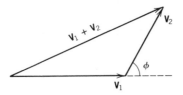

Fig. 11.9.   In ac circuit theory, two voltages should be added vectorially.

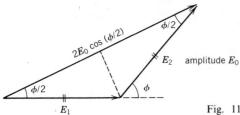

Fig. 11.10. Vectorial addition of two electric fields in the case of double-slit arrangement.

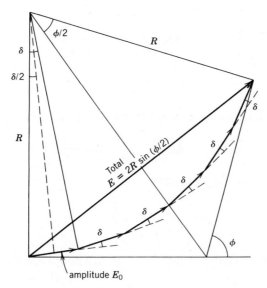

Fig. 11.11. Addition of six electric fields. Notice that the phase difference between any two neighboring fields is δ (const).

arc with a radius R. From Fig. 11.11, we find

$$\frac{E_0/2}{R} = \sin \frac{\delta}{2} \qquad (11.17)$$

$$\phi = 6\delta \qquad (11.18)$$

and

$$E = 2R \sin\left(\frac{\phi}{2}\right). \qquad (11.19)$$

Eliminating R and ϕ, we find

$$E = E_0 \frac{\sin(6\delta/2)}{\sin(\delta/2)}. \qquad (11.20)$$

(Do this.)

A more elegant way to derive this is to use complex variables. Noting $\sin A = \text{Im} (e^{iA})$, where Im indicates the imaginary part, we may write Eq. (11.16) as

$$E = E_0 \, \text{Im} \, [e^{iX}(1 + e^{i\delta} + \cdots + e^{i5\delta})],$$

where $X = kx - \omega t$. Since the amplitude of e^{iX} is 1, the amplitude of E is given by

$$E_0|1 + e^{i\delta} + e^{i2\delta} + e^{i3\delta} + e^{i4\delta} + e^{i5\delta}| = E_0|(1 + e^{i\delta})e^{i2\delta}(e^{-i2\delta} + 1 + e^{i2\delta})|$$

Noting $e^{iA} + e^{-iA} = 2 \cos A$, the amplitude becomes

$$2E_0 \cos \left(\frac{\delta}{2}\right)[1 + 2 \cos (2\delta)]$$

which is identical to Eq. (11.20). [Use $\sin 3A = 3 \sin A - 4 \sin^3 A$ in Eq. (11.20).]
    We can easily generalize this to the case of N light sources as

$$E = E_0 \frac{\sin (N\delta/2)}{\sin (\delta/2)}. \tag{11.21}$$

The case of the double slit corresponds to $N = 2$, and we indeed recover Eq. (11.7)

$$E(N=2) = E_0 \frac{\sin \delta}{\sin (\delta/2)} = 2E_0 \cos \left(\frac{\delta}{2}\right).$$

    The light intensity corresponding to the electric field given by Eq. (11.21) becomes

$$I = I_0 \frac{\sin^2 (N\delta/2)}{\sin^2 (\delta/2)}. \tag{11.22}$$

Here we need some mathematics. We want to know what would happen to the function

$$f(x) = \frac{\sin Nx}{\sin x} \quad (N = \text{integer})$$

if we let $\sin x$ approach zero. Sin $x = 0$ occurs when

$$x = m\pi \quad (m = \text{integer}).$$

Then $\sin Nx = \sin (Nm\pi)$ also becomes zero, and we end up with $0/0$, which is indefinite. Let $x = m\pi + \varepsilon$, with ε a small value. Since

$$\sin (m\pi + \varepsilon) = \sin m\pi \cos \varepsilon + \cos m\pi \sin \varepsilon.$$

$$= \pm \sin \varepsilon,$$

and $\sin N(m\pi + \varepsilon) = \pm \sin N\varepsilon$, we find

$$\lim_{x \to m\pi} f(x) = \lim_{\varepsilon \to 0} \frac{\sin N\varepsilon}{\sin \varepsilon} = \frac{N\varepsilon}{\varepsilon} = N,$$

which is finite.

Then the light intensity I becomes a maximum whenever $\sin(\delta/2)=0$, and its peak value is proportional to N^2. $\sin(\delta/2)=0$ yields

$$\frac{\delta}{2}=0,\ \pm\pi,\ \pm2\pi,\ldots$$

or

$$a\sin\theta=m\lambda \quad (m=\text{integer}). \tag{11.23}$$

The function

$$f(\delta)=\frac{1}{N^2}\frac{\sin^2(N\delta/2)}{\sin^2(\delta/2)} \tag{11.24}$$

indicates the relative intensity (maximum chosen as 1.0) and is plotted in Fig. 11.12 for $N=2$, 5, and 10, as a function of $a/\lambda\sin\theta$. It can be seen that as N

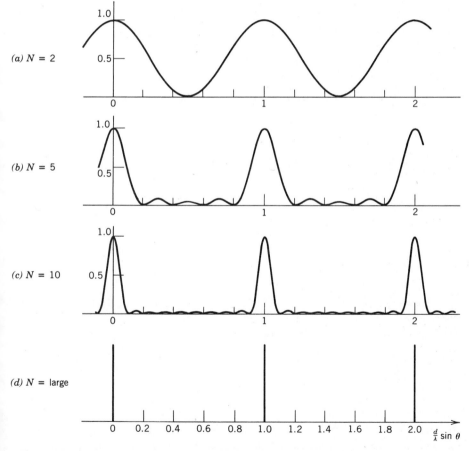

Fig. 11.12. (a) Equation (11.24) plotted as a function of $\delta/2\pi=d\sin\theta/\lambda$ for the case of $N=2$. (This corresponds to the intensity profile in Young's experiment.) (b) $N=5$. (c) $N=10$. (d) N large.

increases, the interference pattern becomes sharper and sharper. You may imagine what profile will result for $N = 100$, say. The profile will be extremely sharp and will simply look like vertical lines located at

$$\frac{d}{\lambda} \sin \theta = 0, \pm 1, \pm 2, \ldots.$$

In optics we have a device called a spectrometer, which can tell us what intensity a particular wavelength has. In other words, a spectrometer can Fourier-analyze light and may be called an optical spectrum analyzer. The principle is the one we have just studied. The number of light sources is usually in the tens of thousands, or more ($N \approx 10^4$!). You may imagine how sharp the interference pattern should be for such a large N. Spectrometers have so-called gratings, a structure similar to the fine grooves on an LP record. (Actually, LP records can be a rough spectrometer. You can easily see color spectrum on the record surface—Problem 14.)

Example 3.   Consider a grating having 5000 grooves per centimeter. Then the separation between grooves (corresponding to the slits) is

$$d = \frac{1}{5000} = 2 \times 10^{-4} \text{ cm.}$$

For red light of $\lambda = 7000 \text{ Å} = 7 \times 10^{-5}$ cm, the sharp peaks appear at

$$\sin \theta_{red} = 0, \pm \frac{\lambda}{a}, \pm 2 \frac{\lambda}{a}, \ldots$$

$$= 0, \pm 0.35, \pm 0.70, (\pm 1.05 \ldots)$$
$$\text{disregard}$$

or

$$\theta_{red} = 0°, \pm 20.5°, \pm 44.4°.$$

For violet light of $\lambda = 4000 \text{ Å} = 4 \times 10^{-5}$ cm, the angles become

$$\theta_{violet} = 0°, \pm 11.5°, \pm 23.6°, \pm 36.9°, \pm 53.1°.$$

All colors fall on $\theta = 0°$ and this angular location is not useful. The next peak in Fig. 11.13 (called order 1) can be used for spectrum analyses. Colors in the wavelength range 4000–7000 Å fall between $\theta = 11.5°$ and $\theta = 20.5°$. For the second order the angular range is between 23.6° and 44.4°, which is also useful. In this example the red light cannot produce a third- (and higher) order peak, and the violet light cannot produce a fifth- (and higher) order peak.

The result [Eq. (11.22)] we obtained can also be applied to the interference pattern formed by an antenna array. In radio communication it is often desired that antennas have sharp directivity. As we have seen in Chapter 10, a single dipole antenna erected vertically radiates uniformly in all horizontal directions, or it has no horizontal directivity. For commercial radio stations this is a

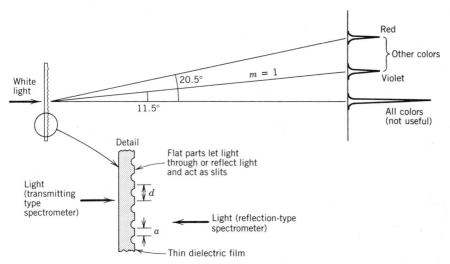

Fig. 11.13.   Grating spectrometer. Light falls normal to the grating. $d = 2 \times 10^{-4}$ cm

desired feature. However, for antenna arrays used for instrumental landing of airplanes, for example, a strong directivity is needed. (As the wavelengths become shorter, the directivity can be realized by a principle completely different from interference. In microwaves, parabolic antennas are used. For such short wavelengths, the concept of geometric optics can be applied and parabolic antennas can be regarded as concave mirrors in geometric optics.)

The wavelengths of electromagnetic waves used in radio communication are of course much longer than optical wavelengths, and the spacing between wave sources (d) can be chosen to be of the order of the wavelength. In fact, the spacing d is chosen at $\lambda/2$ in many applications. Consider an antenna array consisting of four coherent dipole antennas as shown in Fig. 11.14. In Eq. (11.22) we choose $N = 4$, and $d = \lambda/2$. Then the intensity becomes

$$I = I_0 \frac{\sin^2 (2\pi \sin \theta)}{\sin^2 (\pi/2 \sin \theta)}, \tag{11.25}$$

which can be calculated easily by either a programmable calculator or a computer (see Fig. 11.15).

We can intuitively see that the array does not radiate waves along the array itself ($\theta = 90°$) since in this direction, A_1 and A_2 are out of phase ($\theta = \pi$), and A_3 and A_4 are also out of phase. Thus at $\theta = 90°$, the electric field is zero.

Of course each antenna radiates radio waves with a power determined by the current supplied and its radiation resistance. By making an array, the angular distribution of the power can be made so that the power can be radiated in a preferential direction. It is like making a high hill by collecting rocks widely spread. By increasing the number of antennas, the directivity can be made sharper and sharper, exactly like the case of the grating spectrometer.

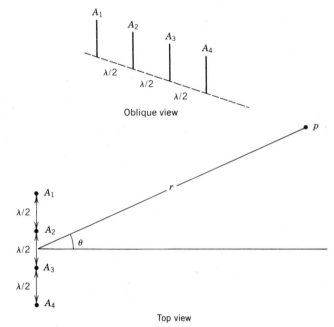

Fig. 11.14. Antenna array for producing strong directivity. Antennas are $\lambda/2$ apart.

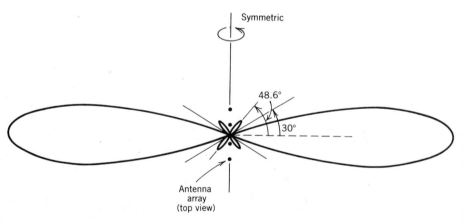

Fig. 11.15. Polar plot of the interference intensity for the antenna arrangement of Fig. 11.14.

Similar array arrangement can be used for receiving antennas when sharp directivity (or high resolving power) is required. In radio astronomy an antenna array consisting of tens of antennas is not unusual.

11.5. Optical Interference in Thin Films

We frequently observe that a motor oil film on a water surface appears colored. Also, good cameras all have lenses coated with certain material (such as MgO_2) to minimize the light reflection from the lenses. Here we analyze the mechanism behind these phenomena.

In Section 9.6 we learned that whenever electromagnetic waves in air try to penetrate into a medium having a characteristic impedance lower than that of air ($\sqrt{\mu_0/\varepsilon_0} = 377 \, \Omega$), the reflected electric field (or voltage) suffers a phase change of π (180°). Such a medium is called a "hard" medium. For example, glass has a dielectric constant of about 2.3 in the frequency range of visible light. Then the velocity of light in glass is about $c/\sqrt{2.3} = 0.67c$, and the characteristic impedance of glass for visible light is about $250 \, \Omega$. This yields the electric field reflection coefficient.

$$\Gamma = \frac{250 - 377}{250 + 377} = -0.20$$

and the power reflection coefficient

$$\Gamma^2 = 0.04,$$

for light incident normal to a flat glass surface. (For oblique incidence the analysis becomes complicated, and we do not consider it here.) This indicates that about 4% of the incident light energy must be reflected at the glass surface, and about 96% can penetrate into the glass. Γ itself is negative, and the reflected electric field must suffer a phase change of π on reflection.

With this knowledge we can now analyze the interference on a thin dielectric film (such as water film, soap film, oil film, etc.) placed in air (see Fig. 11.16).

Consider monochromatic light, of wavelength λ in air falling almost normal to the film. The electric field E_1 is the one reflected at the upper film surface and out of phase with respect to the incident field E_0, since the film is a medium harder than air. The electric field E_2 is due to the reflection at the lower surface, and in phase with respect to E_0, since air is softer than the film. But E_2 travels a

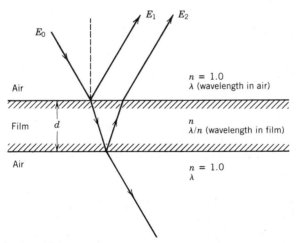

Fig. 11.16.   Reflection of light at both film surfaces causes interference. E_1 changes its sign relative to E_0, but E_2 does not.

longer distance by $2d$ than E_1, and we have to take the path difference into account to calculate the total phase difference between E_1 and E_2. Since the light velocity in the film is smaller than that in air by a factor $\sqrt{\varepsilon/\varepsilon_0} = \sqrt{\kappa}$ with κ the relative dielectric constant, the wavelength in the film becomes shorter by the same factor. We define the *index of refraction* by

$$n = \sqrt{\kappa} = c/c_{\text{film}} \qquad (11.26)$$

which is the ratio between the two light velocities.

The phase difference between E_1 and E_2 resulting from the path difference alone is

$$\frac{2\pi \cdot 2d}{\lambda/n} \quad \text{(rad)}$$

However, since E_1 has a phase difference of π relative to E_0, the net phase difference between E_1 and E_2 is

$$\phi = n\frac{4\pi d}{\lambda} - \pi$$

If this total phase difference is an integer multiple of 2π, or

$$n\frac{4\pi d}{\lambda} - \pi = m \cdot 2\pi,$$

we have constructive interference, or the reflected light is intensified. We may rewrite the preceding equation as

$$2d = (m + \tfrac{1}{2})\frac{\lambda}{n} \qquad (m = 0, 1, 2, 3, \ldots). \qquad (11.27)$$

Destructive interference occurs if

$$2d = \frac{m\lambda}{n}. \qquad (11.28)$$

If the thickness gradually varies as shown in Fig. 11.17 many stripes appear at locations where

$$d = (m + \tfrac{1}{2})\frac{\lambda}{2n} \qquad (m = 0, 1, 2, 3, \ldots) \qquad (11.29)$$

is satisfied. (Also see Newton's rings, Problem 7.)

Coatings on camera lenses work on the same principle, except we now have three media—air, film, and glass (see Fig. 11.18). We assume that $n_g > n_f > 1$ holds, where n_g is the index of refraction of lens glass, and n_f is that of the film. Reflected light now suffers π phase change at *both* surfaces, and we have

$$2d = \frac{m\lambda}{n} \quad \text{for intensification} \qquad (11.30)$$

and

$$2d = (m + \tfrac{1}{2})\frac{\lambda}{n} \quad \text{for destruction.} \qquad (11.31)$$

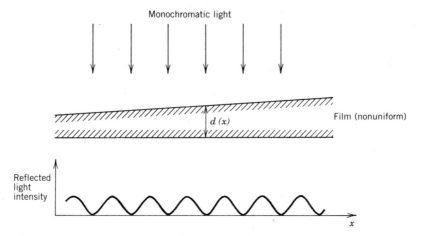

Fig. 11.17.   Dielectric film of nonuniform thickness can cause interference stripes.

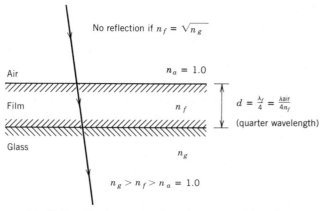

Fig. 11.18.   Coating on glass is used to prevent light reflection.

A particularly important case is the destructive interference, which of course indicates minimum reflected light intensity. The minimum thickness corresponds to $m=0$, and we have

$$d=\frac{\lambda}{4n}=\frac{\text{wavelength in film}}{4}$$

This is the well-known *quarter wavelength coating*, which is routinely applied to high-quality optical devices. It is possible to reduce the reflection to less than 0.1%, depending on the uniformity of coating and the glass surface.

The $\lambda/4$ coating alone, however, cannot completely eliminate reflection. It only gives us a necessary condition. Another condition to be imposed is that the amplitudes of E_1 and E_2 be the same. Then we can have complete destructive interference. Let Γ_1 be the voltage reflection coefficient at the air–film boundary and Γ_2 that of the film–glass boundary. The impedances of air, film, and glass are, $Z_a = 377\,\Omega$, $Z_f = 377/n_f\,\Omega$, and $Z_g = 377/n_g\,\Omega$, respectively. Then

$$\Gamma_1 = \frac{1-n_f}{1+n_f},$$

and

$$\Gamma_2 = \frac{n_f - n_g}{n_f + n_g}.$$

Thus the amplitude of the electric field E_1 is

$$E_1 = |\Gamma_1| E_0$$

and that of E_2 is

$$E_2 = (1 + \Gamma_1)|\Gamma_2| E_0.$$

For complete destructive interference, $E_1 = E_2$. Then for small Γ's, we find

$$\Gamma_1 \simeq \Gamma_2,$$

or, in terms of n's, we have

$$n_f^2 = n_g \times 1,$$

This can be rewritten in terms of the characteristic impedances as

$$Z_f^2 = Z_g Z_a.$$

Although we have derived the condition for complete destructive interference $n_f^2 = n_g n_a$ by assuming small reflection coefficients, this holds for any values of reflection coefficients.

Example 4. Find the thickness of coating and its index of refraction to minimize light reflection on the surface of glass having $n_g = 1.5$. Assume $\lambda = 5000\ \text{Å}$ in air.

From $n_f^2 = n_g$, we find $n_f = \sqrt{1.5} = 1.22$. The coating should be $\lambda_f/4$, where λ_f is the wavelength in the film. Then

$$\tfrac{1}{4}\lambda_f = \frac{1}{4n_f}\lambda_{\text{air}} = \frac{1}{4 \times 1.22} \times 5000\ \text{Å}$$

$$= 1.0 \times 10^3\ \text{Å}.$$

The concept of impedance matching using a quarter wavelength medium can also be applied to transmission line problems. When a load resistance $R\,(\neq Z)$ is to be matched to a transmission line having an impedance Z, one should insert another transmission line which has an impedance $\sqrt{RZ}$ and is a quarter wavelength long, between the load and the transmission lines. The impedance seen by the transmission line is then Z, or matching is achieved. Note that the wavelength λ' is that of waves in the transmission to be inserted, and not necessarily equal to the wavelength in the transmission line to be matched. (See Fig. 11.19.)

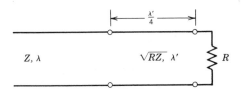

Fig. 11.19. Quarter wavelength impedance transformer.

11.6. Diffraction I (Fraunhofer Diffraction)

We have seen that a narrow opening, such as a slit, can act as a new, light source. Even though a plane wave falls on a narrow slit, the slit emits a cylindrical wave in the region behind it (Fig. 11.20). In other words, light does not always travel along a straight line. Another example is AM radio waves, which can be received even behind a high mountain. AM radio waves can go around the mountain without too much difficulty. TV waves, on the other hand, are difficult to receive. You may intuitively see that short wavelengths tend to travel along straight lines, and long wavelengths suffer stronger bending, which we call diffraction.

The analysis we are going to do for diffraction is very similar to what we did for multislit (grating) structure. As Huygens' principle assures us, we may assume a large number of light sources equally spaced at the slit opening, which has a width a (Fig. 11.21). The difference between this and the multislit case is that here we have no definite spacing between two neighboring light sources.

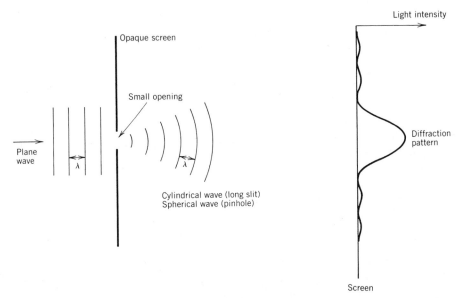

Fig. 11.20. A small opening (pinhole or narrow slit) "diffracts" light. Light does not travel in a straight line.

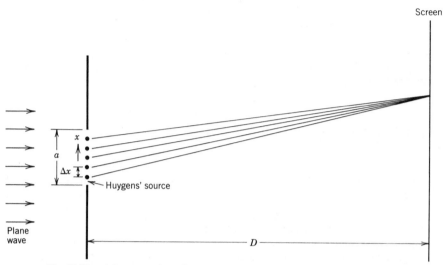

Fig. 11.21.   A large number of coherent light sources to simulate the opening.

Rather, we consider an infinitely large number of light sources, and eventually let $\Delta x \to 0$.

If D is sufficiently large, the phase difference between two neighboring waves is given by

$$\delta = \frac{2\pi}{\lambda} \Delta x \sin \theta, \tag{11.32}$$

a constant along x. The diffraction in this case is called *Fraunhofer diffraction* and is the easiest to analyze. We may substitute the preceding δ into Eq. (11.22) for the multislit case,

$$I = I_0 \frac{\sin^2\left(\frac{\pi}{\lambda} N \Delta x \sin \theta\right)}{\sin^2\left(\frac{\pi}{\lambda} \Delta x \sin \theta\right)} \tag{11.33}$$

However, $N \Delta x = a$, and if we make N very large, Δx becomes very small. Then we may approximate

$$\sin\left(\frac{\pi}{\lambda} \Delta x \sin \theta\right) \simeq \frac{\pi}{\lambda} \Delta x \sin \theta,$$

$$= \frac{1}{N} \frac{\pi}{\lambda} a \sin \theta.$$

Therefore the intensity becomes proportional to a function,

$$I = \frac{\sin^2 \alpha}{\alpha^2}, \qquad \alpha = \frac{\pi}{\lambda} a \sin \theta, \tag{11.34}$$

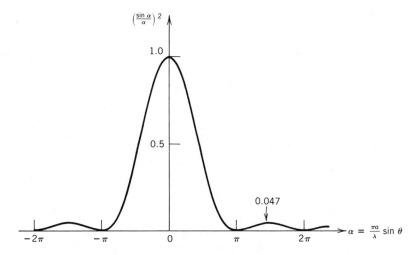

Fig. 11.22.   The function $\sin^2 \alpha/\alpha^2$ plotted versus α. It does not diverge at $\alpha = 0$.

which gives the Fraunhofer, diffraction pattern (as a function of the angular location θ) caused by a single slit with a width a.

The function $\sin^2 \alpha/\alpha^2$ is plotted in Fig. 11.22. The angular spread is approximately given by $\Delta \alpha \simeq \pi$, or

$$\sin \theta = \frac{\lambda}{a}, \tag{11.35}$$

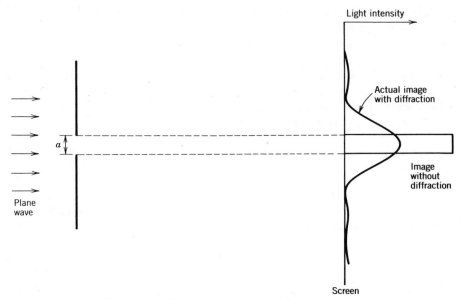

Fig. 11.23.   Diffraction pattern caused by a small opening.

and on a screen a distance D away, the vertical spread Δy is given approximately by

$$\Delta y \simeq D \frac{\lambda}{a} \qquad (11.36)$$

Thus the ideal square-shaped image that would be observed if there were no diffraction actually appears as a blurred image, without clear cut edges, as shown in Fig. 11.23. As the wavelength λ increases the image becomes more widely spread out; that is, long wavelengths suffer stronger diffraction.

11.7. Resolution of Optical Devices

As long as light has wave nature, we cannot avoid diffraction. Diffraction imposes a serious limitation on optical devices such as telescopes, microscopes, the human eye, and so on. Most optical devices have circular lenses or apertures, and we have to analyze the diffraction caused by a circular aperture, rather than a one-dimensional slit. The analysis, however, is complicated, and we do not attempt to do it here. It only introduces a factor of 1.22 in Eq. (11.35), and the angular spread for a circular aperture is given by

$$\sin \theta = 1.22 \frac{\lambda}{a}, \qquad (11.37)$$

with a now the diameter of the aperture.

Consider two light sources separated by an angle β as seen from a lens (Fig. 11.24). Images formed by the lens are inevitably blurred because of diffraction. Their angular spread is given by Eq. (11.37), about each image. If β becomes small, the two images are superposed, and we cannot tell which is which anymore. It is obvious that this critical angle is approximately given by θ itself.

Example 5. Let us take a human eye as an example. We assume it has a 5-mm aperture and wish to find the resolution for green light, $\lambda = 5500$ Å. (The human eye is most sensitive to the color green.)

From Eq. (11.37) (notice $\sin \theta \simeq \theta$ for small θ).

$$\theta \simeq 1.22 \times \frac{5.5 \times 10^{-5} \text{ cm}}{0.5 \text{ cm}}$$

$$= 1.3 \times 10^{-4} \text{ rad}.$$

Thus if the two headlights of a car are 1.5 m apart, a human eye cannot resolve them as two separate light sources at a distance more than

$$D = \frac{1.5 \text{ m}}{1.3 \times 10^{-4}} = 11.5 \text{ km}.$$

This sounds too good. Of course, we have assumed that the resolution is limited by diffraction only, and neglected other effects, such as the finite size of photoreceptors on the retina, aging, and so on.

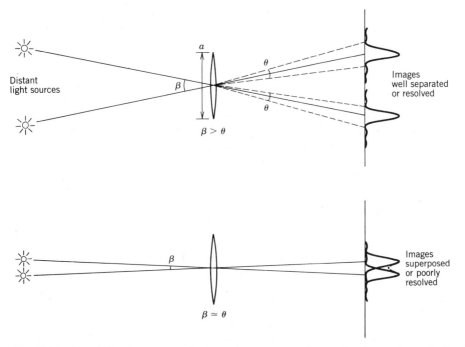

Fig. 11.24.   Two diffraction patterns (or images) get superposed on each other as the angle β becomes smaller. Light intensities are additive when the sources are incoherent.

Equation (11.37) also explains why an electron microscope can "see" better than an optical microscope. The diffraction becomes smaller as the wavelength becomes smaller, and the resolving power improves. According to De Broglie, any object having a momentum p has a wave nature. Its wavelength is

$$\lambda = \frac{h}{p}, \qquad (11.38)$$

where $h = 6.63 \times 10^{-34}$ J·sec is *Planck's constant* ($\hbar \equiv h/2\pi$), which we will learn when studying photoelectric effects. Consider an energetic electron having an energy of $100 \, \text{keV} = 10^5 \times 1.6 \times 10^{-19}$ J. From

$$\tfrac{1}{2}mv^2 = eV,$$

we find $mv = 1.7 \times 10^{-22}$ kg m/sec. Then the wavelength associated with the electron is

$$\lambda = 3.9 \times 10^{-12} \, \text{m}$$
$$= 3.9 \times 10^{-2} \, \text{Å},$$

which is roughly 10^5 times shorter than visible light wavelengths (4000–7000 Å). Thus the diffraction is expected to be small, and the resolution of electron microscopes is expected to be much better than that of optical microscopes.

Another example is astronomical telescopes. The larger the aperture, of course, the more light is collected and the brighter the image becomes. However, a more important benefit is the higher resolving power.

11.8. Diffraction II (Fresnel Diffraction)

In Fraunhofer diffraction we assumed the distance between the slit and the screen is large enough so that the phase varies linearly with x, for $0 \leqslant x \leqslant a$. If the screen is brought closer to the slit, however, this linear phase variation breaks down.

We first consider an obstacle blocking a light beam. If there were no diffraction, the image on the screen would be a sharply defined step function. The actual image, however, is blurred, with a wavy structure, as shown in Fig. 11.25. Light can even go around the obstacle.

In order to find the light intensity at the point P on the screen a distance y above the edge of the obstacle, consider the phase difference between the waves emitted from points A and B. The path difference is

$$\sqrt{D^2 + h^2} - D \simeq D\left(1 + \frac{h^2}{2D^2}\right) - D = \frac{h^2}{2D} \qquad (D \gg h),$$

which is proportional to h^2, in contrast to the previous cases of interference and Fraunhofer diffraction. In these cases, we have assumed that D is almost infinitely large and the phase difference was proportional to h, as shown below.

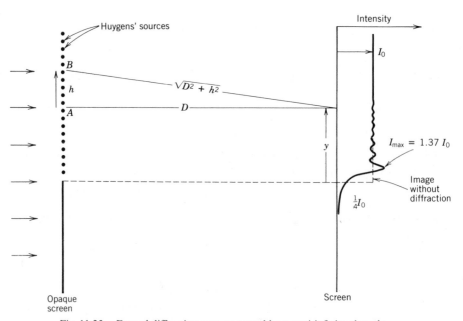

Fig. 11.25.   Fresnel diffraction pattern caused by a semi-infinite obstacle.

The phase difference corresponding to the path difference $h^2/2D$ is

$$\phi(h) = \frac{2\pi}{\lambda} \frac{h^2}{2D} = \frac{\pi h^2}{\lambda D}, \qquad (11.39)$$

in contrast to the case of multislit interference and Fraunhofer diffraction, in which we had

$$\phi(h) = \frac{2\pi}{\lambda} h \sin \theta,$$

independent of D, as noted in Fig. 11.26.

To see this difference more carefully, consider two sources located at h_1 and h_2, as given in Fig. 11.27. The path difference is given by

$$\sqrt{D^2 + h_2^2} - \sqrt{D^2 + h_1^2}$$

$$\simeq D\left(1 + \frac{h_2^2}{2D^2}\right) - D\left(1 + \frac{h_1^2}{2D^2}\right)$$

$$= \frac{1}{2D}(h_2^2 - h_1^2).$$

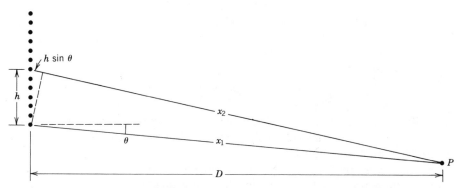

Fig. 11.26.   If the point P is very far away from the light sources, the path difference is $h \sin \theta$ independent of the distance D.

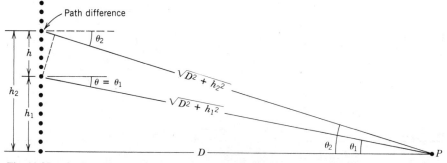

Fig. 11.27.   As D becomes smaller, the difference between the angles θ_1 and θ_2 is not negligible.

Letting $h_2 = h + h_1$, we find

$$\text{Path difference} = \frac{1}{2D}(2hh_1 + h^2).$$

If $D \gg h_1, h$, we may approximate

$$\sin \theta \simeq \frac{h_1}{D}.$$

Then the path difference becomes

$$h \sin \theta + \frac{h^2}{2D}.$$

In the case of multislit interference and Fraunhofer diffraction, we retained only the first term by assuming $D \to \infty$.

What about the amplitude of the electric fields? As we have seen before, the amplitude of cylindrical waves is inversely proportional to the square root of the distance. In the present case, the amplitude of the electric field emitted at A in Fig. 11.25 is

$$E_A \propto \frac{1}{\sqrt{D}}$$

and that of the field emitted at B is

$$E_B \propto \frac{1}{(D^2 + h^2)^{1/4}}$$

Thus if $D^2 \gg h^2$, the amplitude difference is still negligible, and we may assume that the electric fields all have an equal amplitude. Otherwise, the analysis will be insurmountably complicated.

Now we are ready to draw a phase vector diagram in Fig. 11.28 for the Fresnel diffraction. For comparison, the phase diagram for the Fraunhofer

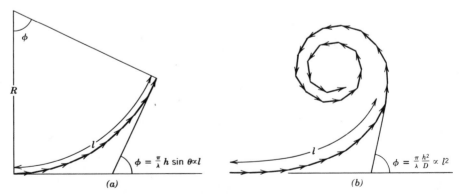

$$\phi = \frac{\pi}{\lambda} h \sin \theta \propto l \qquad \qquad \phi = \frac{\pi}{\lambda}\frac{h^2}{D} \propto l^2$$

(a)   (b)

Fig. 11.28.   Phase diagrams for (a) Fraunhofer and (b) Fresnel diffraction.

diffraction is also shown, which, as we have seen before, forms a circular arc, since the phase difference is proportional to the distance h. In contrast, for the Fresnal diffraction, the phase difference increases more rapidly, being proportional to h^2, and the phase diagram becomes a spiral. Unfortunately, there are no simple mathematical equations to describe the spiral. The only thing we can do is to describe it in terms of a parameter h, noting that the length l along the spiral curve is related to the phase angle ϕ through

$$\phi = \frac{\pi}{\lambda} \frac{h^2}{D}.$$

$$l = \text{const} \times h,$$

where l is the total length along the spiral. The spiral (Fig. 11.29) is known as the Cornu spiral, which can be parametrically described by the so-called Fresnel integrals

$$C(s) = \int_0^s \cos\left(\frac{\pi}{2} s^2\right) ds \qquad (11.40)$$

$$S(s) = \int_0^s \sin\left(\frac{\pi}{2} s^2\right) ds \qquad (11.41)$$

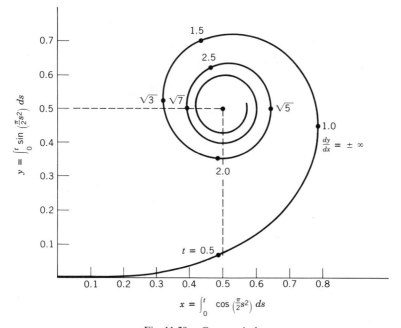

Fig. 11.29. Cornu spiral.

Using the spiral, we can qualitatively discuss how Fresnel diffraction looks, when formed by an obstacle with a sharp edge. We choose the reference electric field E_R (located at the origin O) as the one emitted by the source located at the same height as the observing point on the screen ($h=0$ in Fig. 11.25). Consider an observing point on the screen above the edge, $y=0$. Huygens' light sources below the point A in Fig. 11.25 are in the third quadrant in Fig. 11.30, and those above the point A are in the first quadrant. The amplitude of the electric field at the point P on the screen is then given by the length AP in Fig. 11.30. As y increases from zero, the length AP first increases, then oscillates about the constant length AA', which corresponds to the field amplitude on the screen well above the edge and finally assumes a constant value corresponding to the length AA'. Note that the field amplitude at $y=0$ is just one half of that at $y=\infty$ (unperturbed field), and thus the light intensity at $y=0$ is one quarter of the unperturbed intensity.

If the observing point P' on the screen is below the edge, or behind the obstacle ($y<0$), Huygens' light sources start at $h=-y$ in the first quadrant in Fig. 11.30. The field amplitude is now given by AP', which monotonically decreases as the observing point is lowered, and finally becomes zero.

The light intensity $I(y)\propto E^2(y)$ is qualitatively shown in Fig. 11.31. Fresnel diffraction (Fraunhofer, too) becomes more pronounced as the wavelength increases (Fig. 11.32).

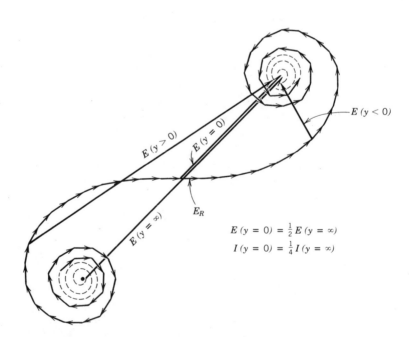

$$E (y = 0) = \tfrac{1}{2} E (y = \infty)$$
$$I (y = 0) = \tfrac{1}{4} I (y = \infty)$$

Fig. 11.30. Method to find the total electric field at various positions y on the screen. Note $E(y=0)=\tfrac{1}{2}E(y=\infty)$.

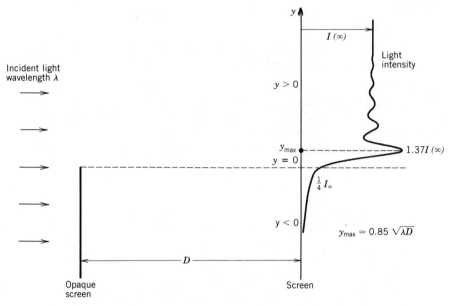

Fig. 11.31. Fresnel intensity ($I(y)$) pattern constructed from the Cornu spiral.

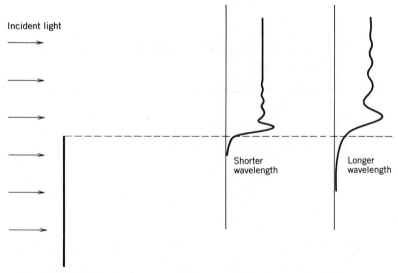

Fig. 11.32. Diffraction becomes more pronounced as the wavelength increases.

Problems

1. In Young's double-slit experiment, a fringe spacing of $\Delta y = 5$ mm is observed. Assuming the slit separation $d = 0.1$ mm and the slit-screen distance $D = 1$ m, find the wavelength.

   (*Answer:* 5000 Å.)

2. In Young's double-slit experiment, one slit is covered with a thin mica film. Discuss what changes should result.

3. What would happen if the whole apparatus of Young's double-slit experiment is immersed in water having an index of refraction of $n = 1.3$?

4. In the diagram (Fig. 11.33) S_1 and S_2 are two coherent wave sources, which radiate waves spherically in every direction. The field amplitude thus has $1/r$ dependence. Assuming that the field vector is normal to the page, find the wave intensity along the x axis as a function of x. (*Hint:* At P, the amplitudes of the fields emitted by S_1 and S_2 are not equal.)

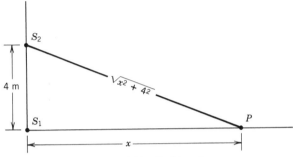

Fig. 11.33.   Problem 4.

5. A dielectric film of MgO_2 is used to prevent reflection of ruby laser light of $\lambda = 7000$ Å from a lens. Assuming that the film has an index of refraction $n = 1.4$, find the minimum thickness to be coated.

6. *Directional coupler.* In microwave waveguide circuits, a device called a directional coupler is frequently used, when it is desired that microwave energy be branched off into another waveguide system. It consists of two waveguides joined together (Fig. 11.34). Through the wall, two holes $\lambda/4$ apart are drilled. Explain why no microwaves can exist in the region III.

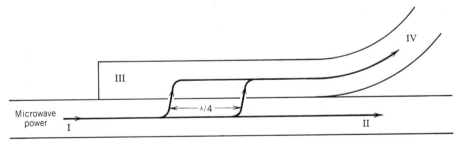

Fig. 11.34.   Problem 6.

7. *Newton's rings.* A planoconvex lens rests on a flat glass surface. Light of wavelength λ falls normal to the plane surface (Fig. 11.35).
   (a) Find the spacing d as a function r, the radial position. Assume R (curvature radius) $\gg r$.

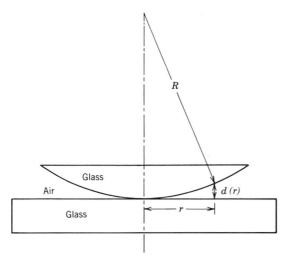

Fig. 11.35. Problem 7.

(b) Show that bright interference fringes are located at the positions given by

$$r=[(m+\tfrac{1}{2})R\lambda]^{1/2},$$

where m is an integer.

8. A spy satellite is said to be able to resolve two points on the earth a distance 50 cm apart. Assuming the satellite is 200 km high and $\lambda = 5000$ Å, find the minimum diameter of the telescope carried by the satellite.

(*Answer:* 24 cm.)

9. In single-slit diffraction what would happen if the slit opening is doubled?

(*Answer:* The peak intensity is quadrupled, and the angular spread becomes one half. Check energy conservation (i.e., explain why the peak intensity is quadrupled even though the amount of light energy passing through the slit is only doubled).

10. *Diffraction in double-slit experiment.* Assume each slit has an opening a in Young's double-slit experiment (Fig. 11.36). Then, in addition to the interference, we expect diffraction due to the finite aperture. Show that the intensity on the screen is given by

$$I = I_0 \cos^2 \beta \left(\frac{\sin \alpha}{\alpha}\right)^2,$$

where

$$\alpha = \frac{\pi a}{\lambda} \sin \theta$$

$$\beta = \frac{\pi d}{\lambda} \sin \theta.$$

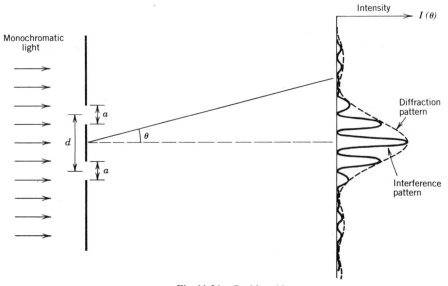

Fig. 11.36.   Problem 10.

11. In Problem 10, assuming $d = 2a$, plot the light intensity as a function of θ.

12. Explain why AM radio can be received better than FM radio in mountain areas.

(*Note:* You may wonder why TV stations do not use lower frequencies since lower frequencies can be diffracted more and received better. The reason is simply that it is impossible. When we say an AM radio wave has a frequency, say, of 540 kHz, what we actually mean is that the radio station uses a frequency band 520 kHz $< f <$ 560 kHz, where ± 20 kHz corresponds to the maximum audio frequency. For TV signals the required frequency *band* is of the order of 4 MHz! Thus the carrier frequency must be much higher than this, and in fact all TV stations have carrier frequencies around 100 MHz. Now we see why communication engineers are so interested in using laser light for communication. By increasing the carrier frequency, more channels can be allocated and more information can be transmitted. Laser communication is already in use, in which the development of high-quality optical fibers has played the major role.)

13. The schematic diagram of a spectrometer is shown in Fig. 11.37. Light enters the entrance slit, is reflected by a spherical mirror ($M1$), hits a grating (G), is reflected by another mirror ($M2$) again, and finally emerges from an exit slit. The grating is rotatable and a desired wavelength can thus be chosen. The distance between the grating and the slit corresponds to the slit-screen distance D in the text.

(a)   It is desired that the *dispersion* ($\Delta\lambda/\Delta y$) be 10 Å/mm at the exit slit.

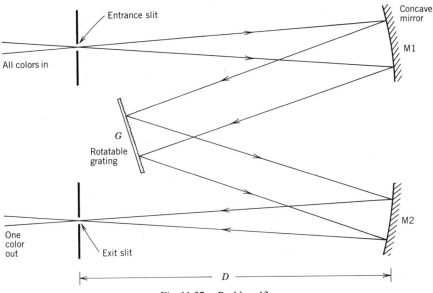

Fig. 11.37. Problem 13.

Assuming $\lambda = 5000$ Å, $D = 1.3$ m, and m (order) $\doteq 1$, find the number of grooves (in 1 cm) of the grating.

(*Answer:* 7700/cm.)

(b) Assuming that the grating has a width of 10 cm, estimate the halfwidth of the intensity profile at $\lambda = 5000$ Å.

14. Hold horizontally an LP record close to your eyes and let the record reflect light coming from a distant light source. You will see color spectrum. Explain.

15. Look at a distant point light source (a flashlight will do) through a nylon stocking. You will find several colored rings around the light source. Can you explain? (Rings around the moon on a foggy night are caused by the same principle. They are due to a diffraction by small water drops.)

16. Consider a glass surface coated with a quarter-wavelength-thick dielectric film, as in Fig. 11.18. Show that the effective impedance for light incident on the coated glass is given by Z_f^2/Z_g, where Z_f is the characteristic impedance of film and Z_g is that of glass.

17. Light of wavelength 6000 Å (in air) falls normally on a plastic film that has a permittivity $\varepsilon = 4\varepsilon_0$ (Fig. 11.38).

(a) If the film is to be a quarter wavelength thick, what should the thickness be?

(b) Calculate how much (in %) energy is reflected and transmitted.

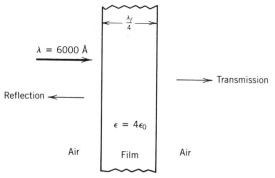

Fig. 11.38.   Problem 17.

(*Answer:*   750 Å, reflection 36%, transmission 64%.) (*Hint:*   Use the result of Problem 16.)

18. (a) Calculate the radiation electric field from a half-wave dipole antenna (Fig. 11.39). Assume that the current is distributed on the antenna as $I = I_0 \cos\left[(2\pi/\lambda)Z\right]$.

  (*Answer:* $E_\theta \propto I_0/r \cos\left[(\pi/2) \cos\theta\right]/\sin\theta$).

  (b) Sketch the radiation field pattern as a function of θ.

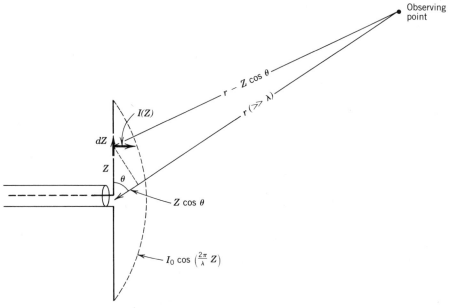

Fig. 11.39.   Problem 18.

19. (a) Using the expression for the radiation electric field due to an oscillating dipole, Eq. (10.24), show that the amplitude of the radiation electric field due to a small oscillating current loop (Fig. 11.40) is

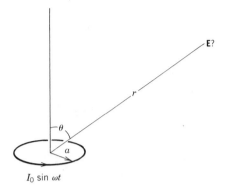

$I_0 \sin \omega t$

Fig. 11.40. Problem 19.

given by

$$E = \frac{\mu_0 M c}{4\pi} \frac{1}{r} \left(\frac{\omega}{c}\right)^2 \sin \theta$$

where $M = I_0 \pi a^2$ is the magnetic dipole moment.

(b) Calculate the total power radiated by the magnetic dipole.
(*Answer:* $\omega^4 M^2 / 12\pi\varepsilon_0 c^5$.)

CHAPTER 12

Geometrical Optics

12.1. Introduction

In Chapter 11 we saw that the diffraction pattern strongly depends on the aperture size a. A small circular aperture essentially acts as a point light source, and no clear image can be formed on the screen behind the hole. As the aperture size increases, the diffraction becomes less significant and the image becomes clear.

Geometrical optics is one branch of optics in which we neglect interference and diffraction. We assume that a light beam propagates along a straight line in a uniform medium, neglecting the increase in the angular spread that is inevitable as the beam propagates. On encountering a foreign medium (from air to glass, for example), the light beam changes its direction, being reflected and refracted, but both the reflected and refracted beams again travel along straight lines. Geometrical optics can greatly simplify the analysis of conventional optical devices, such as mirrors and lenses, as long as the medium (e.g., lens glass) is uniform. Another requirement is that the boundary between two different media (air and glass, for example) be smooth. Otherwise reflection and refraction become random and the light beam is scattered.

12.2. Reflection and Refraction

Suppose a light beam falls obliquely on the flat surface of glass with an index of refraction n_g. The light speed in glass is thus c/n_g, or the wavelength in glass is shorter than that in air by a factor of n_g/n_a, where n_a is the index of refraction of air ($n_a \simeq 1$) (Fig. 12.1).

You may know that the reflection angle θ'_1 is equal to the incident angle θ_1, just like the case of an elastic ball hitting a heavy wall. Here we prove this and also find the relationship between the incident angle θ_1 and refraction angle θ_2. Before doing this, we need some mathematics. As we have seen, a plane wave (Fig. 12.2) propagating in the x direction is described by

$$E(x, t) = E_0 \sin (kx - \omega t), \qquad (12.1)$$

where $k = 2\pi/\lambda$ is the wavenumber. How can we describe the wave propagating

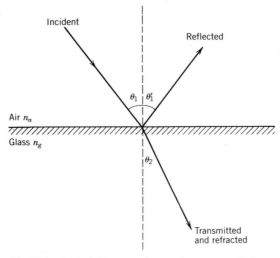

Fig. 12.1.   Light falling on a glass surface at an angle θ_1.

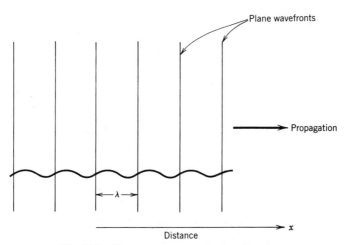

Fig. 12.2.   Plane wave propagating in x direction.

obliquely as shown in Fig. 12.3? It is obvious that the wave is now written as

$$E(x, y, t) = E_0 \sin (ks - \omega t), \tag{12.2}$$

where $s = \sqrt{x^2 + y^2}$ is the distance along the straight line OS. The wave number k is still defined by $k = 2\pi/\lambda$. Now we rewrite ks as (note that $\cos^2 \theta + \sin^2 \theta = 1$)

$$ks = k \cos \theta \, s \cos \theta + k \sin \theta \, s \sin \theta.$$

However, $s \cos \theta = x$, and $s \sin \theta = y$ from Fig. 12.4. Thus

$$ks = k \cos \theta \, x + k \sin \theta \, y. \tag{12.3}$$

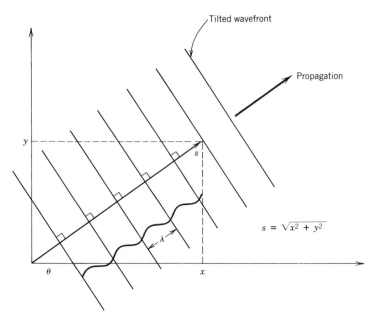

Fig. 12.3.   Plane wave propagating in an oblique direction.

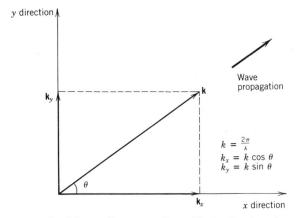

Fig. 12.4.   The wavenumber k is actually a vector directed in the direction of wave propagation.

It is now clear that k is actually a vector directed along the direction of wave propagation. Its magnitude is given by $2\pi/\lambda$, and its x component by $k \cos \theta$, and y component by $k \sin \theta$. We can write ks as

$$ks = \mathbf{k} \cdot \mathbf{r}, \quad \mathbf{r} = \mathbf{x} + \mathbf{y} \tag{12.4}$$

and a wave propagating in an arbitrary direction is conveniently written as

$$E = E_0 \sin (\mathbf{k} \cdot \mathbf{r} - \omega t). \tag{12.5}$$

For the problem of reflection and refraction, we assign $\mathbf{k}_1$, $\mathbf{k}_1'$, and $\mathbf{k}_2$ for the incident, reflected, and refracted waves, respectively (Fig. 12.5). The incident wave is actually composed of two waves, one propagating in the negative y direction, and another propagating in the positive x direction (Fig. 12.6). The component propagating along the x direction never hits the boundary, and the x component of k should not change. Thus

$$k_{1x}' = k_{1x}.$$

The vertical component k_{1y} should simply change its sign on reflection. Thus

$$k_{1y}' = -k_{1y}.$$

It is then obvious that the angles θ_1 and θ_1' must be the same.

For refraction the x components k_{1x} and k_{2x} are the same, as seen from Fig. 12.7. The number of waves contained in a given distance along the x axis is the same in both air and glass. Thus

$$k_1 \sin \theta_1 = k_2 \sin \theta_2. \tag{12.6}$$

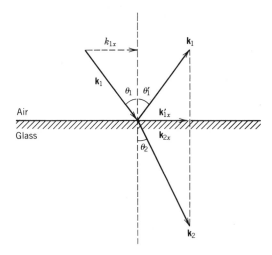

Fig. 12.5. $\mathbf{k}$ diagram corresponding to Fig. 12.1. Note that $k_{1x} = k_{1x}' = k_{2x}$.

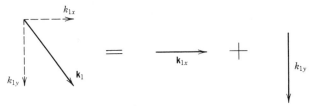

Fig. 12.6. Decomposition of the incident wavenumber $\mathbf{k}_1$.

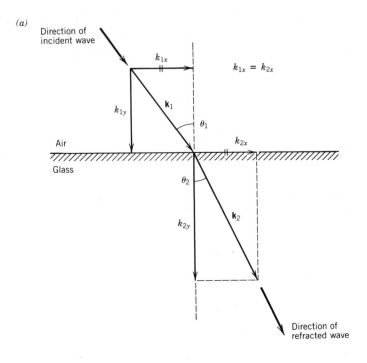

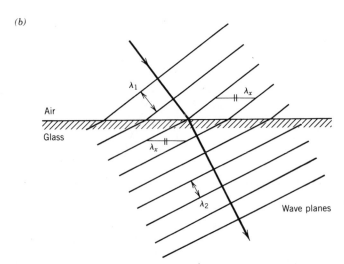

Fig. 12.7. (a) Change in the vector **k** on refraction. The component parallel to the boundary (k_x) remains unchanged. (b) Illustration of refraction by wave planes. Note that the wavelength along $x(\lambda_x)$ does not change.

244

However, the light velocity in glass is c/n_g, or

$$\frac{\omega}{k_2} = \frac{c}{n_g}$$

Thus, recalling $\omega/k_1 = c/n_a$, we find

$$\frac{\sin \theta_1}{\sin \theta_2} = \frac{k_2}{k_1} = \frac{n_g}{n_a} \qquad (12.7)$$

This is known as *Snell's law*.

There is a more rigorous way to derive the Snell's law using proper boundary conditions for electric and magnetic fields associated with light, but let us be satisfied here with the preceding somewhat qualitative derivation.

The physical meaning behind refraction is that light (or wave) is bent toward a region of lower phase velocity, as we briefly saw before (Chapter 9). Since the light velocity in glass is smaller than that in air, light is pulled by glass and k_{2y} is indeed larger than k_{1y}. The reflection, on the other hand, is caused by impedance mismatching, which we also have studied. Glass has a lower characteristic impedance than air, and some energy must be reflected (about 4% for ordinary glass if light falls almost normally on glass).

As an example of light refraction in a nonuniform medium, let us take a look at the bending of sunlight in the earth's atmosphere. The air density gradually becomes lower with height (Fig. 12.8). The velocity of light in air is not exactly equal to c, but slightly lower, depending on the air density. Thus the light beam is bent toward the lower phase velocity region, where the air density is higher. This example is opposite to the case of mirage discussed earlier (Chapter 9, Problem 21).

The index of refraction n of a given material (say, glass) is not a constant, but weakly depends on the wavelength λ. Shown in Fig. 12.9 is the index of refraction of flint glass, crown glass and quartz as functions of visible wave-

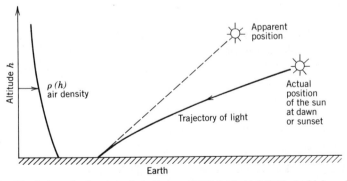

Fig. 12.8. The index of refraction of air (1 atm, 20°C) is about 1.0003. At higher altitudes, air density decreases and so does the index of refraction. Thus the velocity of light becomes a function of the altitude, and light is refracted or bent toward the slower velocity region.

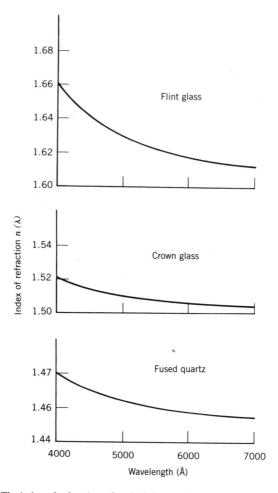

Fig. 12.9.   The index of refraction of typical glass and quartz as functions of wavelength.

lengths, 4000–7000 Å. It can be seen that blue light is refracted more than red light, and this explains the prism spectrograph. This wavelength dependence of the index of refraction is undesirable for lenses and causes chromatic aberration, as we will see later.

Example 1.   Monochromatic (single color or wavelength) light is incident on 45° prism at an angle 40°. Taking $n = 1.5$ for glass, find the total refraction angle α (see Fig. 12.10).

For the refraction at the first surface, Snell's law gives

$$\frac{\sin 40°}{\sin \theta_1} = 1.5.$$

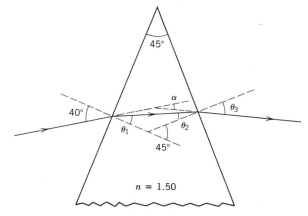

Fig. 12.10.   Example 1.

Then

$$\theta_1 = \sin^{-1}\left(\frac{\sin 40°}{1.5}\right) = 25.4°.$$

Since

$\theta_1 + \theta_2 = 45°$, we find $\theta_2 = 19.6°$.

Applying again Snell's law at the exit surface,

$$\frac{\sin \theta_3}{\sin \theta_2} = 1.5,$$

we find

$$\theta_3 = \sin^{-1}(\sin \theta_2 \times 1.5) = 30.2°.$$

Then the total refraction angle is

$$\alpha = (40° - \theta_1) + (\theta_3 - \theta_2)$$

$$= 25.2°.$$

Example 2.   Repeat Example 1 by assuming another wavelength for which $n = 1.46$ (see Fig. 12.11).

$$\theta_1 = \sin^{-1}\frac{\sin 40°}{1.46} = 26.1°$$

$$\theta_2 = 45° - \theta_1 = 18.9°$$

$$\theta_3 = \sin^{-1}(\sin \theta_2 \times 1.46) = 28.2°.$$

Then

$$\alpha = 40° - \theta_1 + \theta_3 - \theta_2 = 23.2°.$$

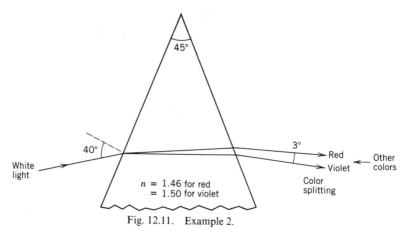

Fig. 12.11. Example 2.

12.3. Total Reflection

Consider a light beam emerging from the water surface into air. Water has an index of refraction of about 1.3. Then from Snell's law (see Fig. 12.12) we can write:

$$\frac{\sin \theta_2}{\sin \theta_1} = \frac{1}{n}.$$

As the incident angle θ_2 increases, the refraction angle θ_1 increases too, and finally reaches 90°. The incident angle at which this occurs is

$$\sin \theta_2 = \frac{1}{n} = \frac{1}{1.3}$$

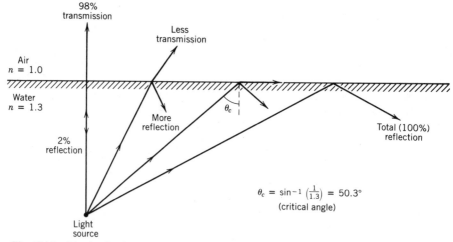

Fig. 12.12. Total reflection occurs when the incident angle θ_2 is larger than the critical angle θ_c.

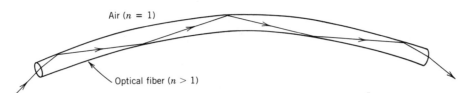

Fig. 12.13.   Hair-thin optical fiber can guide light that is totally reflected at the surface. The fiber is a dielectric waveguide for light.

or $\theta_2 = 50.3°$ (for water). At angles larger than this, light is completely reflected at the boundary, and no light will come out of water. The angle is called the *critical angle for total reflection.*

The optical fiber uses this principle of total reflection, as shown in Fig. 12.13. The light beam can be bent effectively along the fiber. Again for light not to be scattered at the surface, the fiber surface must be extremely smooth, without lumps. Recent development of high-quality optical fibers has enabled using laser beams for communication. The important aspects of optical fibers for communication are

1. Small light attenuation and scattering.
2. Low dispersion.

Otherwise signals are strongly damped and/or dispersed, that is, the signal waveform sent out by a transmitter is severely deformed.

Example 3.   For what incident angles does total reflection occur at the vertical surface? Take $n = 1.3$ (see Fig. 12.14).

From Snell's law,

$$\frac{\sin \theta}{\sin \theta_1} = n = 1.3.$$

Since $\theta_2 = 90° - \theta_1$ and the critical angle θ_2 is given by

$$\sin \theta_2 = \cos \theta_1 = \frac{1}{n} = \frac{1}{1.3},$$

we find

$$\sin \theta = 1.3 \times \sin \theta_1 = 1.3\sqrt{1 - \cos^2 \theta_1}$$
$$= 1.3 \times \sqrt{1 - 1/1.3^2} = 0.83$$

or

$$\theta = 56°.$$

Total reflection at the vertical surface occurs if the angle θ is less than 56°.

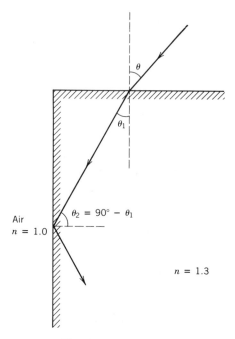

Fig. 12.14. Example 3.

12.4. Reflection at Spherical Surfaces (Mirrors)

Light reflection at a spherical surface works in the same way as that at the flat surface. The incident angle θ is now with respect to the curvature radius R of the surface, as shown in Fig. 12.15. Let us place a point light source 0 at a distance o from the mirror. We want to find out where a light beam leaving the source at an angle α to the axis finally hits the axis again. Let this distance from the mirror be i. Using geometry, we find

$$\alpha + 2\theta = \beta, \qquad \alpha + \theta = \gamma$$

or

$$\alpha + \beta = 2\gamma. \tag{12.8}$$

Here we make a major assumption that will be used throughout this section. That is, we assume that all the angles—α, β, θ—are small. When measured in radians, this means

$$\alpha, \beta, \gamma \ll 1 \quad (\text{rad}),$$

This assumption can alternatively be put as $h, \delta \ll o, i, R$ where h is the height of the point where the light beam hits the mirror and δ is the deviation from the mirror center.

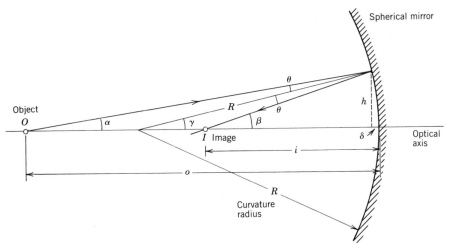

Fig. 12.15.   Reflection at a spherical surface.

If α, β, γ are small, we may approximate

$$\alpha \simeq \tan \alpha, \; \beta \simeq \tan \beta, \; \gamma \simeq \tan \gamma,$$

since the Taylor expansion for $\tan x$ is given by (Chapter 3)

$$\tan x = x + \frac{1}{3} x^3 + \frac{2}{15} x^5, \ldots.$$

But $\tan \alpha = h/(o - \delta) \simeq h/o$ ($\delta \ll o$), $\tan \beta = h/i$ and $\tan \gamma = h/R$. Substituting these into Eq. (12.8), we find

$$\frac{1}{o} + \frac{1}{i} = \frac{2}{R}, \tag{12.9}$$

which is the mirror formula. We define the focal length of the mirror as the image distance formed when the light source is placed far away, $o \to \infty$, $f \equiv R/2$. Using this *focal length*, the mirror formula becomes (Fig. 12.16)

$$\frac{1}{o} + \frac{1}{i} = \frac{1}{f} \tag{12.10}$$

Equation (12.10) can be generalized if we properly interpret the sign ($+$ or $-$) of all the quantities, o, i, and f. The sign convention we adopt is

$o, i > 0$ for object and image in front of the mirror

$o, i < 0$ for object and image behind the mirror

$f > 0$ for concave mirrors ($R > 0$)

$f < 0$ for convex mirrors ($R < 0$).

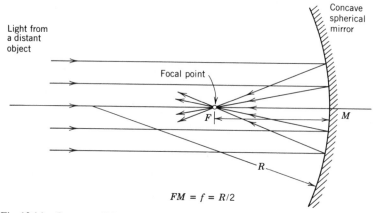

Fig. 12.16.   Rays parallel to the axis are all converged (or focused) at the focal point.

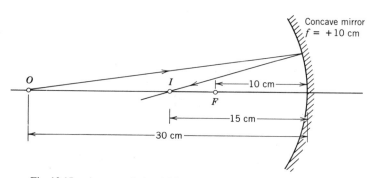

Fig. 12.17.   An example in which $o=30$ cm, $f=10$ cm, and $i=15$ cm.

Let us work on the example in Fig. 12.17. If an object is placed at $o=+30$ cm, the image is formed at $i=+15$ cm, as we easily find from Eq. (12.10). In high school you must have learned the method of ray tracing, assuming a finite height for the object. This works in the same way as the formula we obtained, since in Fig. 12.18 we have

$$i=\frac{R}{2}+\frac{h'}{\tan 2\theta}, \qquad h'=\frac{i}{o}h$$

and

$$\tan\theta=\frac{h}{R}.$$

Again assuming a small θ so that $\tan\theta\simeq\theta$, we readily find

$$\frac{1}{o}+\frac{1}{i}=\frac{2}{R}.$$

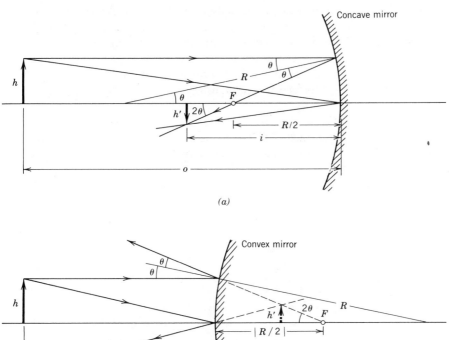

Fig. 12.18. Finding the image location by ray tracing. (a) Concave mirror. (b) Convex mirror.

The ratio between the heights h' and h may be called the magnification m. We define

$$m = -\frac{i}{o} \qquad (12.11)$$

so that m becomes positive for an *erect image* (arrows in the same direction) and negative for an *inverted image* (arrows in the opposite direction). In the preceding sample,

$$m = -\frac{15}{30} = -0.5,$$

which tells us that the image size is one-half of the object size and inverted.

There is one more complication. Consider now $o = +5$ cm in the preceding. example. If we directly use the formula

$$\frac{1}{5} + \frac{1}{i} = \frac{1}{10},$$

we find $i = -10$ cm ($m = +2.0$), which tells us that the image is formed behind the mirror ($i < 0$) (Fig. 12.19). But light can never go into the region behind the mirror, and the image in this case is not formed by real light beam intersecting with the axis. The image is formed by the beam extended toward the region behind the mirror (broken line in the figure). Such an image is called a *virtual image*, and the image formed by real light rays is called a *real image*.

Example 4.   *Virtual Object.*   Consider a light beam falling on a concave mirror as shown in Fig. 12.20. To find the location of the image, using Eq. (12.10), we have to put $o = -30$ cm, since the beam intersects with the axis behind the mirror. From

$$\frac{1}{-30} - \frac{1}{i} = \frac{1}{20}$$

we find $i = 12$ cm.

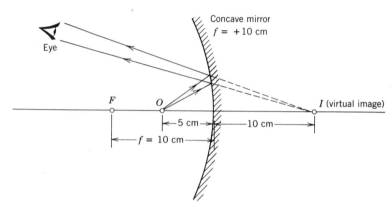

Fig. 12.19.   Example of virtual image, $i < 0$.

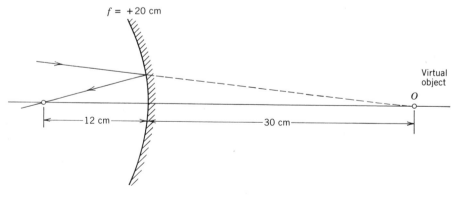

Fig. 12.20.   Example 4.

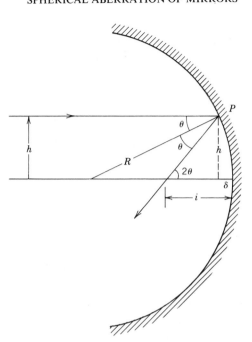

Fig. 12.21.   Ray tracing without assuming $h \ll R$.

12.5. Spherical Aberration of Mirrors

In deriving the mirror formula, we assumed that h, the vertical distance between P and the axis, is small compared with the curvature radius R. Under this condition the light beams parallel to the axis are all focused at $i = R/2$, the focal distance. Let us remove this assumption here and find where a parallel beam hits the axis after reflection. The maximum value of h is obviously R.

In Fig. 12.21 we observe that

$$i = \delta + \frac{h}{\tan 2\theta} = \delta + h \frac{\cos 2\theta}{\sin 2\theta}$$

$$R^2 = (R - \delta)^2 + h^2$$

$$\sin \theta = \frac{h}{R}.$$

Eliminating δ and θ, we find

$$i = R \left[1 - \frac{1}{2\sqrt{1 - (h/R)^2}} \right]$$

If $h \ll R$, we indeed recover $i = R/2 = f$. As h increases, the image distance i becomes smaller and at $h = (\sqrt{3}R/2)$, i becomes zero. The variation of the focal distance with the height h is called spherical aberration and cannot be

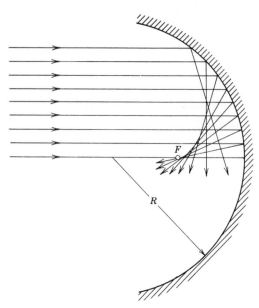

Fig. 12.22.   Incident rays parallel to the axis are not focused if the mirror aperture is large (spherical aberration of mirror).

avoided as long as the reflector surface is spherical (Fig. 12.22). However, if h is small, or if the opening aperture is kept small, the aberration can practically be neglected.

Spherical aberration can be avoided for light beams parallel to the axis by using a parabolic mirror (Problem 19). Microwave antennas usually have this parabolic shape to eliminate spherical aberration.

12.6. Refraction at Spherical Surfaces

Refraction at spherical surfaces provides us with a basis for understanding how lenses work. Here again we assume that all angles α, β, γ, θ are small or the height h is much less than the curvature radius R. This is necessary if the spherical aberration is to be negligible. In Fig. 12.23, we have

$$\gamma = \beta + \theta_2$$

$$\theta_1 = \alpha + \gamma$$

However, Snell's law requires

$$\frac{\sin \theta_1}{\sin \theta_2} = n$$

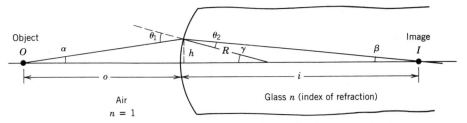

Fig. 12.23. Refraction at a spherical surface. R is the curvature radius of the surface.

where n is the index of refraction of glass. Then for small θ_1 and θ_2, we have

$$\frac{\sin \theta_1}{\sin \theta_2} \simeq \frac{\theta_1}{\theta_2} = \frac{\alpha + \gamma}{\gamma - \beta} = n,$$

or

$$\alpha + n\beta = (n-1)\gamma.$$

Using

$$\alpha \simeq \frac{h}{o}, \qquad \beta \simeq \frac{h}{i}, \qquad \text{and} \qquad \gamma \simeq \frac{h}{R},$$

we obtain

$$\frac{1}{o} + \frac{n}{i} = (n-1)\frac{1}{R}. \qquad (12.13)$$

The sign convention we adopt here is as follows:

$i > 0$ for image in glass or behind the surface

$i < 0$ for image in air or in front of the surface,

in contrast to the case of mirrors. For the curvature radius R,

$R > 0$ for a convex surface as seen from the object.

$R < 0$ for a concave surface as seen from the object.

As an example, consider a concave glass surface shown in Fig. 12.24. Since R is negative, we have

$$\frac{1}{20} + \frac{1.5}{i} = (1.5 - 1)\frac{1}{-10}$$

or

$$i = -15 \text{ cm},$$

which indicates that the image is formed in front of the surface.

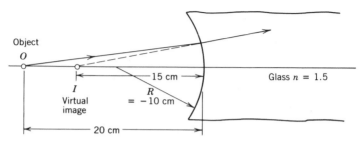

Fig. 12.24. Example of refraction at a concave spherical surface.

12.7. Lenses

The formula we obtained enables us to analyze lenses that have two spherical surfaces with curvature radii R_1 and R_2. These surfaces must have a common optical axis. Otherwise the lens cannot form clear images. For the lens shown in Fig. 12.25, $R_1 > 0$ and $R_2 < 0$ according to the sign convention for curvature radii.

At the first surface, we have

$$\frac{1}{o} + \frac{n}{i'} = (n-1)\frac{1}{R_1} \tag{12.14}$$

where i' is the location of the image if the second surface is absent. At the second surface,

$$\frac{1}{i} - \frac{n}{i'-l} = (n-1)\frac{1}{-R_2} \quad (R_2 < 0). \tag{12.15}$$

This may need some explanation. The distance i can be regarded as an object distance and the image is formed at $i'-l$ in *front* of the surface (Fig. 12.26). (Compare this with Fig. 12.23.) Thus when using the *refraction* formula [Eq. (12.13)], we have to substitute $-(i'-l) < 0$. Similarly, since $R_2 < 0$, and the "object I" sees a convex surface, the effective curvature radius must be positive, $-R_2 (>0)$.

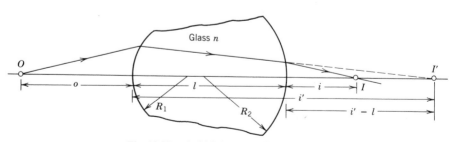

Fig. 12.25. A thick lens used for analysis.

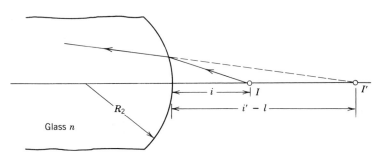

Fig. 12.26. Rays can be reversed—explanation of Eq. (12.15).

Now we neglect l compared with the distance i', that is, we assume that the lens is very thin, $l \to 0$. Adding the two equations, we find

$$\frac{1}{o} + \frac{1}{i} = (n-1)\left(\frac{1}{R_1} - \frac{1}{R_2}\right), \tag{12.16}$$

where the focal length f is defined by

$$\frac{1}{f} = (n-1)\left(\frac{1}{R_1} - \frac{1}{R_2}\right). \tag{12.17}$$

If $f > 0$, the lens is a *converging lens*, and if $f < 0$, the lens is a *diverging lens*. Four fundamental lenses are illustrated in Fig. 12.27. The lens formula

$$\frac{1}{o} + \frac{1}{i} = \frac{1}{f} \tag{12.18}$$

is identical in form to the mirror formula, except for the sign convention for the image distance i. The magnification m is still defined by

$$m = -\frac{i}{o}$$

and can be interpreted in exactly the same way as the magnification for mirrors.

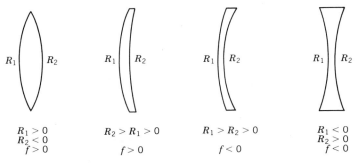

Fig. 12.27. Four fundamental lenses.

Example 5. Find the location and nature (real or virtual, erect or inverted) of the final image formed (Fig. 12.28).

For the first convergent lens, the lens formula gives

$$\frac{1}{50} + \frac{1}{i} = \frac{1}{30}$$

or

$$i = 75 \text{ cm}.$$

The magnification due to the first lens is

$$m_1 = -\frac{75}{50} = -1.5.$$

For the second lens we have -35 cm object distance (virtual!). Then

$$-\frac{1}{35} + \frac{1}{i'} = -\frac{1}{10}$$

or $i' = -17.5$ cm, and $m_2 = -0.5$. Thus the final image is formed at 17.5 cm to the left of the divergent lens, and the final magnification is

$$m = m_1 m_2 = 0.75.$$

Since $m > 0$, the image is erect. The image is virtual since the ray does not intersect with the axis after going through the second (final) lens.

In the preceding example let the separation between the lenses approach zero. Then the virtual object distance for the divergent lens becomes -75 cm, and the final image is formed at i' determined from

$$\frac{1}{50} + \frac{1}{i'} = \frac{1}{30} - \frac{1}{10}.$$

In general, if two lenses of focal lengths f_1 and f_2 are attached together, the compound lenses form a lens with an effective focal length determined from

$$\frac{1}{f_{\text{eff}}} = \frac{1}{f_1} + \frac{1}{f_2}. \tag{12.19}$$

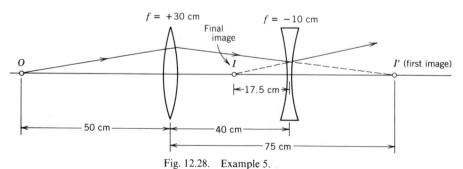

Fig. 12.28. Example 5.

Good cameras or any high-quality optical devices all have compound lenses to eliminate chromatic aberration, which we study next.

12.8. Chromatic Aberration

Besides the spherical aberration, which is common to both mirrors and lenses, lenses have another aberration problem. As we have seen before, the index of refraction n depends on wavelength. Blue light is refracted more than red light, and color splitting takes place (Fig. 12.29). This is extremely undesirable for lenses, since we want lenses to form images of blue light and red light at the same location. Otherwise images become blurred.

The focal length for blue light is given by

$$\frac{1}{f_B} = (n_B - 1)\left(\frac{1}{R_1} - \frac{1}{R_2}\right)$$

and for red light   (12.20)

$$\frac{1}{f_R} = (n_R - 1)\left(\frac{1}{R_1} - \frac{1}{R_2}\right),$$

where n_B is the index of refraction at blue wavelength, 5000 Å, and n_R is that at red wavelength, 6500 Å. If $n_B = n_R$, we have no problems, but all glass materials used for lenses have dispersion, that is, the light velocity depends on wavelengths.

There is an ingenious way to eliminate the chromatic aberration. Consider a compound lens (Fig. 12.30). Since the focal length of a compound lens is given by

$$\frac{1}{f} = \frac{1}{f_1} + \frac{1}{f_2},$$

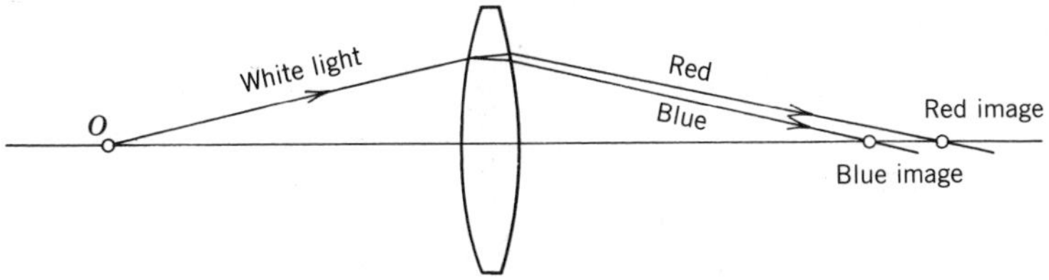

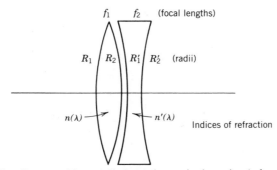

Fig. 12.29.   Prism action of lens to cause chromatic aberration.

Fig. 12.30.   Compound lens to eliminate chromatic aberration (achromatic lens).

the focal length for blue light becomes

$$\frac{1}{f_B} = \frac{1}{f_{1B}} + \frac{1}{f_{2B}} = (n_B - 1)\left(\frac{1}{R_1} - \frac{1}{R_2}\right) + (n'_B - 1)\left(\frac{1}{R'_1} - \frac{1}{R'_2}\right)$$

and that for red light is

$$\frac{1}{f_R} = (n_R - 1)\left(\frac{1}{R_1} - \frac{1}{R_2}\right) + (n'_R - 1)\left(\frac{1}{R'_1} - \frac{1}{R'_2}\right).$$

Thus if $f_B = f_R$, or

$$(n_B - n_R)\left(\frac{1}{R_1} - \frac{1}{R_2}\right) = (n'_R - n'_B)\left(\frac{1}{R'_1} - \frac{1}{R'_2}\right) \tag{12.21}$$

the chromatic aberration can be eliminated. Since $n_B > n_R$, the compound lens must be composed of one convergent lens and one divergent lens.

Example 6. Crown glass has $n_B = 1.510$, $n_R = 1.505$ and flint glass has $n_B = 1.630$, $n_R = 1.615$. Design an achromatic compound lens having a focal length of 50 cm. You may assume the compound lens consists of two lenses attached together having a common curvature radius as shown (Fig. 12.31).

From Eq. (12.21), we have

$$0.005\left(\frac{1}{R_1} - \frac{1}{R_2}\right) = -0.015\left(\frac{1}{R'_1} - \frac{1}{R'_2}\right)$$

where $R_2 = R'_1$ and $R'_2 = \infty$. Then

$$\frac{1}{R_1} - \frac{1}{R_2} = -3\frac{1}{R_2}$$

or

$$R_2 = -2R_1 \tag{A}$$

From the formula for a compound lens, we have

$$\frac{1}{50} = 0.51\left(\frac{1}{R_1} - \frac{1}{R_2}\right) + 0.63\left(\frac{1}{R_2} - \frac{1}{\infty}\right)$$

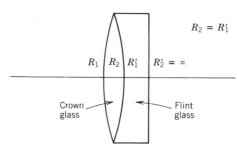

Fig. 12.31. Example 6.

or

$$\frac{1}{50} = 0.51 \frac{1}{R_1} + 0.12 \frac{1}{R_2} \qquad \text{(B)}$$

Solving (A) and (B) for R_1 and R_2, we find

$$R_1 = 22.5 \text{ cm} \quad \text{and} \quad R_2 = -45 \text{ cm}.$$

You should check that the compound lens indeed has a focal length of 50 cm.

12.9. Optical Instruments

Magnifying Glass.

A convergent lens can create a virtual image with a magnification larger than unity. This can be seen from the formula

$$\frac{1}{o} + \frac{1}{i} = \frac{1}{f},$$

which yields

$$m = -\frac{i}{o} = \frac{f}{f - o}.$$

The magnification becomes large as the object distance o approaches the focal length. In practice, however, the image must be formed at a point where the human eye, which is also an optical instrument can form a distinct image without effort. For a normal person this distance is about 25 cm, or $i = -25$ cm, since the image must be virtual. Also, the object should be placed near the focal point of the lens, and we can put $o \simeq f$. Then the magnification of a magnifying glass is (Fig. 12.32)

$$m = -\frac{i}{o} \simeq \frac{25}{f} \quad (>0). \qquad (12.22)$$

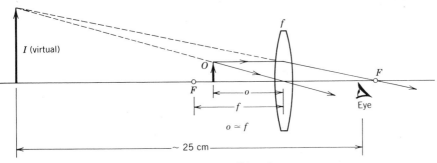

Fig. 12.32. A magnifying glass.

Microscope.

The microscope uses two convergent lenses called objective and eyepiece, as shown in Fig. 12.33. The objective lens has a focal length much shorter than that of the eyepiece. The objective forms a real image by placing an object slightly beyond the focal point, $o \gtrsim f_1$. The image is formed at a point close to the focal point of the eyepiece, or practically at the length of the microscope L, since f_1 and f_2 are usually much smaller than the length of the microscope. Then the magnification of the objective is

$$m_1 = -\frac{L}{f_1} \quad (<0)$$

The rest is the same as that in the magnifying glass. The magnification of the eyepiece is

$$m_2 = \frac{25}{f_2} \quad (>0)\,(f_2 \text{ in cm})$$

and the total magnification is

$$m = m_1 m_2 = -\frac{25 \times L}{f_1 f_2} \quad \text{(all in cm)},\qquad (12.23)$$

which is negative. The image of the microscope is virtual and inverted.

The maximum useful magnification of optical microscopes is at most 600. This limitation is caused by the diffractive nature of light. (See Section 11.6.)

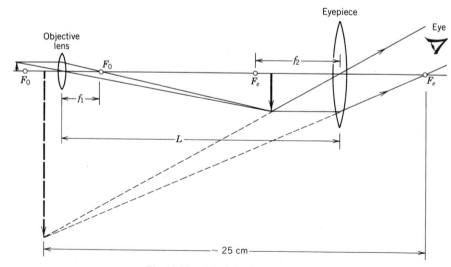

Fig. 12.33.   Principle of a microscope.

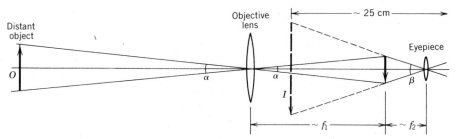

Fig. 12.34.   Principle of a telescope. The eyepiece can be a diverging lens. In this case the image is erect.

Telescope.

In the telescope the objective lens has a longer focal length than the eyepiece, in contrast to the microscope. Since the object is far away from the telescope the image of the objective is formed at the focal point, which is then magnified by the eyepiece (Fig. 12.34). The magnification is given by

$$m \simeq -\frac{\beta}{\alpha} \simeq -\frac{f_1}{f_2}.$$

where α and β are the angles subtended by the object O and the image I. The eyepiece can be either convergent ($f_2 > 0$) or divergent ($f_2 < 0$). Again the maximum magnification of optical telescope is limited by diffraction. As we have seen in Chapter 11, the resolution of a lens having a diameter a is

$$\alpha \simeq 1.2 \frac{\lambda}{a}.$$

The resolution of the human eye was (Example 5),

$$\beta \simeq 1.2 \frac{\lambda}{a_e},$$

where a_e is the aperture diameter of the human eye, $a_e \simeq 5$ mm. Then the maximum magnification of the telescope is

$$|m_{max}| \simeq \frac{\beta}{\alpha} = \frac{a}{a_e}$$

which is proportional to the telescope diameter. If $\alpha = 1$ m, this value becomes 200.

12.10. Physical Meaning of Focusing

Spherical surfaces for reflection and refraction are characterized by focal lengths. Consider a spherical reflector shown in Fig. 12.35. We saw that two

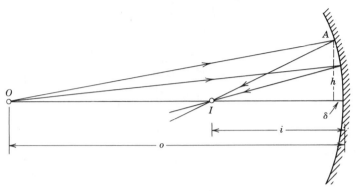

Fig. 12.35. Time required for light to travel from O to I is independent of α, or h, provided h is sufficiently small.

rays leaving the object are focused at the same image position, irrespective of the height h, provided h is much smaller than the curvature radius R, or provided the spherical aberration is negligible. Here it will be shown that the time required for light to travel along either path is the same, which is the physical meaning of focusing.

In Fig. 12.35 the length OA is

$$OA = \sqrt{(o-\delta)^2 + h^2} \simeq o - \delta + \frac{1}{2}\frac{h^2}{o}$$

and the length IA is

$$IA = \sqrt{(i-\delta)^2 + h^2} \simeq i - \delta + \frac{1}{2}\frac{h^2}{i},$$

where we have made use of binomial expansion, noting $\delta, h \ll o, i$.

The quantity δ can be found from Fig. 12.36. For the triangle OAB, we have

$$R = \sqrt{(R-\delta)^2 + h^2}$$

or neglecting δ^2

$$\delta = \frac{h^2}{2R}.$$

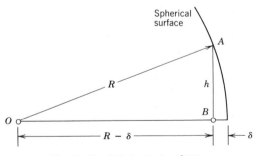

Fig. 12.36. If $h, \delta \ll R, \delta \simeq h^2/2R$.

Then the time required for light to travel from O to I in Fig. 12.35 is

$$t = \frac{OA}{c} + \frac{AI}{c}$$

$$= \frac{1}{c}(o + i - 2\delta) + \frac{h^2}{2c}\left(\frac{1}{o} + \frac{1}{i}\right)$$

$$= \frac{1}{c}(o + i) + \frac{h^2}{2c}\left(\frac{1}{o} + \frac{1}{i} - \frac{2}{R}\right).$$

However, the mirror formula was

$$\frac{1}{o} + \frac{1}{i} = \frac{2}{R},$$

and we see that the time is independent of the height h, or the light path.

Similarly, it can be shown that in the case of lenses, the light propagation time is independent of the path as long as spherical aberration is negligible. Consider a convergent lens shown in Fig. 12.37. The geometrical distance along the path 1 is obviously longer than that along the path 2. However, the *optical distance* defined by the sum of

$$n \times \text{geometrical distance}$$

in each medium can be shown to be independent of which path light takes.

Example 7. Recalling Eq. (12.13), show that the light propagation time along the path OAI (see Fig. 12.38) does not depend on the height h and is

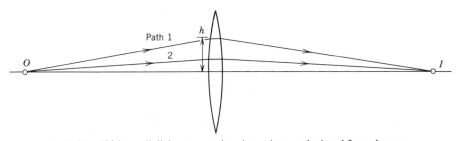

Fig. 12.37.   If h is small, light propagation times along paths 1 and 2 are the same.

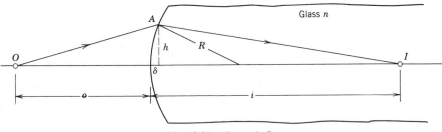

Fig. 12.38.   Example 7.

equal to $(1/c)(o+n\times i)$.

$$OA=\sqrt{(o+\delta)^2+h^2}\simeq o+\delta+\frac{h^2}{2o}$$

$$AI=\sqrt{(i-\delta)^2+h^2}\simeq i-\delta+\frac{h^2}{2i},$$

where $\delta=h^2/2R$, as in the case of the spherical mirror. Then the propagation time along the path OAI is

$$\frac{OA}{c}+\frac{AI}{c/n}=\frac{1}{c}(o+ni)+\frac{h^2}{2c}\left[\frac{1}{o}+\frac{n}{i}+(1-n)\frac{1}{R}\right].$$

The quantity in the brackets [ ] is zero, from Eq. (12.13). Therefore the optical path $OA+nAI$ is independent of the height h, as long as spherical aberration is negligible.

Problems

1. Light beam falls on the water–glass boundary at an angle of 30° (Fig. 12.39). Find the refraction angle, θ.

   (*Answer:*   35.2°.)

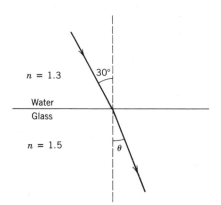

Fig. 12.39.   Problem 1.

2. Explain how heat haze is created on a hot day.

3. Show that the apparent depth of a water pool if looked down normally is d/n, where d is the true depth and n is the index of refraction of water.

4. A glass cube ($n=1.5$) has a spot at the center. Assuming that the cube has an edge of 2 cm, find what parts of the cube face must be covered so that the spot cannot be seen at all.

   (*Answer:*   A circle around the face center with a radius 0.89 cm.)

5. *Fermat's principle.* Show that the time required for light to travel from A to B is minimum when the angles θ_1 and θ_2 satisfy the Snell's law (Fig. 12.40). (*Hint:* Find the time τ as a function of x. Calculate $d\tau/dx$. $d\tau/dx = 0$ yields Snell's law.)

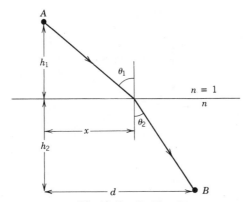

Fig. 12.40.   Problem 5.

6. Show that a mirror having a vertical size half of a person's height is sufficient for the person to view his whole body.

7. A spherical mirror has a focal length of $+20$ cm. An object is placed at 10 cm in front of the mirror. Find the image distance, magnification, and nature (real or virtual, erect or inverted) of the image.
   (*Answer:* -20 cm, $m = 2$, virtual, erect.)

8. Repeat Problem 7 for a mirror having $f = -20$ cm.
   (*Answer:* -6.7 cm, $m = 0.67$, virtual, erect.)

9. When an object is placed at 60 cm in front of a mirror, a magnification of -0.5 results. Find the focal length of the mirror and the image distance.
   (*Answer:* $f = 20$ cm, $i = 30$ cm.)

10. Does the image of a cubic object appear to be cubic when formed by a mirror?
    (*Answer:*   No. The dimension along the axis appears as m^2, where m is the magnification in the vertical direction.)

11. Locate the position of the final image and find the magnification (Fig. 12.41).
    (*Answer:*   50 cm in front of the lens; $m = +1.0$.)

12. Repeat Problem 11 by replacing the lens by a diverging lens having $f = -20$ cm.
    (*Answer:*   11.6 cm to the right of the lens; $m = -0.04$.)

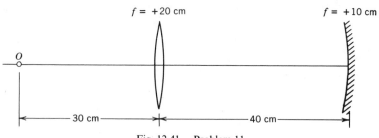

Fig. 12.41.   Problem 11.

13. Complete the table for four lenses—*a*, *b*, *c*, and *d*. The lengths are in centimeters.

	a	*b*	*c*	*d*
f	+10			+40
o	+20	+10	+120	
i				
m		0.5	2	−0.6
Real?			No	
Erect?		Yes		

14. Find the location of the image (Fig. 12.42).
    (*Answer:*   15.8 cm right of the second surface.)

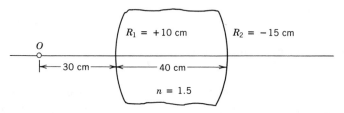

Fig. 12.42.   Problem 14.

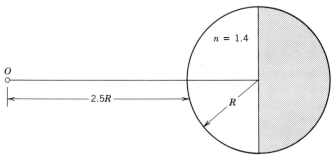

Fig. 12.43.   Problem 15.

15. A solid glass sphere of radius R and index of refraction $n=1.4$ is silvered over one hemisphere. An object is placed at a distance from the sphere as shown in Fig. 12.43. Find the position of the final image. Neglect spherical aberration.

(*Answer:* 0.1R in front of the silvered surface.)

16. Repeat Example 6 for a compound lens of focal length 20 cm. Take $R'_2 = \infty$.

17. An SLR (single lens reflex) camera can select lenses for different purposes.

(a) If a 50-mm lens is attached, what is the minimum object distance that the camera can focus? Assume that the maximum lens–film distance 6.0 cm.

(b) The 50-mm lens is replaced with a telephoto lens of focal length 800 mm. What is the magnification of an object far away relative to that of a 50-mm lens?

(c) Repeat (a) when a lens of focal length 20 mm is attached in front of the 50-mm lens.

(*Answer:* 30 cm, 16, 19 cm.)

18. (a) Does the eyepiece of a telescope have to be a convergent lens?

(*Answer:* No.)

(b) Does the eyepiece of a microscope have to be a convergent lens?

(*Answer:* Yes.)

19. If a mirror surface is parabolic, rays parallel to the axis are all converged at a same point irrespective of the height. Prove this (see Fig. 12.44).

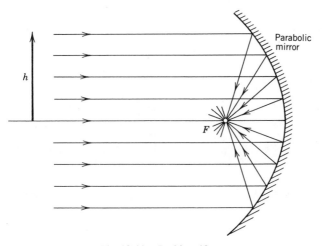

Fig. 12.44. Problem 19.

20. Derive Eq. (12.12).

21. Reflection and refraction of electromagnetic waves (light included) at a medium discontinuity can be analyzed with the aid of Maxwell's equations. One requirement resulting from the Maxwell equations is that the tangential component of electric field be continuous across the boundary. (We have already used this for the case of normal incidence in Chapter 9 where the electric fields were always tangential to the boundary surface.) For the incident wave, there are two possible field configurations depending on the orientation of the electric field, the field normal to the "incident plane" (Fig. 12.45a) and tangential to the incident plane (Fig. 12.45b).

(a)   For the configuration in Fig. 12.45a, show that the reflected electric field is given by

$$E_r = -\frac{\sin(\theta_i - \theta_r)}{\sin(\theta_i + \theta_r)} E_i$$

where E_i is the incident electric field. Does this reduce to the known result of normal incidence, $\theta_i \to 0$, $\theta_r \to 0$?

(b)   For the configuration in Fig. 12.45b, show that the reflected electric field is given by

$$E_r = \frac{\tan(\theta_i - \theta_r)}{\tan(\theta_i + \theta_r)} E_i$$

Note: For case (b), the reflected field becomes zero when

$$\theta_i + \theta_r = \frac{\pi}{2} \quad \left(\tan\frac{\pi}{2} = \infty\right)$$

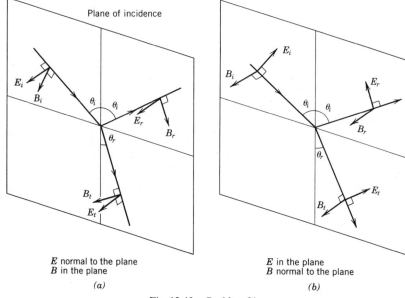

E normal to the plane
B in the plane

(a)

E in the plane
B normal to the plane

(b)

Fig. 12.45.   Problem 21.

or when $\tan \theta_i = n$. This particular incident angle is called the *Brewster angle* and has important applications in polarization of light. For typical glass having an index refraction of $n = 1.5$, the Brewster angle becomes 56.3°.

Hint: Apply the boundary condition for the electric field and energy conservation principle.

CHAPTER 13

Fourier Analyses and Laplace Transformation

13.1. Introduction

In Chapter 2 we learned that the wave equation

$$\frac{\partial^2 f}{\partial t^2} = c_w^2 \frac{\partial^2 f}{\partial x^2}$$

can be satisfied by any function f as long as f depends on x and t in the form $x \pm c_w t$. Although we have studied many kinds of waves (mechanical and electromagnetic) in terms of harmonic waves such as $\xi_0 \sin(kx - \omega t)$, waves do not have to be harmonic. In fact, waves we experience in daily life are hardly harmonic. For example, radio waves emitted by a radio station are not pure sine waves. If they were, we would hear nothing, as we will see. However, studying waves in terms of harmonic waves is extremely important, because no matter how complicated a waveform is, it can be approximated by a collection of many harmonic waves. This technique is known as Fourier analysis, which is an extremely important mathematical tool in engineering and physics. We will also learn the method of Laplace transform, which is closely related to Fourier analysis and can significantly simplify solving differential equations.

13.2. Sum of Harmonic Functions

Let us see how a sum of several harmonic functions creates an anharmonic function. We consider the following function

$$f(x) = \sin x + \tfrac{1}{3} \sin 3x + \tfrac{1}{5} \sin 5x + \cdots \tag{13.1}$$

In Fig. 13.1, $\sin x + \tfrac{1}{3}\sin 3x$, $\sin x + \tfrac{1}{3}\sin 3x + \tfrac{1}{5}\sin 5x$, $\sin x + \tfrac{1}{3}\sin 3x + \tfrac{1}{5}\sin 5x + \tfrac{1}{7}\sin 7x$ are shown. We can expect that as the number of terms in the sum increase, the function approaches a square function. Or said the other way around, we expect that the square function can be expressed in terms of a sum of many harmonic (sinusoidal) functions (Example 2).

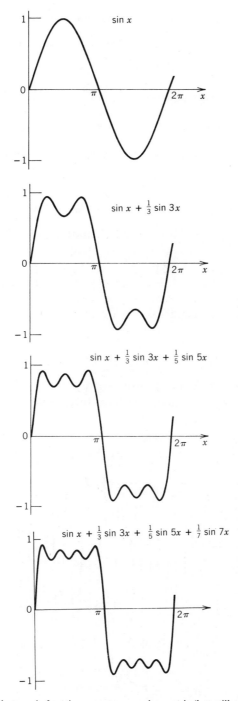

Fig. 13.1. Sum of harmonic functions creates an anharmonic (but still periodic) function.

Consider another example, which is essentially the wave of AM (*Amplitude Modulation*) radio. Let us consider the following function:

$$A \sin \omega t + \alpha \sin (\omega + \Delta\omega)t + \alpha \sin (\omega - \Delta\omega)t, \qquad (13.2)$$

which is a sum of three harmonic waves. Using

$$\sin \alpha + \sin \beta = 2 \sin \frac{\alpha + \beta}{2} \cos \frac{\alpha - \beta}{2},$$

we can rewrite Eq. (13.2) as

$$A \sin \omega t + 2\alpha \sin \omega t \cos \Delta\omega t = A \sin \omega t \left[1 + \frac{2\alpha}{A} \cos \Delta\omega t \right], \qquad (13.3)$$

which is shown in Fig. 13.2 for the case of

$$2\alpha/A = 0.1 \qquad \text{and} \qquad \Delta\omega = 0.1\omega.$$

This is the amplitude modulation of the harmonic carrier wave $A \sin \omega t$. The smaller frequency $\Delta\omega$ corresponds to the audio frequency, which is at most $2\pi \times 20$ kHz.

Take a radio station having a frequency of 540 kHz. Centered at this frequency, the station occupies a frequency band from $540 - 20$ kHz to $540 + 20$ kHz in order to transmit audio signals. Each radio station should be allotted this 40-kHz band centered at each carrier frequency.

Example 1. Estimate the frequency band required for a TV station.

Video (picture) signals are transmitted by amplitude modulation of the carrier wave. In American standards, 30 picture frames are transmitted every

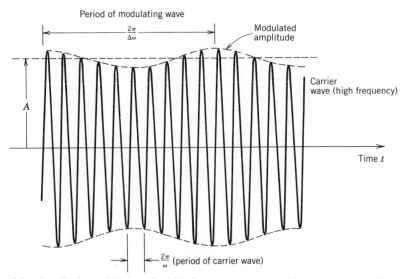

Fig. 13.2.   Amplitude modulation of a high-frequency wave (carrier wave) by a low-frequency wave.

second. Each frame consists of 525 horizontal scannings. Therefore if the resolution in the horizontal direction is to be of the same order as that in the vertical direction (the latter corresponding to the spacing between two neighboring lines), the time resolution must be of the order of

$$\frac{1}{30 \times (525)^2} \simeq 0.1 \; \mu\text{sec}.$$

The frequency corresponding to this pulse is about $1/0.1 \; \mu\text{sec} = 10 \; \text{MHz}$.

In American standards the frequency band allotted for video signals is about 4 MHz, since only half is sufficient in practice because of the special transmission method (Single Side Band). A narrower band would blur the picture because of poor resolution. A wider band does not improve resolution since the vertical resolution is more or less fixed by the distance between adjacent scanning lines.

13.3. Fourier Series

Consider an arbitrary periodic function $f(t)$ with a period T (Fig. 13.3):

$$f(t) = f(t + T). \tag{13.4}$$

Since both $\sin(2\pi nt/T)$ and $\cos(2\pi nt/T)$ (n = integer) are also periodic functions, we put

$$f(t) = a_0 + a_1 \cos\left(\frac{2\pi t}{T}\right) + a_2 \cos\left(\frac{4\pi t}{T}\right) + \cdots$$

$$+ b_1 \sin\left(\frac{2\pi t}{T}\right) + b_2 \sin\left(\frac{4\pi t}{T}\right) + \cdots$$

$$= a_0 + \sum_{n=1}^{\infty} a_n \cos\left(\frac{2\pi nt}{T}\right) + \sum_{n=1}^{\infty} b_n \sin\left(\frac{2\pi nt}{T}\right), \tag{13.5}$$

where the coefficients a_n and b_n are unknown coefficients to be determined for a given function $f(t)$.

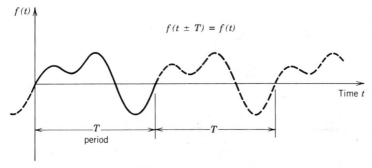

Fig. 13.3.   A periodic function of an arbitrary shape with a period T.

It is convenient to introduce a new variable $x = 2\pi t/T$. Then

$$f(x) = a_0 + \sum_{n=1}^{\infty} a_n \cos(nx) + \sum_{n=1}^{\infty} b_n \sin(nx) \tag{13.6}$$

and the period of the function becomes 2π. To find a_n and b_n, let us recall

$$\int_0^{2\pi} \cos mx \cos nx \, dx = \begin{cases} \pi, & m = m \\ 0, & m \neq n \end{cases}$$

$$\int_0^{2\pi} \sin mx \cos nx \, dx = 0$$

$$\int_0^{2\pi} \sin mx \sin nx \, dx = \begin{cases} \pi, & m = n \\ 0, & m \neq n. \end{cases}$$

These can be proved easily by noting (see Appendix B)

$$\cos mx \cos nx = \tfrac{1}{2}[\cos(m+n)x + \cos(m-n)x]$$

$$\sin mx \sin nx = \tfrac{1}{2}[-\cos(m+n)x + \cos(m-n)x]$$

$$\sin mx \cos nx = \tfrac{1}{2}[\sin(m+n)x + \sin(m-n)x]$$

and integrating over the interval from 0 to 2π. Then multiplying Eq. (13.6) by $\cos nx$ and integrating the result from 0 to 2π, we find

$$a_n = \frac{1}{\pi} \int_0^{2\pi} f(x) \cos nx \, dx \qquad (n = 1, 2, 3 \ldots). \tag{13.7}$$

Similarly, to find b_n, we multiply Eq. (13.6) by $\sin nx$ and integrate

$$b_n = \frac{1}{\pi} \int_0^{2\pi} f(x) \sin nx \, dx \qquad (n = 1, 2, 3 \ldots). \tag{13.8}$$

The coefficient a_0 can be found from

$$a_0 = \frac{1}{2\pi} \int_0^{2\pi} f(x) \, dx, \tag{13.9}$$

that is a_0 corresponds to the average value of the function $f(x)$.

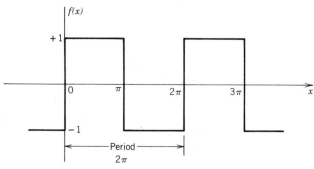

Fig. 13.4.   Example 2.

Example 2. Find the Fourier series of the square wave shown in Fig. 13.4.

The average of $f(x)$ is zero. Then $a_0 = 0$. Also, the function $f(x)$ is odd, or symmetric about the origin 0. Then the even components $a_n = 0$.

$$b_n = \frac{1}{\pi} \int_0^{2\pi} f(x) \sin nx \, dx$$

$$= \frac{1}{\pi} \left[\int_0^{\pi} \sin nx \, dx - \int_{\pi}^{2\pi} \sin nx \, dx \right]$$

$$= \frac{1}{\pi} \left[\frac{1}{n}(-\cos nx) \Big|_0^{\pi} + \frac{1}{n}(\cos nx) \Big|_{\pi}^{2\pi} \right]$$

$$= \frac{1}{n\pi} [2 - 2 \cos n\pi]$$

Then the even terms (b_2, b_4, b_6, . . .) are zero, and odd terms are given by

$$b_1 = \frac{4}{\pi}, \quad b_3 = \frac{4}{3\pi}, \quad b_5 = \frac{4}{5\pi}, \ldots$$

Therefore the square wave train can be Fourier expanded as

$$f(x) = \frac{4}{\pi} [\sin x + \tfrac{1}{3} \sin 3x + \tfrac{1}{5} \sin 5x + \cdots]$$

We previously guessed (Fig. 13.1) that the function

$$\sin x + \tfrac{1}{3} \sin 3x + \tfrac{1}{5} \sin 5x + \cdots$$

would approach a square wave function as the number of terms is increased. This has now been proved.

Example 3. Find the Fourier series for the sawtooth function shown in Fig. 13.5.

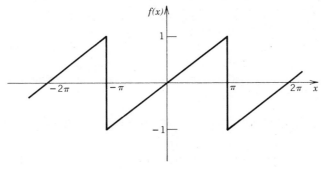

Fig. 13.5. Example 3.

The average of $f(x)$ is zero. Then $a_0=0$. Again the function is odd, and $a_n=0$. The Fourier coefficient can be evaluated over an arbitrary period irrespective of location. Here it is convenient to choose the period from $-\pi$ to π.

$$b_n=\frac{1}{\pi}\int_{-\pi}^{\pi}\frac{x}{\pi}\sin nx\,dx$$

$$=\frac{1}{\pi^2}\left[-\frac{1}{n}x\cos nx\Big|_{-\pi}^{\pi}+\frac{1}{n}\int_{-\pi}^{\pi}\cos nx\,dx\right]$$

$$=\frac{1}{\pi^2}\left[-\frac{\pi}{n}\cos(n\pi)-\frac{\pi}{n}\cos(-n\pi)\right]$$

$$=\frac{2}{\pi}\frac{(-1)^{n-1}}{n}\qquad(n=1,2,3,\ldots).$$

Therefore

$$f(x)=\frac{2}{\pi}[\sin x-\tfrac{1}{2}\sin 2x+\tfrac{1}{3}\sin 3x-\cdots].$$

Note we have made use of integration by parts,

$$\int x\sin x\,dx=-x\cos x+\int\cos x\,dx=-x\cos x+\sin x.$$

13.4. Fourier Spectrum

We have seen that any periodic function can be expanded in terms of harmonic functions. The lowest frequency is determined by the period T, as $\omega_0=2\pi/T$, and higher harmonics have frequencies of integer multiples of the fundamental frequency. Although we discussed only Fourier expansion of time-varying functions, exactly the same procedure can be applied to any functions that are periodic in spatial coordinates.

The amplitude of harmonics, a_n and b_n, can be plotted as a function of harmonic frequency. For example, the square wave function we studied in Example 2 can be characterized by the harmonic amplitudes shown in Fig. 13.6. Such a plot in the frequency coordinate is called a frequency spectrum and enables us to visualize the distribution of harmonics. Spectrum analyzers frequently used in communication research can directly display frequency spectrums on a cathode-ray-tube screen. Also, the grating optical spectrometer (Chapter 11) can be regarded as a spectrum analyzer that can perform Fourier analysis in the wavelength domain.

Example 4. Find the frequency spectrum of the amplitude-modulated wave given by

$$f(t)=A(1+a\cos\omega_a t)\sin\omega_c t,\qquad a<1,$$

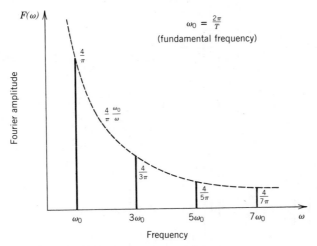

Fig. 13.6. Fourier spectrum of the square wave.

where ω_a is the audio frequency and ω_c is the carrier frequency.
    The function $f(t)$ can be rewritten as

$$A \sin \omega_c t + \tfrac{1}{2}aA \left[\sin (\omega_c - \omega_a)t + \sin (\omega_c + \omega_a)t\right].$$

Therefore we obtain the spectrum shown in Fig. 13.7.

Example 5.   Find the Fourier spectrum of a single pulse shown in Fig. 13.8a.

The single pulse is not a periodic function and you may wonder how to Fourier-analyze such a function. There is a trick. We assume a pulse train with a period T (Fig. 13.8b), Fourier-analyze it, and eventually make $T \to \infty$. Consider

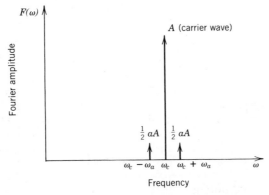

Fig. 13.7.   Example 4.

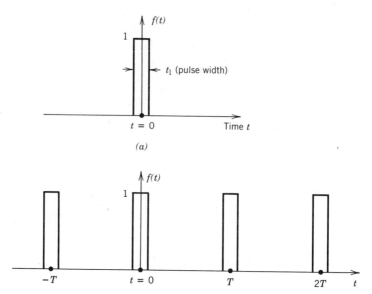

Fig. 13.8.   Example 5.

a pulse train shown above. Since the function is even, $b_n = 0$.

$$a_0 = \text{average of } f(t) = \frac{t_1}{T}$$

$$a_n = \frac{2}{T} \int_0^T f(t) \cos n\omega_0 t \, dt$$

where $\omega_0 = 2\pi/T$ is the fundamental frequency. Since

$$f(t) = 1 \quad \left(0 < t < \frac{t_1}{2}\right)$$

$$= 0 \quad \left(\frac{t_1}{2} < t < T - \frac{t_1}{2}\right)$$

$$= 1 \quad \left(T - \frac{t_1}{2} < t < T\right),$$

we find

$$a_n = \frac{2}{T}\left[\int_0^{t_1/2} \cos n\omega_0 t \, dt + \int_{T-(t_1/2)}^T \cos n\omega_0 t \, dt\right]$$

$$= \frac{2}{T}\frac{1}{n\omega_0}\left[\sin\left(n\omega_0 \frac{t_1}{2}\right) - \sin n\omega_0\left(T - \frac{t_1}{2}\right)\right].$$

Recalling $\omega_0 = 2\pi/T$, and noting

$$\sin(\alpha - \beta) = \sin \alpha \cos \beta - \cos \alpha \sin \beta,$$

we find

$$a_n = \frac{2}{n\pi} \sin\left(n\pi \frac{t_1}{T}\right).$$

This spectrum is shown in Fig. 13.9 for the case $T = 10t_1$. As T is increased, the spacing between adjacent lines becomes narrower, and in the limit of $T \to \infty$, we have a *continuous spectrum*, rather than *discrete spectrum* (Fig. 13.10).

The preceding example indicates that even nonperiodic functions can be Fourier-analyzed. If a function is periodic, its Fourier spectrum is discrete. If a function is nonperiodic, its Fourier spectrum is, in general, continuous.

The preceding example also reveals an interesting fact, about the relationship between time and frequency. The characteristic frequency band width for the single pulse with a width t_1 is about $2\pi/t_1$ as seen from Fig. 13.10. Then

$$\text{Pulse width} \times \text{band width} \simeq 2\pi \text{ (const)}, \qquad (13.10)$$

which indicates that if we want to transmit a narrow pulse, we have to use a

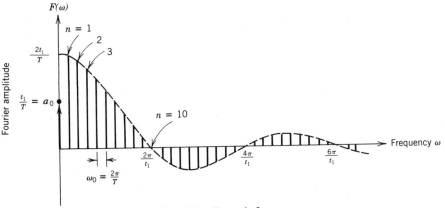

Fig. 13.9.   Example 5.

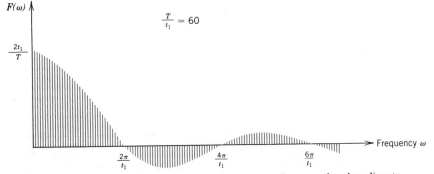

Fig. 13.10.   As T increases, the spectrum becomes continuous rather than discrete.

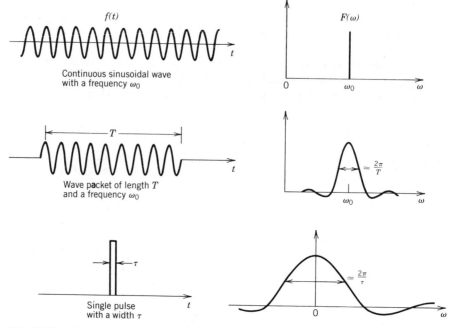

Fig. 13.11.   Fourier spectra for several functions. Note that a smaller time spread results as a wider frequency spread.

wider-frequency band. In Fig. 13.11 the Fourier spectra for some waveforms are schematically shown.

13.5. Operator Method

Harmonic waves are usually described by

$$A \cos (kx - \omega t) \quad \text{or} \quad A \sin (kx - \omega t)$$

as we have seen in previous chapters. Since

$$\cos \alpha = \text{Re } e^{i\alpha}$$

$$\sin \alpha = \text{Im } e^{i\alpha}, \tag{13.11}$$

it is more convenient to use the exponential function

$$A e^{i(kx - \omega t)} \tag{13.12}$$

to describe a wave. When we need a cosine wave, we take its real part, and for a sine wave, we take the imaginary part.

Writing a wave in the form Eq. (13.12) greatly simplifies wave analysis. Since

$$\frac{\partial}{\partial x} e^{i(kx - \omega t)} = ike^{i(kx - \omega t)}$$

$$\frac{\partial}{\partial t} e^{i(kx - \omega t)} = -i\omega e^{i(kx - \omega t)},$$

we can simply replace $\partial/\partial x$ by ik and $\partial/\partial t$ by $-i\omega$. For example, the wave equation

$$\frac{\partial^2 \xi}{\partial t^2} = c_w^2 \frac{\partial^2 \xi}{\partial x^2}$$

can readily be converted into

$$(-i\omega)^2 \xi = (ik)^2 c_w^2 \xi$$

or

$$\omega^2 = c_w^2 k^2,$$

which immediately yields the dispersion relation.

Example 6. Find the natural oscillation frequency and the damping coefficient of an *LCR* circuit. (See p. 18.)

The differential equation to describe the charge on the capacitor was (Chapter 1)

$$\frac{d^2 q}{dt^2} + \frac{R}{L}\frac{dq}{dt} + \frac{1}{LC} q = 0.$$

Let us assume a solution of the form $q_0 e^{-i\omega t}$, where ω may be a complex number. Using $d/dt = -i\omega$, we obtain

$$-\omega^2 q - i\omega \frac{R}{L} q + \frac{1}{LC} q = 0$$

or

$$\omega^2 + i\frac{R}{L}\omega - \frac{1}{LC} = 0.$$

Solving for ω we find

$$\omega = \frac{1}{2}\left[-i\frac{R}{L} \pm \sqrt{\frac{4}{LC} - \left(\frac{R}{L}\right)^2} \right]. \tag{13.13}$$

The complex frequency indicates that the oscillation damps as

$$e^{\gamma t},$$

where

$$\gamma = \text{Im } \omega = -\frac{R}{2L} (<0). \tag{13.14}$$

The oscillation frequency is

$$\text{Re}\,\omega = \sqrt{\frac{1}{LC}-\left(\frac{R}{2L}\right)^2}\qquad\left[\frac{1}{LC}>\left(\frac{R}{2L}\right)^2\right]. \tag{13.15}$$

You should check that these results are consistent with Chapter 1.

Electromagnetic waves can propagate without damping if the medium is loss-free as in ideal vacuum. In practice, all dielectric materials have finite conductivity although it is usually much smaller than that in metals. In Chapter 9 we saw that a lossy medium has finite resistance in parallel with the capacitors (Fig. 13.12). Let the conductance per unit element be $G/\Delta x$. The characteristic impedance is then

$$Z=\sqrt{\frac{-i\omega L/\Delta x}{(-i\omega C+G)/\Delta x}}. \tag{13.16}$$

In the limit of $G\to 0$, this reduces to the lossless case. In general, the characteristic impedance should be given by

$$Z=\sqrt{\frac{\text{reactance per unit length}}{\text{susceptance per unit length}}}=\sqrt{\frac{(-i\omega L+R)/\Delta x}{(-i\omega C+G)/\Delta x}}. \tag{13.17}$$

For waves in unbounded space such as in air, the impedance becomes

$$Z=\sqrt{\frac{-i\omega\mu_0}{-i\omega\varepsilon+\sigma}}, \tag{13.18}$$

where σ is conductivity ($1/\Omega\text{m}$). In the presence of loss, the impedance becomes complex.

The propagation velocity in the presence of the finite conductivity can be found as follows. Kirchhoff's voltage and current theorems yield (Fig. 13.12).

$$V(x)=L\frac{\partial i}{\partial t}+V(x+\Delta x) \tag{13.19}$$

$$i(x)=C\frac{\partial V}{\partial t}+GV+i(x+\Delta x), \tag{13.20}$$

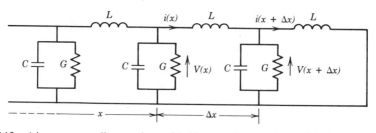

Fig. 13.12.   A lossy wave medium can be modeled by a conductance in parallel with a capacitance.

from which we obtain for the current i

$$\frac{\partial^2 i}{\partial x^2} = \frac{LC}{(\Delta x)^2} \frac{\partial^2 i}{\partial t^2} + \frac{LG}{(\Delta x)^2} \frac{\partial i}{\partial t}. \tag{13.21}$$

Substituting $L/\Delta x = \mu_0$, $C/\Delta x = \varepsilon$, and $G/\Delta x = \sigma$, we get

$$\frac{\partial^2 i}{\partial x^2} = \varepsilon\mu_0 \frac{\partial^2 i}{\partial t^2} + \mu_0\sigma \frac{\partial i}{\partial t}. \tag{13.22}$$

This is not the conventional wave equation because of the presence of the term having only the first partial derivative.

Let us apply the operator method to find the propagation velocity. Replacing $\partial/\partial x$ by ik and $\partial/\partial t$ by $-i\omega$, we find

$$k^2 = \varepsilon\mu_0\omega^2 + i\omega\mu_0\sigma. \tag{13.23}$$

Then the propagation velocity becomes

$$\frac{\omega}{k} = \frac{1}{\sqrt{\varepsilon\mu_0}} \frac{1}{\sqrt{1 + \dfrac{i\sigma}{\omega\varepsilon}}}, \tag{13.24}$$

which is complex! The complex velocity should be interpreted as follows. Since

$$k = \sqrt{\varepsilon\mu_0}\,\omega\left(1 + \frac{i\sigma}{\omega\varepsilon}\right)^{1/2} \tag{13.25}$$

is complex for a real ω, the function e^{ikx} exponentially damps, or the wave amplitude damps as the wave propagates. The damping length (e-folding length) is given by the imaginary part of k as

$$\delta = \frac{1}{\text{Im } k}. \tag{13.26}$$

Since

$$1 + i\frac{\sigma}{\omega\varepsilon} = \sqrt{1 + \left(\frac{\sigma}{\omega\varepsilon}\right)^2}\, e^{i\theta}$$

$$\theta = \tan^{-1}\frac{\sigma}{\omega\varepsilon},$$

Im k can be found as

$$\text{Im } k = \omega\sqrt{\varepsilon\mu_0}\left[1 + \left(\frac{\sigma}{\omega\varepsilon}\right)^2\right]^{1/4} \sin\frac{\theta}{2} \tag{13.27}$$

and the damping length becomes

$$\delta = \frac{1}{\omega\sqrt{\varepsilon\mu_0}}\left[1 + \left(\frac{\sigma}{\omega\varepsilon}\right)^2\right]^{-1/4} \frac{1}{\sin\theta/2}. \tag{13.28}$$

This quantity is often called the skin depth, which is a measure of how deep an electromagnetic wave can penetrate with normal incidence.

The propagation phase velocity can be found from

$$\frac{\omega}{\operatorname{Re} k} = \frac{1}{\sqrt{\varepsilon\mu_0}} \left[1 + \left(\frac{\sigma}{\omega\varepsilon}\right)^2 \right]^{-1/4} \frac{1}{\cos \theta/2} \tag{13.29}$$

Example 7.   Calculate the skin depth in soil for 1-MHz (AM radio range) electromagnetic waves. Assume that the conductivity of soil is 10^{-2} 1/Ωm.

From Eq. (13.28)

$$\delta = \frac{3 \times 10^8}{2\pi \times 10^6} \left[1 + \left(\frac{10^{-2}}{2\pi \times 10^6 \times 8.85 \times 10^{-12}}\right)^2 \right]^{-1/4} \frac{1}{\sin \theta/2} = \frac{3.6}{\sin \theta/2},$$

where

$$\theta = \tan^{-1} \left(\frac{10^{-2}}{2\pi \times 10^6 \times 8.85 \times 10^{-12}}\right) = \tan^{-1} (180)$$

$$\simeq \frac{\pi}{2} \text{ rad.}$$

Then $\delta = 5.0$ m, which is much less than the wavelength $\lambda = 300$ m. Therefore the earth should be regarded as a good conductor.

In the presence of loss (finite conductivity), the electromagnetic waves necessarily become dispersive since, as Eq. (13.29) indicates, the phase velocity depends on the frequency ω. Dissipative medium is thus necessarily dispersive too. However, the inverse is not always true.

13.6. Laplace Transform

In solving differential equations, we usually guess a possible solution and substitute it back into the differential equation to see if the solution is correct or not. In most cases we depend on our own experience and skills.

Let us consider a simple case

$$\frac{d^2 f}{dx^2} - 4f = 0 \tag{13.30}$$

It can be observed that $Ae^{-2x} + Be^{2x}$ is a solution to the differential equation. To obtain the exponential solution, we have assumed a solution of the form e^{Dx}, which yields

$$(D^2 - 4)f = 0 \tag{13.31}$$

and

$$D = \pm 2. \tag{13.32}$$

This is already one significant step to convert the differential equation into an algebraic equation.

The Laplace transform exactly does this. It can convert a differential equation into a mere algebraic equation. What is more, the Laplace transform can automatically take into account initial conditions (or boundary conditions) in a natural manner.

The Laplace transform of a function $f(t)$ is defined by

$$F(s) = \int_0^\infty e^{-st} f(t) \, dt. \tag{13.33}$$

This has an interesting property. The Laplace transform of df/dt is

$$\int_0^\infty e^{-st} \frac{df}{dt} \, dt = e^{-st} f \Big|_0^\infty + s \int_0^\infty e^{-st} f \, dt$$

$$= -f(0) + sF(s), \tag{13.34}$$

where $f(0)$ is the initial ($t=0$) value of the function $f(t)$. Similarly,

$$\int_0^\infty e^{-st} \frac{d^2 f}{dt^2} \, dt = s^2 F(s) - (s f(0) + f'(0)), \tag{13.35}$$

where $f'(0)$ is the initial value of the derivative df/dt. Therefore, differentiation in time corresponds to multiplying by s. Using these properties, we can convert the following differential equation

$$\frac{d^2 f}{dt^2} + A \frac{df}{dt} + Bf = 0$$

as

$$s^2 F(s) + AsF(s) + BF(s) = sf(0) + f'(0) + Asf(0). \tag{13.36}$$

This can be easily solved for $F(s)$. To find $f(t)$, we can inverse Laplace transform $F(s)$ into $f(t)$. You can find several books in the library that have Laplace transforms of most fundamental functions. These Laplace transform tables can be used like logarithmic tables. Some examples follow.

	$f(t)$	$F(s)$
(1)	1 (step function)	$1/s$
(2)	t^n (n integer)	$\dfrac{n!}{s^{n+1}}$
(3)	e^{at}	$\dfrac{1}{s-a}$
(4)	$e^{at} f(t)$	$F(s-a)$

$f(t)$	$F(s)$
(5) $\sin \omega t$	$\dfrac{\omega}{s^2+\omega^2}$
(6) $\cos \omega t$	$\dfrac{s}{s^2+\omega^2}$
(7) $\delta(t)$ (delta function)	1
(8) $\ln t$	$-\dfrac{1}{s}(\ln s+0.577)$

The delta function in row (7) needs some explanation. It is extremely peculiar but very useful. There are several ways to define the delta function. Here we use the following definition (Fig. 13.13)

$$\delta(t)=\begin{array}{l}\lim\limits_{\Delta t\to 0}\dfrac{1}{\Delta t}\quad (0<t<\Delta t)\\[2mm]=0\qquad\quad\ (\text{otherwise})\end{array} \qquad (13.37)$$

The area under the delta function is unity as can be easily seen. Also, for any function $f(t)$,

$$\int_{-\infty}^{\infty}\delta(t)f(t)\,dt=f(0). \qquad (13.38)$$

In engineering, the delta function is given other terminology, namely, the impulse function. A short disturbance given to mechanical or electrical systems can be well approximated by the delta function.

Example 8. A capacitor charged to q_0 is suddenly connected to a resistor R (Fig. 13.14). Find the charge $q(t)$.

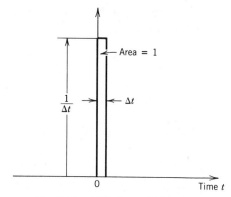

Fig. 13.13.   Definition of δ function.

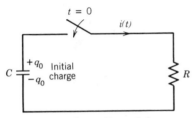

Fig. 13.14. Example 8.

The differential equation for the charge q is

$$\frac{dq}{dt}+\frac{1}{RC}q=0.$$

By Laplace transforming, we obtain

$$sQ(s)-q_0+\frac{1}{RC}Q(s)=0,$$

where q_0 is the initial charge. Solving for $Q(s)$, we find

$$Q(s)=\frac{q_0}{s+1/RC}.$$

Using row (3) in the table, the inverse Laplace transform $q(t)$ can be found as

$$q(t)=q_0e^{-t/RC},$$

which is the familiar exponential damping.

Example 9.   Solve the differential equation for the *LCR* circuit in Example 6. Assume that the capacitor has an initial charge q_0 and

$$\frac{1}{LC}>\left(\frac{R}{2L}\right)^2$$

is satisfied.

The differential equation

$$\frac{d^2q}{dt^2}+\frac{R}{L}\frac{dq}{dt}+\frac{1}{LC}q=0$$

can be Laplace transformed as

$$s^2Q-(sq_0+q'(0))+\frac{R}{L}[sQ-q_0]+\frac{1}{LC}Q=0,$$

$q'(0)$ is the initial current, which is zero.* Then

$$Q(s)=\frac{(s+R/L)q_0}{s^2+(R/L)s+1/LC}$$

$$=\frac{[s+R/2L]q_0+(R/2L)q_0}{(s+R/2L)^2+1/LC-(R/2L)^2}.$$

*Recall that the inductor cannot instantly acquire a current.

Using (4), (5), and (6) in the table, we find

$$q(t) = q_0 e^{-(R/2L)t}\left[\cos \omega t + \frac{R}{2\omega L}\sin \omega t\right],$$

where

$$\omega = \sqrt{\frac{1}{LC} - \left(\frac{R}{2L}\right)^2}.$$

You should check that the solution is consistent with that of Problem 12, Chapter 1, and compare the two methods.

Example 10. An ac voltage source $V_0 \sin \omega t$ is suddenly connected to an RL circuit as shown in Fig. 13.15. Find the current $i(t)$.

Kirchhoff's voltage theorem yields

$$L\frac{di}{dt} + Ri = V_0 \sin \omega t,$$

which can be Laplace transformed as

$$sLI(s) + RI(s) = V_0 \frac{\omega}{s^2 + \omega^2}$$

since $i(0) = 0$. Then

$$I(s) = \frac{\omega V_0}{(sL + R)(s^2 + \omega^2)}$$

$$= \frac{\omega V_0}{L}\frac{1}{\omega^2 + (R/L)^2}\left[\frac{1}{s + R/L} - \frac{s}{s^2 + \omega^2} + \frac{R/L}{s^2 + \omega^2}\right].$$

The current $i(t)$ becomes

$$i(t) = \frac{\omega V_0}{L}\frac{1}{\omega^2 + (R/L)^2}\left[e^{-(R/L)t} + \frac{R}{\omega L}\sin \omega t - \cos \omega t\right].$$

After many time constants, the current becomes harmonic and consistent with the result obtained from ac circuit theory.

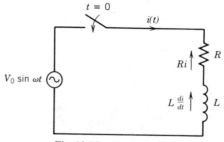

Fig. 13.15.   Example 10.

Problems

1. High-tone instruments (violin, flute, etc.) can play fast, but low-tone instruments (bass, tuba, etc.) cannot. Explain why in terms of Fourier spectra.

2. Estimate the Fourier spectrum of the wave packet shown in Fig. 13.16. (*Answer:* $\Delta\omega \simeq 2\pi \times 10^5$ sec^{-1} centered at $\omega = 2\pi \times 10^6$ sec^{-1}.)

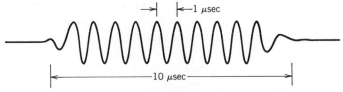

Fig. 13.16.   Problem 2.

3. Calculate the Fourier series of the triangle wave shown in Fig. 13.17.

$$\left(Answer: \quad f(t) = 0.5 + (4/\pi^2)\left(\cos \omega_0 t + \frac{1}{3^2}\cos 3\omega_0 t + \frac{1}{5^2}\cos 5\omega_0 t + \cdots \right) \right.$$

$$\left. \omega_0 = \pi \times 10^3 \text{ sec}^{-1}. \right)$$

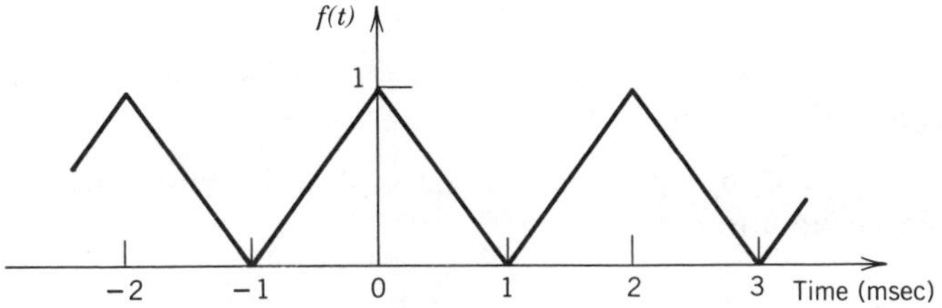

f(t)

1

−2   −1   0   1   2   3   Time (msec)

Fig. 13.17.   Problem 3.

4. Since $\cos(n\omega_0 t)$ is an even function of n, the Fourier series in Example 5 may be written as

$$f(t) = \sum_{n=-\infty}^{\infty} a_n \cos(n\omega_0 t)$$

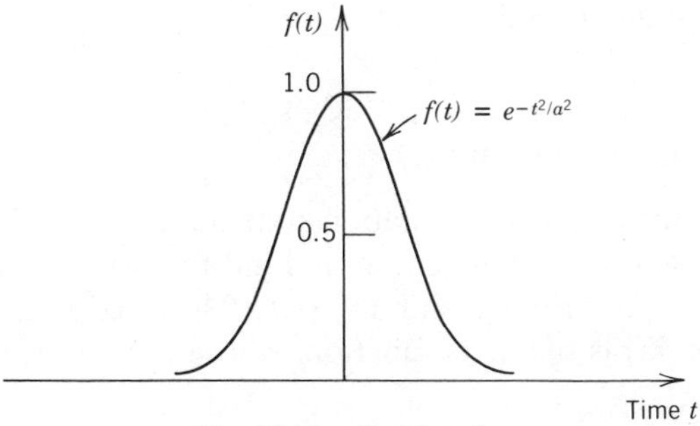

f(t)

1.0

$f(t) = e^{-t^2/a^2}$

0.5

Time t

Fig. 13.18.   Problem 4.

where

$$a_n = \frac{1}{\pi n} \sin \left(\frac{n\omega_0}{2} t_1 \right), \quad \omega_0 = \frac{2\pi}{T}$$

In the limit of $T \to \infty$, the summation over n can be replaced by integration $(n\omega_0 \to \omega)$

$$\lim_{\omega_0 \to 0} \sum_{n=\infty}^{\infty} a_n \cos(n\omega_0 t) \to \int_{-\infty}^{\infty} \frac{F(\omega)}{2\pi} \cos(\omega t) \, d\omega$$

(a) Show that $F(\omega)$ for the single pulse in Example 5 becomes

$$F(\omega) = \frac{2}{\omega} \sin \left(\frac{\omega}{2} t_1 \right)$$

$F(\omega)$ is called the Fourier transform of $f(t)$ and in general can be calculated from

$$F(\omega) = \int_{-\infty}^{\infty} f(t) e^{i\omega t} \, dt$$

The inverse Fourier transform yields $f(t)$,

$$f(t) = \frac{1}{2\pi} \int_{-\infty}^{\infty} F(\omega) e^{-i\omega t} \, d\omega$$

(b) Find the Fourier transform of $f(t) = e^{-t^2/a^2}$ (see Fig. 13.18).
(*Answer:* $F(\omega) = \sqrt{\pi} a e^{-a^2 \omega^2/4}$.)

5. A certain wave in a plasma (ionized gas) is described by the following differential equation

$$\frac{\partial^2}{\partial t^2} \left(\frac{\partial^2}{\partial x^2} - k_D^2 \right) \xi = -\omega_{pi}^2 \frac{\partial^2 \xi}{\partial x^2},$$

where k_D and ω_{pi} are constants. Find the dispersion relation and plot ω/ω_{pi} as a function of k/k_D.
(*Answer:* $\omega^2/\omega_{pi}^2 = k^2/(k^2 + k_D^2)$.)

6. A 50-Ω coaxial cable has polyethylene dielectric that has $\sigma/\omega\varepsilon \simeq 2 \times 10^{-4}$ at 1 MHz. Assuming $\varepsilon = 2.3\varepsilon_0$, calculate the e-folding damping length at 1 MHz.

(*Answer:* 315 km. Dielectric loss is negligible. In practice, the loss in conductors is more important.)

7. The mass in a mass–spring oscillation system is suddenly hit by a hammer and instantly acquires a momentum p. Find the solution for the displacement from the equilibrium position. [*Hint:* The force given by the hammer is $p\delta(t)$, where $\delta(t)$ is the delta function, whose Laplace transform is 1.]
(*Answer:* $x(t) = (p/\sqrt{Mk}) \sin \omega t$, $\omega = \sqrt{k/M}$.)

CHAPTER 14

Particle Nature of Light

14.1. Introduction

Young's experiment (see Chapter 11) clearly demonstrated that light is a wave phenomena since interference can be explained only in terms of superposition of waves. In 1887, Hertz (the discoverer of electromagnetic waves) found that electrons can be released from a metal surface illuminated by light. Many people tried to explain this photoelectric effect in terms of classical theories, involving the wave nature of light, but no one was successful. Einstein introduced the concept of the photon to explain the photoelectric effect. Light behaves as if it were a collection of photon particles in the photoelectric effect. His theory could explain every aspect of the photoelectric effect. However, it should be emphasized that light does behave as a wave too. Both wave and particle nature can coexist in light.

14.2. Photoelectric Effect and Einstein's Photon Theory

The experimental arrangement to study the photoelectric effect is shown in Fig. 14.1. If electrons are emitted from the metal surface, they can be collected by the anode and the ammeter deflects. The experimental results can be summarized as follows:

1. Electrons can be emitted only if the frequency of light is higher than a certain frequency called the *cutoff frequency*. The cutoff frequency depends on the metal that is used.
2. When released from the metal surface, electrons have a kinetic energy that increases with the light frequency. The energy does not depend on the light intensity.

Wave nature of light fails to explain the presence of a cutoff frequency. One would think that electrons are emitted when they acquire sufficient energy by absorbing light energy. Then it is immaterial which frequency light has. By the same token, the wave nature of light is not consistent with the observation that the photoelectron energy does not depend on the light intensity.

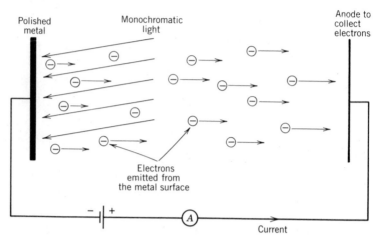

Fig. 14.1.   Arrangement for photoelectric experiment.

Einstein proposed that monochromatic light of frequency v is a collection of photons each having an energy hv (J) where $h = 6.63 \times 10^{-34}$ J·sec is Planck's constant. He postulated that electrons are emitted, being liberated by photons, rather than gradually absorbing light energy. Electrons, however, are bound to the metal and some energy has to be given to the electrons before they are released. This binding energy is called the work function W (J). Therefore, the difference $hv - W$ would appear as the kinetic energy of the electron, or

$$\tfrac{1}{2}mv^2 = hv - W. \tag{14.1}$$

This relationship is illustrated in Fig. 14.2. If the photon energy hv is less than the work function W, the electron cannot climb up the hill and cannot be released from the metal surface. This explains the cutoff frequency and the dependence of the kinetic energy on the frequency.

Example 1.   Sodium (Na) has a work function $W = 2.9 \times 10^{-19}$ J. Find the cutoff frequency v_c for photoelectric emission.

$$v_c = \frac{W}{h} = \frac{2.9 \times 10^{-19} \text{ J}}{6.63 \times 10^{-34} \text{ J·sec}}$$

$$= 4.4 \times 10^{14} \text{ Hz}$$

The corresponding wavelength is

$$\lambda_c = \frac{c}{v_c} = 6.9 \times 10^{-7} \text{ m}$$

$$= 6900 \text{ Å (red)}.$$

Therefore visible light (4000–7000 Å) can cause photoelectric emission from sodium.

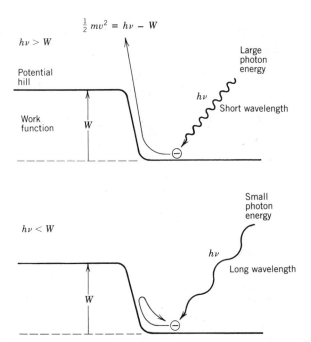

Fig. 14.2.   Electron absorbs the photon energy $h\nu$. If the energy is larger than the work function (potential energy), the electron is released from the metal.

Example 2.   A metal having a work function of 2.3 eV is illuminated by ultraviolet radiation of $\lambda = 3000$ Å. Calculate the maximum kinetic energy of photoelectrons emitted.

$$\tfrac{1}{2}mv^2 = h\nu - W,$$

where

$$h\nu = \frac{hc}{\lambda}$$

$$= \frac{6.63 \times 10^{-34} \times 3 \times 10^8}{3 \times 10^{-7}}$$

$$= 6.63 \times 10^{-19} \text{ J} = 4.1 \text{ eV}.$$

Then

$$\tfrac{1}{2}mv^2 = 4.1 - 2.3 = 1.8 \text{ eV}$$

$$= 3.0 \times 10^{-13} \text{ J}.$$

In Chapter 10 we learned that accelerated or decelerated charges can create electromagnetic radiation. For example, if an energetic electron hits a hard metal surface (such as tungsten), X rays can be created. In this case the electron experiences tremendously large deceleration and its energy can be converted into that of the X ray. This process may be called an inverse photoelectric effect. A more common terminology is *Bremsstrahlung*, which is a German word for *Bremse* (braking, deceleration) and *Strahlung* (radiation).

14.3. Hydrogen Atom

Neon signs use a gas discharge to create various colors. Incandescent lamps can create light for illumination. In this section, you will be introduced to quantum theory of radiation in order to understand the mechanism of radiation from atoms.

If the radiation spectrum from a hydrogen gas is carefully analyzed, the spectrum is composed of many discrete lines rather than a continuous spectrum. As we have seen in the previous section, if the electron in the hydrogen atom somehow loses a certain amount of energy, that energy is expected to be released as electromagnetic radiation.

The simplest picture of the hydrogen atom is that one electron is revolving around a proton. Let the distance between the electron and proton be r. Then the potential energy of the electron–proton system is

$$U = - \frac{e^2}{4\pi\varepsilon_0 r} \quad \text{(J)}. \tag{14.2}$$

The Coulomb force keeps the electron from flying away, balancing the centrifugal force.

$$\frac{mv^2}{r} = \frac{e^2}{4\pi\varepsilon_0 r^2}. \tag{14.3}$$

Then the kinetic energy of the electron is

$$\tfrac{1}{2}mv^2 = \frac{e^2}{8\pi\varepsilon_0 r} \quad \text{(J)}. \tag{14.4}$$

The total energy E of the atom is found from $U + \tfrac{1}{2}mv^2$,

$$E = - \frac{e^2}{8\pi\varepsilon_0 r} \quad \text{(J)}. \tag{14.5}$$

The question is: What is the radius r? According to classical dynamics, the radius r can be arbitrary, and each hydrogen atom could have a different radius and a different energy. That this is not the case is clearly evident from the discrete spectrum of radiation from hydrogen gas since if the radius is arbitrary, so is the energy, and we would expect the radiation spectrum to be continuous. We conclude that the radius must be discrete to explain the discrete spectrum.

In 1913 Niels Bohr (a Danish physicist) proposed that the angular momentum of electrons in atoms is discrete. For the case of the hydrogen atom the angular momentum of the electron is

$$mvr = \sqrt{\frac{me^2 r}{4\pi\varepsilon_0}} \tag{14.6}$$

He postulated that

$$mvr = n\frac{h}{2\pi} \qquad (n = 1, 2, 3, \ldots), \tag{14.7}$$

where h is the Planck's constant. Then the radius r is given by

$$r = n^2 \frac{\varepsilon_0 h^2}{\pi m e^2} = n^2 \times 5.3 \times 10^{-11} \quad \text{(m)} \tag{14.8}$$

and the energy becomes

$$E_n = -\frac{me^4}{8\varepsilon_0^2 h^2}\frac{1}{n^2} \quad \text{(J)}$$

$$= -13.6\frac{1}{n^2} \quad \text{(eV)}, \tag{14.9}$$

which is shown in Fig. 14.3.

The state of $n = 1$ is the lowest energy level that the ordinary hydrogen atom has. The electron in this bottom state can be brought to higher energy levels by electrical or optical disturbances (this process is called *excitation*). When this excited electron falls down to a lower state, it releases an energy that appears as radiation. Since the energy levels are discrete (Fig. 14.4), the frequencies of radiation are also discrete,

$$h\nu = \Delta E = 13.6 \left(\frac{1}{n^2} - \frac{1}{m^2} \right) \quad \text{(eV)}.$$

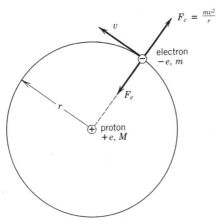

Fig. 14.3. A model of hydrogen atom: $F_e = e^2/4\pi\varepsilon_0 r^2$ (Coulomb force) and $F_c = mv^2/r$ (centrifugal force).

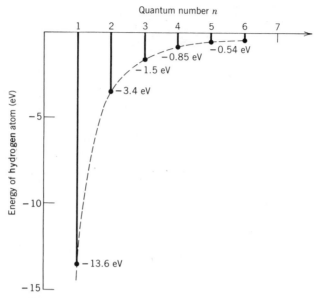

Fig. 14.4.   Discrete allowed energy states of a hydrogen atom.

Example 3.   The electron in a hydrogen atom is excited to the second energy level. When it falls down to the bottom level, what radiation (in frequency) should appear?

$$h\nu = \Delta E = 13.6 \left(1 - \tfrac{1}{4}\right)$$

$$= 10.2 \text{ eV} = 1.63 \times 10^{-18} \text{ J}$$

$$\therefore \; \nu = \frac{\Delta E}{h} = \frac{1.63 \times 10^{-18}}{6.63 \times 10^{-34}} = 2.46 \times 10^{15} \text{ Hz}.$$

The wavelength is

$$\lambda = \frac{c}{\nu} = 1.2 \times 10^{-7} \text{ m}$$
$$= 1200 \text{ Å},$$

which is in the ultraviolet range.

Example 4.   How much energy is required to completely free the electron in a hydrogen atom or to ionize the atom?

A free electron corresponds to $r = \infty$, or $n = \infty$. Then,

$$E = 13.6 \text{ eV} \left(1 - \frac{1}{\infty}\right)$$

$$= 13.6 \text{ eV}.$$

This is the ionization potential of the hydrogen atom.

14.4. De Broglie Wave

In Chapters 3 and 9 we saw that waves, both mechanical and electro-magnetic, carry momentum as well as energy. The relationship between them was

$$\text{Momentum} = \frac{\text{wave energy}}{\text{wave velocity}}. \qquad (14.10)$$

Since a photon carries an energy $h\nu$ at a velocity c, we expect that the momentum associated with a photon is

$$p = \frac{h\nu}{c} = \frac{h}{\lambda}. \qquad (14.11)$$

This is indeed correct and a more rigorous proof can be given by quantum mechanics.

Here we reverse the argument. We ask: Can we assign a wavelength λ defined through

$$\lambda = \frac{h}{p} \qquad (14.12)$$

to any object having a momentum p? De Broglie argued that we can; that is, any physical object should have both particle *and* wave nature. In Chapter 11 we saw that electron microscopes can have much higher resolving power than conventional optical microscopes because the wavelength associated with energetic electrons is much shorter than that of visible light. For example, the momentum of a 100-keV electron is 1.7×10^{-22} kg m/sec and the De Broglie wavelength is 3.9×10^{-12} m. This is about 10^5 times shorter than the wave-length of visible light ($4-7 \times 10^{-7}$ m). If you recall that the resolving power is inversely proportional to the wavelength, it can be understood why electron microscopes can "see" better than optical microscopes.

Problems

1. It is sometimes convenient to consider light (or electromagnetic waves) as a collection of photons. Estimate the number of photons emitted every second from a 100-W light bulb. Assume that the light is monochromatic and has a wavelength 5500 Å.

2. A laser beam ($\lambda = 6000$ Å) has a power density of 50 W/cm^2. What is the photon density associated with the beam?
   (*Answer:* 5.0×10^9/cm^3.)

3. Aluminum has a work function of 4.2 eV.
   (a) What is the cutoff frequency and wavelength for photoelectric emission?

(b) Ultraviolet light of $\lambda = 1500$ Å falls on an aluminum surface. What is the maximum kinetic energy of emitted electrons?

(*Answer:* 1.0×10^{15} Hz, 2960 Å, 4.1 eV.)

4. A hydrogen atom is excited from $n=1$ to $n=4$ energy level.
(a) What is the energy at each level?
(b) What energy is required for the excitation?
(c) If the electron falls down to $n=1$ level again, what is the wavelength of emitted radiation?

(*Answer:* -13.6 eV, -0.85 eV, 12.8 eV, 970 Å.)

5. Find the electron orbit radii and energy levels of a singly ionized helium ion which has one electron revolving around an α particle (two protons and two neutrons).

(*Answer:* $2.7 \times 10^{-11} n^2$ (m), $-54/n^2$ (eV).)

6. Calculate the De Broglie wavelength of a 200-keV electron.

(*Answer:* 2.7×10^{-12} m.)

7. Referring to the result of Example 2, Chapter 10, answer the following questions.
(a) What is the acceleration acting on the electron in a hydrogen atom at the ground state?
(b) What is the approximate time constant for the electron to lose energy by radiation?

(*Answer:* 9.0×10^{22} m/sec², 4.7×10^{-11} sec. This indicates that the classical model of a hydrogen atom consisting of a discrete electron revolving around a proton is unstable against radiation.)

CHAPTER 15

Nonlinear Waves

15.1. Introduction

Up to now, we have studied linear waves and oscillations. This is a drastic simplification to nature. Nonlinearity is a more common state, and in addition, its study makes life more interesting. Compare a drive through a flat region with that of a mountain. In Chapter 1 we encountered oscillatory phenomena and examined, in particular, the pendulum (Fig. 15.1). It was shown there that the equation of oscillation could be written as

$$\frac{d^2\theta}{dt^2} = -\frac{g}{l}\sin\theta \tag{15.1}$$

This is a nonlinear equation! Only in the very small region where $\theta \ll 1$ was it possible to write

$$\frac{d^2\theta}{dt^2} = -\frac{g}{l}\theta \tag{15.2}$$

and obtain the simple oscillatory solution given there.

Equation (15.1) is difficult as it stands, so let us relax the requirement that $\theta \ll 1$ to be $\theta < 1$. From Chapter 3, we can write

$$\frac{d^2\theta}{dt^2} \sim \frac{g}{l}\left(\theta - \varepsilon\frac{\theta^3}{3!}\right) \tag{15.3}$$

We have introduced a parameter ε ("bookkeeping parameter") to indicate clearly the nonlinear term. Equation (15.3) is still formidable as it stands, but there are techniques available to solve it. We shall illustrate a powerful *perturbation* technique here, although there are others.

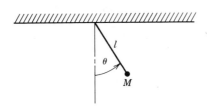

Fig. 15.1. Pendulum having a mass M and a length l.

303

The philosophy of the technique is to let

$$\theta = \theta^{(0)} + \varepsilon\theta^{(1)} + \varepsilon^2\theta^{(2)} + \cdots \tag{15.4}$$

where $\theta^{(0)} \gg \theta^{(1)} \gg \theta^{(2)} \gg \cdots$ and ε is the same bookkeeping parameter. Substitute (15.4) in (15.3).

$$\frac{d^2}{dt^2}\left\{\theta^{(0)} + \varepsilon\theta^{(1)} + \varepsilon^2\theta^{(2)} + \cdots\right\} =$$

$$-\frac{g}{l}\left\{[\theta^{(0)} + \varepsilon\theta^{(1)} + \varepsilon^2\theta^{(2)} + \cdots]\right.$$

$$\left.-\frac{\varepsilon}{3!}[\theta^{(0)} + \varepsilon\theta^{(1)} + \varepsilon^2\theta^{(2)} + \cdots]^3\right\}. \tag{15.5}$$

We shall collect terms of this equation in orders of ε.

$$\varepsilon^0: \quad \frac{d^2\theta^{(0)}}{dt^2} = \frac{-g}{l}\theta^{(0)} \tag{15.6}$$

$$\varepsilon^1: \quad \frac{d^2\theta^{(1)}}{dt^2} = \frac{-g}{l}\left\{\theta^{(1)} - \frac{1}{3!}[(\theta^{(0)})^3]\right\}, \tag{15.7}$$

$$\varepsilon^2: \quad \frac{d^2\theta^{(2)}}{dt^2} = \frac{-g}{l}\left\{\theta^{(2)} - \frac{1}{3!}[3(\theta^{(0)})^2\theta^{(1)}]\right\}, \tag{15.8}$$

and so on. We could add more terms but would soon be lost in a sea of algebra. The solution of Eq. (15.6) can be immediately written down as

$$\theta^{(0)} = A\ \sin\ \omega t + B\ \cos\ \omega t$$

where $\omega = \sqrt{g/l}$. If we now substitute $\theta^{(0)}$ in (15.7), we can obtain an equation for $\theta^{(1)}$. To simplify, let $B=0$ in $\theta^{(0)}$. Hence

$$\frac{d^2\theta^{(1)}}{dt^2} + \frac{g}{l}\theta^{(1)} = \frac{-1}{3!}\frac{g}{l}[A\ \sin\ \omega t]^3$$

$$= \frac{1}{3!}\frac{g}{l}A^3\left[\frac{-3}{4}\sin\ \omega t + \frac{1}{4}\sin\ 3\ \omega t\right]$$

Recall that $\omega = \sqrt{g/l}$ and note that there is a forcing term that oscillates at the *same* frequency as the terms on the left-hand side of the equation. The driving term that oscillates at $3\omega = 3\sqrt{g/l}$ creates no problem as far as the natural oscillation at the frequency $\omega = \sqrt{g/l}$. The solution for $\theta^{(1)}$ is straightforward and we write

$$\theta^{(1)} = \hat{A}\ \sin\ \omega t + \hat{B}\ \cos\ \omega t + \frac{1}{3!}\frac{gA^3}{8l}\sin\ 3\omega t$$

$$-\frac{1}{3!}\frac{gA^3}{8l}\omega t\ \cos\ \omega t \tag{15.9}$$

Note that as time increases, the last term in Eq. (15.9) increases and $\theta^{(1)}$ may eventually become comparable with or larger than $\theta^{(0)}$, a violation of the assumption given after (15.4). This is a *secular* term. A similar result would ensue for the terms $\theta^{(2)}$, $\theta^{(3)}$, and so on. Let us suggest a technique to get out of this dilemma, that is, develop a secular-free perturbation expansion. The technique is also called a *multiple time scale* perturbation expansion and it has found extensive use in the study of nonlinear oscillations.

In this technique, we define two (multiple) time scales given by t and $\tau = \varepsilon t$, where the same ordering parameter has been employed. Hence from (15.3), we write using the chain rule

$$\frac{d}{dt} = \frac{\partial}{\partial t} + \frac{\partial \tau}{\partial t}\frac{\partial}{\partial \tau} = \frac{\partial}{\partial t} + \varepsilon \frac{\partial}{\partial \tau}.$$

We use the partial differential notation and treat t and τ as *independent* variables. Hence we obtain from Eq. (15.3)

$$\left(\frac{\partial^2}{\partial t^2} + 2\varepsilon\frac{\partial}{\partial t}\frac{\partial}{\partial \tau} + \varepsilon^2 \frac{\partial^2}{\partial \tau^2}\right)\theta = \frac{-g}{l}\left(\theta - \varepsilon\frac{\theta^3}{3!}\right)$$

Substitute (15.4) into (15.3) and equate powers of ε

$$\varepsilon^{(0)}: \quad \frac{\partial^2\theta^{(0)}}{\partial t^2} = -\frac{g}{l}\theta^{(0)} \tag{15.10}$$

$$\varepsilon^{(1)}: \quad \frac{\partial^2\theta^{(1)}}{\partial t^2} + 2\frac{\partial}{\partial \tau}\frac{\partial\theta^{(0)}}{\partial t} = -\frac{g}{l}\left\{\theta^{(1)} - \frac{1}{3!}(\theta^{(0)})^3\right\}, \tag{15.11}$$

and so on.

The solution of (15.10) is as before

$$\theta^{(0)} = \tfrac{1}{2}(A\varepsilon^{i\omega t} + A^*\varepsilon^{-i\omega t}), \text{ where } \omega = \sqrt{g/l},$$

and we use the exponential notation for convenience except that the constant of integration A depends on the slow time variable $\tau = \varepsilon t$, that is, $A = A(\tau)$. The "$*$" indicates the complex conjugate. Substitute this solution in (15.11)

$$\frac{\partial^2\theta^{(1)}}{\partial t^2} + \omega^2\theta^{(1)} = \frac{\omega^2}{3!}[A^3 e^{i\omega t} + 3A^2 A^* e^{i\omega t} + 3AA^{*2}e^{-i\omega t} + A^{*3}e^{-i3\omega t}]$$

$$- 2i\left[\omega\frac{\partial A}{\partial \tau}e^{i\omega t} - \omega\frac{\partial A^*}{\partial \tau}e^{-i\omega t}\right].$$

We shall choose the slow time dependence so the terms multiplying $e^{\pm i\omega t}$ are set equal to zero. We eliminate the secularity causing terms in the equation by examining the slow time dependence of the solution. Hence

$$\frac{\partial A}{\partial \tau} + i\frac{\omega}{16}|A|^2 A = 0$$

$$\frac{\partial A^*}{\partial \tau} - i\frac{\omega}{16}|A|^2 A^* = 0$$

The solution of the first of these

$$A = A_0 e^{-i(\omega/16)|A|^2\tau}. \tag{15.12}$$

Hence

$$\theta^{(0)} = \tfrac{1}{2}\{A_0 e^{i\omega(1-|A|^2/16)t} + A_0^* e^{i\omega(1-|A|^2/16)t}\}, \tag{15.13}$$

and we note that the deviation of the resonant frequency depends on the square of the amplitude of the signal. The oscillation is essentially "detuned," with the detuning being proportional to the signal's power.

One could conjecture at this stage that this "detuning process" could have profound implications if there were a coupling between two oscillatory systems, one with a natural frequency of oscillation ω_1 being much higher than the second, that is, $\omega_1 \gg \omega_2$. If the first system is excited at a frequency $\omega \sim \omega_1$, the system may "detune" enough such that the difference is approximately ω_2 and the high-frequency signal will "parametrically" excite the signal at ω_2. The coupling between electrons and ions in a plasma is an example that exhibits such behavior.

Our first encounter with nonlinearity has had a dramatic effect. We shall see that it will produce several new features, such as shocks and solitons for wave propagation. New mathematical tools will be required and some will be introduced here.

15.2. Nonlinear Wave Equations

The derivation of nonlinear wave equations is usually quite straightforward. Just the simple equation of continuity

$$\frac{\partial n}{\partial t} + \frac{\partial (nv)}{\partial x} = 0$$

possesses a product of two terms and we encounter a nonlinearity. If a velocity depends on position x, the equation of motion for fluids

$$\frac{mdv(x, t)}{dt} = m\left(\frac{\partial v}{\partial t} + \frac{\partial v}{\partial x}\frac{\partial x}{\partial t}\right)$$

$$= m\left(\frac{\partial v}{\partial t} + v\frac{\partial v}{\partial x}\right) = \text{force}$$

also contains a nonlinear term. It is not difficult to find nonlinearity almost anywhere in nature. The problem arises when one tries to solve the nonlinear equation that has been derived to describe the phenomena. One can use the perturbation approach outlined in the previous section, use other techniques that are detailed in other books that just treat nonlinear waves, or find numerical solutions. Another approach is to be able to transform the equation of interest to a nonlinear equation that possesses known solutions. We illustrate this first with a slight generalization of Problem 3 in Chapter 9.

The problem will be to derive a nonlinear wave equation for electromagnetic wave propagation on a nonlinear-dispersive transmission line shown in Fig. 15.2. In this transmission line C_N is a nonlinear capacitor such as a "VARICAP" or a reverse-biased p–n junction diode, the capacitance of which depends on the voltage applied across it. Following the procedure outlined in Chapter 9 (and Problem 3), one can write the set of partial differential equations for the currents and voltages as

$$\frac{\partial I}{\partial x} + \frac{\partial \rho(V)}{\partial t} = 0, \tag{15.14}$$

$$\frac{\partial V}{\partial x} + L \frac{\partial I'}{\partial t} = 0, \tag{15.15}$$

$$\frac{\partial^2 V}{\partial x \partial t} + \frac{1}{C_S}(I - I') = 0, \tag{15.16}$$

where the current through the nonlinear capacitor is given by $\partial \rho(V)/\partial t$. From (15.14)–(15.16), we can eliminate I and I' and write

$$C_S \frac{\partial^4 V}{\partial x^2 \partial t^2} + \frac{1}{L}\frac{\partial^2 V}{\partial x^2} - \frac{\partial^2 \rho(V)}{\partial t^2} = 0. \tag{15.17}$$

The dependence of $\rho(V)$ (ρ is a charge density, Coulomb/length) must be specified before we can proceed. The simplest choice is to expand $\rho(V)$ in a Taylor series as

$$\rho(V) \sim C_0 V - C_N V^2.$$

Therefore (15.17) can be written as

$$\frac{1}{L}\frac{\partial^2 V}{\partial x^2} - C_0 \frac{\partial^2 V}{\partial t^2} + C_S \frac{\partial^4 V}{\partial x^2 \partial t^2} + C_N \frac{\partial^2 V^2}{\partial t^2} = 0. \tag{15.18}$$

We can recognize some features of Eq. (15.18). The first two terms are identical to the linear wave equation studied in Chapter 10. As such, its velocity of propagation is given by $c_w = 1/\sqrt{LC_0}$ if the other two terms were absent. The third term accounts for the dispersion introduced by the capacitors C_S, and the fourth term is the nonlinear term. If the nonlinear term can be neglected

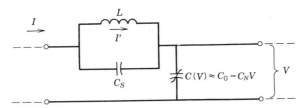

Fig. 15.2. A typical section for a distributed nonlinear-dispersive transmission line. The nonlinear capacitor $C(V)$ could be a reverse-biased p–n diode. The units are: L=henry/meter, C_S=farad-meter and $C(V)$=farad/meter.

and $V \sim V_0 e^{i(kx-\omega t)}$, one derives the dispersion relation

$$\frac{\omega}{k} = \frac{1}{\sqrt{LC_s k^2 + LC_0}} \sim \frac{1}{\sqrt{LC_0}} \left(1 - \frac{C_s}{2C_0} k^2 \right), \qquad (15.19)$$

which is illustrated in Fig. 15.3.

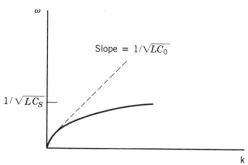

Fig. 15.3.   Dispersion curve for the linearized version of the transmission line shown in Fig. 15.2.

Note that in the long-wave length limit ($k \equiv 2\pi/\lambda \to 0$), the velocity approaches the linear velocity $c_w \to 1/\sqrt{LC_0}$. In the shortwave length limit ($k \equiv 2\pi/\lambda \to \infty$), there is an upper cutoff frequency $1/\sqrt{LC_s}$, which is the resonant frequency of the "tank" circuit in the series arm. For frequencies above this value, the wave damps.

To derive a well-known nonlinear wave equation, we shall introduce a technique entitled the *reductive perturbation technique*. We shall define two new variables

$$\xi = \varepsilon^{1/2} \left[x - \frac{t}{\sqrt{LC_0}} \right],$$

$$\tau = \varepsilon^{3/2} \frac{t}{\sqrt{LC_0}}, \qquad (15.20)$$

where ε is a bookkeeping parameter. The first states that we shall examine deviations in the wave frame. The ordering of the variables $\varepsilon^{1/2}$ and $\varepsilon^{3/2}$ reflects the dispersion given in the weakly dispersive limit (small k) in Eq. (15.19). Using the chain rule for differentiation, we write:

$$\frac{\partial}{\partial t} = \frac{\partial \xi}{\partial t} \frac{\partial}{\partial \xi} + \frac{\partial \tau}{\partial t} \frac{\partial}{\partial \tau}$$

$$= \frac{-\varepsilon^{1/2}}{\sqrt{LC_0}} \frac{\partial}{\partial \xi} + \varepsilon^{3/2} \frac{\partial}{\partial \tau}$$

$$\frac{\partial}{\partial x} = \frac{\partial \xi}{\partial x} \frac{\partial}{\partial \xi} + \frac{\partial \tau}{\partial x} \frac{\partial}{\partial \tau}$$

$$= \varepsilon^{1/2} \frac{\partial}{\partial \xi} + 0.$$

The voltage will be expressed in a perturbation series

$$V = \varepsilon u^{(1)} + \varepsilon^2 u^{(2)} + \cdots.$$

(15.21)

Collecting all terms that have similar powers in (15.18), which is now transformed to the new variables ξ and τ, we find to lowest order $(\theta(\varepsilon^3))$ that

$$\sqrt{LC_0}\frac{\partial u^{(1)}}{\partial \tau} + \frac{C_N}{C_0}u^{(1)}\frac{\partial u^{(1)}}{\partial \xi} + \frac{C_S}{2C_0}\frac{\partial^3 u^{(1)}}{\partial \xi^3} = 0$$

(15.22)

This equation is the first of an extensive series of what are called nonlinear evolution equations and was studied first by Korteweg and de Vries to describe water wave pulses that propagate along shallow canals. In their honor, it is now known as the Korteweg de Vries (KdV) equation. Although it was derived at the end of the nineteenth century, it is still being extensively studied as it admits solutions which describe a topic of current interest, namely solitons. More about that later.

Another approach to derive the KdV equation is suggested from quantum mechanics. In this approach one considers frequencies and wave numbers in the dispersion relation to correspond to the operators: $\omega \to i(\partial/\partial t)$ and $k \to -i(\partial/\partial x)$ and let the dispersion equation operate on a wave function ψ. Here we look at the long-wave length expansion given in Eq. (15.19) and write

$$\omega\psi \sim \left(\frac{k}{\sqrt{LC_0}} - \frac{C_S}{2C_0}\frac{1}{\sqrt{LC_0}}k^3\right)\psi$$

$$-\frac{1}{i}\frac{\partial\psi}{\partial t} = \frac{1}{i\sqrt{LC_0}}\frac{\partial\psi}{\partial x} + \frac{1}{i}\frac{C_S}{2C_0\sqrt{LC_0}}\frac{\partial^3\psi}{\partial x^3}$$

$$\sqrt{LC_0}\frac{\partial\psi}{\partial t} + \frac{\partial\psi}{\partial x} + \frac{C_S}{2C_0}\frac{\partial^3\psi}{\partial x^3} = 0.$$

(15.23)

Let us transform to the wave frame defined by

$$\xi = x - \frac{1}{\sqrt{LC_0}}t$$

$$\tau = t$$

(15.24)

and Eq. (15.23) becomes

$$\sqrt{LC_0}\frac{\partial\psi}{\partial \tau} + \frac{C_S}{2C_0}\frac{\partial^3\psi}{\partial \xi^3} = 0$$

(15.25)

One recognizes this as the linear part of the KdV equation (15.22).

In Eq. (15.24) we have used the fact that the velocity was a constant. This can be extended to the nonlinear case by just asserting that the velocity depends on the amplitude, say

$$\frac{1}{\sqrt{LC_0}} \to \frac{1}{\sqrt{LC_0}}(1 + \alpha\psi)$$

Therefore Eq. (15.25) becomes

$$\sqrt{LC_0}\,\frac{\partial \psi}{\partial \tau} + \frac{C_N}{C_0}\,\psi\,\frac{\partial \psi}{\partial \xi} + \frac{C_S}{2C_0}\,\frac{\partial^3 \psi}{\partial \xi^3} = 0 \tag{15.26}$$

where $\alpha = -C_N/C_0$.

The second nonlinear wave equation that has received considerable attention is the nonlinear Schrödinger (NLS) equation. One can derive it from first principles using the idea that a force can exist in a medium (say, a dielectric or a plasma) that is given by $\bar{F} \propto -\nabla \phi^2$, where ϕ is the amplitude of the modulation of a plane wave $\psi(x, t)$, where

$$\psi(x, t) = \phi(x, t)e^{i[k_0 x - \omega_0 t]}. \tag{15.27}$$

To find this force F, which is called the ponderomotive force or Miller force, we must examine charged particle motion in a spatially inhomogeneous oscillating electric field $E(x, t)$, where

$$E(x, t) = E_0(x) \cos \omega t. \tag{15.28}$$

Assume that $E(x > 0) > E(x < 0)$. During the first half cycle, a charged particle will be brought into a region of larger electric field and receive a stronger push to the left. During the second half cycle, the particle will be in a region of *weaker* electric field and the particle will receive less of a push to the right. In Fig. 15.4 the particle motion in (a) a spatially homogeneous and (b) a spatially inhomogeneous field are illustrated. In Fig. 15.4b, the dashed line indicates an average displacement of the particle's motion which we will designate as x_0. Its instantaneous value is given by $x = x_0 + x_1$. Hence the equation of motion is given by

$$m\,\frac{d^2 x}{dt^2} = qE(x, t)$$

or

$$m\,\frac{d^2 x_0}{dt^2} + m\,\frac{d^2 x_1}{dt^2} \sim q\left[E_0(x_0) + \frac{\partial E_0}{\partial x}\bigg|_{x_0} x_1\right]\cos \omega t. \tag{15.29}$$

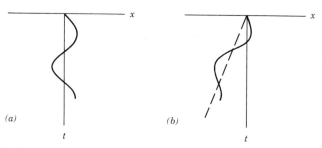

Fig. 15.4. Particle motion in (a) a spatially homogeneous and (b) a spatially inhomogeneous field.

If we time-average this equation over one period defined by $2\pi/\omega$, we obtain

$$m\frac{d^2x_0}{dt^2}=q\left.\frac{\partial E_0}{\partial x}\right|_{x_0}\langle x_1\cos\omega t\rangle,\tag{15.30}$$

where $\langle\ \rangle$ indicates a time average

$$\langle x_1\cos\omega t\rangle\equiv\int_0^{2\pi/\omega}x_1\cos\omega t\,dt.$$

To find x_1 we use (15.29) with the approximation

$$\frac{d^2x_1}{dt^2}\gg\frac{d^2x_0}{dt^2}\quad\text{and}\quad E_0(x_0)\gg\left.\frac{\partial E_0}{\partial x_0}\right|_{x_0}x_1$$

and write

$$m\frac{d^2x_1}{dt^2}\approx qE_0(x_0)\cos\omega t$$

Hence $x_1=-(qE_0(x_0)/m\omega^2)\cos\omega t$. The expression for the time average becomes

$$-\frac{1}{2}\frac{qE_0(x_0)}{m\omega^2}$$

and (15.30) is written as

$$m\frac{d^2x_0}{dt^2}=-\frac{1}{4}\frac{q^2}{m\omega^2}\frac{\partial(E_0(x_0))^2}{\partial x_0}=\text{force}\tag{15.31}$$

An examination of Fig. 15.4b indicates there are two time scales for the particle's trajectory, a fast time scale for the particle's oscillatory motion and a slow time scale governing its drift motion. This illustrates the "multiple-time scale" discussed in the previous section.

To derive the NLS equation, it is most convenient to assert that the dispersion relation depends on this ponderomotive force and write

$$\omega=\omega(k,|\phi|^2).\tag{15.32}$$

In Eq. (15.27), ω_0 and k_0 correspond to the frequency and wave number of the carrier wave. Therefore Eq. (15.27) can be written as

$$\omega-\omega_0\sim\left.\frac{\partial\omega}{\partial k}\right|_{k_0}(k-k_0)+\frac{1}{2}\left.\frac{\partial^2\omega}{\partial k^2}\right|_{k_0}(k-k_0)^2+\left.\frac{\partial\omega}{\partial|\phi|^2}\right|_{|\phi_0|^2}(|\phi|^2-|\phi_0|^2)\tag{15.33}$$

The terms $(\omega-\omega_0)$ and $(k-k_0)$ are replaced by the operators $i\,\partial/\partial t$ and $-i\,\partial/\partial x$, respectively, and Eq. (15.33) operates on the amplitude ϕ. Therefore

$$i\left(\frac{\partial\phi}{\partial t}+\left.\frac{\partial\omega}{\partial k}\right|_{k_0}\frac{\partial\phi}{\partial x}\right)+\frac{1}{2}\left.\frac{\partial^2\omega}{\partial k^2}\right|_{k_0}\frac{\partial^2\phi}{\partial x^2}-\left.\frac{\partial\omega}{\partial|\phi|^2}\right|_{|\phi_0|^2}(|\phi|^2-|\phi_0|^2)\phi=0.\tag{15.34}$$

One notes that a wave ϕ will move with the group velocity $\equiv \partial\omega/\partial k|_{k_0}$. The substitution

$$\xi = x - \frac{\partial\omega}{\partial k}\bigg|_{k_0} t, \qquad \tau = t$$

transforms (15.34) to the NLS equation

$$i\frac{\partial\phi}{\partial\tau} + \frac{1}{2}\frac{\partial^2\omega}{\partial k^2}\bigg|_{k_0}\frac{\partial^2\phi}{\partial\xi^2} - \frac{\partial\omega}{\partial|\phi|^2}\bigg|_{|\phi_0|^2}(|\phi|^2 - |\phi_0|^2)\phi = 0. \qquad (15.35)$$

The KdV and NLS equations are two nonlinear wave equations that fall into the general category of nonlinear evolution equations that are currently topics of much current interest among soliton aficionados. Details of solitons will be presented later. We shall just examine solutions of the KdV and NLS equations here.

To find the solution of the KdV equation, it is convenient to look for a "wave of permanant profile, "that is, a wave solution that does not change its shape as the wave propagates. To do this let $\zeta = \xi - (1/\sqrt{LC_0})\tau$ and Eq. (15.22) becomes

$$-\frac{du^{(1)}}{d\zeta} + \frac{C_n}{C_0}u^{(1)}\frac{du^{(1)}}{d\zeta} + \frac{C_S}{2C_0}\frac{d^3u^{(1)}}{d\zeta^3} = 0, \qquad (15.36)$$

whose first integral is

$$-U + \frac{C_N}{2C_0}U^2 + \frac{C_S}{2C_0}\frac{d^2U}{d\zeta^2} = k_1, \qquad (15.37)$$

where the superscript 1 is dropped for convenience, that is, $U = u^{(1)}$. We shall look for a pulse solution that satisfies $U \to 0$, $dU/d\zeta \to 0$ and $d^2U/d\zeta^2 \to 0$ as $|\zeta| \to \infty$. Hence $k_1 = 0$. The integral of Eq. (15.37) is found by multiplying Eq. (15.37) by $(dU/d\zeta)d\zeta$ and integrating

$$-\frac{U^2}{2} + \frac{C_N}{6C_0}U^3 + \frac{C_S}{4C_0}\left(\frac{dU}{d\zeta}\right)^2 = k_2,$$

where once again $k_2 = 0$. The integral of this equation is

$$U = U_0 \operatorname{sech}^2\left[\sqrt{\frac{C_N U_0}{6C_S}}\left(\zeta - \frac{C_N U_0}{3C_0}\tau\right)\right].$$

In laboratory coordinates it is written as

$$U = U_0 \operatorname{sech}^2\left[\sqrt{\frac{C_N U_0}{6C_S}}\left(x - \left(1 + \frac{C_N U_0}{3C_0}\right)\frac{t}{\sqrt{LC_0}}\right)\right]. \qquad (15.38)$$

The velocity c of this "solitary wave" (so called because it is a single pulse shaped object as shown in Fig. 15.5)

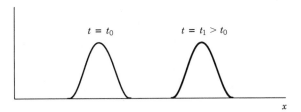

Fig. 15.5.   Propagation of a solitary wave.

is given by

$$c=\left[1+\frac{C_N}{3C_0}U_0\right]\frac{1}{\sqrt{LC_0}},\tag{15.39}$$

which indicates that its velocity increases with amplitude. The half-width of this pulse can be computed from $U/U_0=1/2=\text{sech}^2\,[.88]$. Hence

$$t\big|_{\text{half}}=.88\times2\sqrt{\frac{6C_SLC_0}{C_N}}\frac{1}{\sqrt{U_0}}\tag{15.40}$$

The product of the pulse's amplitude times the square of its half-width is a constant.

To find a solution of the NLS equation (15.35), we rewrite it as

$$i\frac{\partial\phi}{\partial\tau}+a\frac{\partial^2\phi}{\partial\xi^2}+b\phi|\phi|^2=0,\tag{15.41}$$

where

$$a=\frac{1}{2}\frac{\partial^2\omega}{\partial k^2}\bigg|_{k_0},\qquad b=-\frac{\partial\omega}{\partial|\phi|^2}\bigg|_{|\phi_0|^2}$$

and we let $|\phi_0|^2=0$. We look for a solution of the form

$$\phi(\xi,\tau)=e^{i\Omega\tau}f(\xi).\tag{15.42}$$

Substitute (15.42) in (15.41)

$$-\Omega f+a\frac{d^2f}{d\xi^2}+bf^3=0\tag{15.53}$$

The first integral of (15.43) is

$$-\Omega\frac{f^2}{2}+\frac{a}{2}\left(\frac{df}{d\xi}\right)^2+\frac{bf^4}{4}=\text{const},\tag{15.44}$$

where the constant of integration is set equal to zero, as in the KdV equation. The second integral of (15.44) is

$$f=\sqrt{\frac{2\Omega}{b}}\,\text{sech}\left[\frac{\sqrt{\Omega a}}{b}\xi\right].\tag{15.45}$$

Combine (15.42) and (15.45)

$$\phi(\xi, \tau) = \sqrt{\frac{2\Omega}{b}} \operatorname{sech} \left[\frac{\sqrt{\Omega a}}{b} \xi \right] e^{i(\Omega/b)\tau}. \tag{15.46}$$

We shall encounter the solutions given by Eqs. (15.38) and (15.46) later when the topic of solitons is discussed.

15.3. Characteristics

We have seen in the previous section that nature has provided several examples of nonlinear waves that included effects of dispersion. As such there arose solutions that tended to balance the effects of nonlinearity with the dispersion, and pulselike waves were found as solutions to the KdV and the NLS equations. You may have wondered, "What would have happened if the dispersive term were somehow absent from the equation?" Let us answer this by looking at two wave equations,

$$\frac{\partial u}{\partial t} + c \frac{\partial u}{\partial x} = 0 \tag{15.47}$$

and

$$\frac{\partial u}{\partial t} + u \frac{\partial u}{\partial x} = 0. \tag{15.48}$$

In Eq. (15.47), c is a constant and therefore we have a linear equation. The most general solution is $u(x, t) = f(x - ct)$, as we have already noted. Let us assume an initial condition of

$$u(x, 0) = \begin{cases} 1 - x^2 & \text{for } |x| \leqslant 1 \\ 0 & \text{for } |x| > 1 \end{cases} \tag{15.49}$$

Hence at any later time, we can write

$$u(x, t) = \begin{cases} 1 - (x - ct)^2 & \text{for } |x - ct| \leqslant 1 \\ 0 & \text{for } |x - ct| > 1 \end{cases} \tag{15.50}$$

This, of course, is a wave of permanent profile, and we illustrate it in Fig. 15.6.

To solve (15.48) with the same initial condition that we used to solve (15.47), we must solve (15.48) along the *characteristics* defined by solving the characteristic equations. A first-order equation

$$P(x, t, u) \frac{\partial u}{\partial x} + Q(x, t, u) \frac{\partial u}{\partial t} = S(x, t, u) \tag{15.51}$$

can be solved along a characteristic curve defined by

$$\frac{dx}{P} = \frac{dt}{Q} = \frac{du}{S}$$

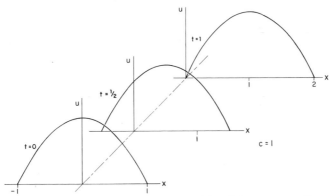

Fig. 15.6.   Propagation of a pulse governed by a linear wave equation.

From (15.48), we write

$$\frac{dx}{u} = \frac{dt}{1} = \frac{du}{0},$$

so that along the characteristic

$$\frac{dx}{dt} = u$$

u is conserved. Applying this to the linear equation (15.47), we would obtain $\partial x/\partial t = c$. Therefore the general solution of (15.48) is

$$u(x, t) = f(x - ut), \tag{15.52}$$

where f is determined by an initial condition.

Before specifying an initial condition, let us contrast the characteristics obtained from the linear equation (15.47), $dx/dt = c$; with the nonlinear equation (15.48), $dx/dt = u$. This is illustrated in Fig. 15.7. In Fig. 15.7a the characteristics are all parallel, which implies that all parts of the wave travel with the same velocity c. In Fig. 15.7b, the characteristic curves travel with different velocities, parts with a higher value of u move faster. This is borne out by writing the

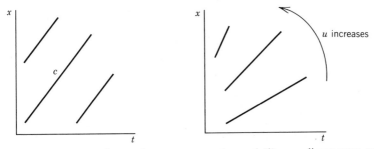

Fig. 15.7.   Characteristics of (a) a linear wave equation and (b) a nonlinear wave equation.

solution for (15.48) as

$$u(x, t) = \begin{cases} 1-(x-ut)^2 & \text{for } |x-ut| \leqslant 1 \\ 0 & \text{for } |x-ut| > 0 \end{cases} \tag{15.53}$$

Solving (15.53) explicitly for u, we obtain,

$$u(x, t) = \begin{cases} \dfrac{1}{2t^2} [(2xt-1) \pm (1-4xt+4t^2)^{1/2}] & \text{for } |x-ut| \leqslant 1 \\ 0 & \text{for } |x-ut| > 1 \end{cases} \tag{15.54}$$

It is instructive to look at the characteristics posed by the initial conditions given in (15.49). They are shown in Fig. 15.8. From this figure we note that all characteristics that start $(x > -1, t=0)$ will intersect characteristics starting $(x > 1, t=0)$ at some time or other. An intersection of two characteristics implies two values of u, a physically impossible situtation. Hence we have to introduce a discontinuity, which in fluid dynamics is called a *shock*.

In Fig. 15.9 we sketch the evolution of the pulse. Note that $u(x, t)$ can have multiple values at some time $T > 0$ when $\partial u / \partial x$ changes sign from $\partial u / \partial x|_{t=0} < 0$ to $\partial u / \partial x|_{t=T} > 0$. From (15.54), we compute

$$\frac{\partial u}{\partial x}\bigg|_{x=1} = \frac{1}{2t^2}\left[2t \pm \frac{1}{2}\frac{-4t}{(1-4xt+4t^2)^{1/2}}\right]\bigg|_{x=1}$$

$$= \frac{1}{2t^2}\left[2t \mp \frac{2t}{1-2t}\right]$$

and note that $\partial u / \partial x$ will change sign (through infinity) at a time $T = \frac{1}{2}$.

This example can be generalized for an initial condition of

$$u(x, 0) = \begin{cases} f(x) & \text{for } |x| \leqslant a \\ 0 & \text{for } |x| > a, \end{cases} \tag{15.55}$$

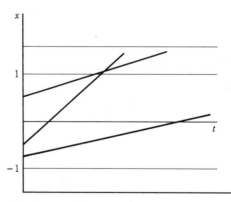

Fig. 15.8.   Characteristics for a nonlinear wave equation with the initial conditions specified by Eq. (15.49).

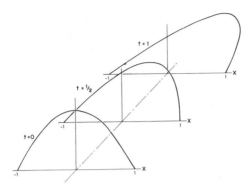

Fig. 15.9.   Propagation of a pulse governed by a nonlinear wave equation.

where $f(x)$ is a continuously differentiable function with positive and negative slopes. Analogous to (15.53), we write

$$u(x, t) = \begin{cases} f(x-ut) & \text{for } |x-ut| \leqslant a \\ 0 & \text{for } |x-ut| > a. \end{cases} \tag{15.56}$$

Define the wave variable ξ as $\xi = x - ut$ and compute from (15.56)

$$\frac{\partial u}{\partial x} = \frac{\partial f}{\partial \xi} \frac{\partial \xi}{\partial x}$$

$$= \frac{\partial f}{\partial \xi} \left(1 - \frac{\partial u}{\partial x} t \right).$$

Solving for $\partial u / \partial x$, we obtain

$$\frac{\partial u}{\partial x} = \frac{\partial f / \partial \xi}{1 + t \partial f / \partial \xi}. \tag{15.57}$$

Equation (15.57) expresses the slope of the profile of u in terms of the initial profile at ξ, where x is the position at time t of the point that was initially at ξ. If $\partial f / \partial \xi < 0$, $\partial u / \partial x$ is infinite at a time $T = 1/(-\partial f / \partial x)$. Therefore if the profile initially had a negative slope at some point ξ, then for $t > T = (1/-\partial f / \partial \xi)_{min}$, the solution is no longer single-valued in the neighborhood of the point $\xi_0 + Tf(\xi_0)$, where ξ_0 is the point where $(1/-\partial f / \partial \xi)$ attains its minimum value.

To show this we examine (15.57) at a time t, where

$$t = T + \varepsilon = \left(\frac{1}{-\partial f / \partial \xi} \right)_{min} + \varepsilon$$

where ε is small and $(\partial f / \partial \xi)_{min}$ occurs at $\xi = \xi_0$. Therefore from (15.57)

$$\frac{\partial u}{\partial x}(x_0, T + \varepsilon) = \left(\frac{\partial f/\partial \xi}{1 + t\,\partial f/\partial \xi}\right)_{\xi = \xi_0,\, t = T + \varepsilon}$$

$$= \frac{\partial f/\partial \xi|_{\xi_0}}{1 + \dfrac{\partial f}{\partial \xi}\bigg|_{\xi_0}\left[\dfrac{1}{-\dfrac{\partial f}{\partial \xi}\bigg|_{\xi_0}} + \varepsilon\right]}$$

$$= \frac{1}{\varepsilon}$$

This implies

$$\frac{\partial u}{\partial x}\bigg|_{T-0} \to -\infty \quad \text{and} \quad \frac{\partial u}{\partial x}\bigg|_{T+0} \to \infty$$

Therefore we see that a linear wave travels without a change of profile. The role of nonlinearity progressively deforms the profile as it propagates. After some time $t > T$, the wave contains a jump discontinuity. As we have noted in the previous section, dispersion can enter and cause a wave with a constant profile to propagate even if nonlinearity is included.

15.4. Self-similarity

We have examined one technique to solve nonlinear wave equations that uses characteristics. If we put ourselves in the frame that is moving with the wave, the detailed processes in that frame may actually be diffusing in space. One may be able to transform the equation to a nonlinear extension of the diffusion equation

$$D\frac{\partial^2 n}{\partial x^2} - \frac{\partial n}{\partial t} = 0, \tag{15.58}$$

where D is the diffusion equation, which, as we shall see later, may depend on n, x, and/or t. We recognize this as the linear part of the NLS equation, where the term $i = \sqrt{-1}$ has been incorporated into D. The procedure that we shall outline in this section is an attempt to answer the question, "Can the variables n, x, and t be combined in some manner such that the partial differential equation (15.58) (or any PDE) could be transformed to an ordinary differential equation (ODE) that may be easier to solve?" For the wave equation we could set $\zeta = x - ct$ and look for waves of permanent profile. We shall seek other solutions here.

In particular, we shall seek solutions that leave (15.58) invariant if the variables are replaced by

$$n = \bar{n}(x, t, n, \varepsilon) \sim n + \tilde{n}(x, t, n) + O(\varepsilon^2)$$

$$x = \bar{x}(x, t, n, \varepsilon) \sim x + \tilde{x}(x, t, n) + O(\varepsilon^2) \tag{15.59}$$

$$t = \bar{t}(x, t, n, \varepsilon) \sim t + \tilde{t}(x, t, n) + O(\varepsilon^2),$$

where ε is a parameter. To transform (15.58), one must apply the chain rule, collect all terms with like derivatives of n, and require that the transformation leaves (15.58) invariant, that is, it looks like (15.58) with barred variables replacing unbarred variables. This will yield a large set of simultaneous equations to be solved. The solution of this set can be found and is called the invariants of the transformation group. The term group comes from algebra and was discussed in the nineteenth century by Sophus Lie. The invariants are the self-similar variables.

The above paragraph was rather formidable and has probably hidden the procedure from the reader. It is included only to let you know that there is some firm mathematical foundation for what we shall do. We can ride on the shoulders of the giants who have preceded us and who will carry us through the dense forest of algebra. Here we shall just choose the simplest linear group, which is defined as

$$G = \begin{cases} n = a^\alpha \bar{n} \\ x = a^\beta \bar{x} \\ t = a^\gamma \bar{t} \end{cases} \tag{15.60}$$

where a is a parameter and α, β, and γ will be determined from invariance requirements. Several significant problems can be treated with this group and it will allow us to illustrate the procedure. In particular, a function $F(y)$ is said to be "constant conformally invariant" (CCI) if $F(y) = f(a)F(\bar{y})$, where $f(a)$ is some function of the parameter a.

Substituting (15.59) into (15.58), we write

$$a^{\alpha - 2\beta} D \frac{\partial^2 \bar{n}}{\partial \bar{x}^2} - a^{\alpha - \gamma} \frac{\partial \bar{n}}{\partial \bar{t}} = 0 \tag{15.61}$$

For (15.60) to be CCI under the transformation group G, one requires

$$\alpha - 2\beta = \alpha - \gamma$$

or $\gamma = 2\beta$. We shall defer until later the further specification of these constants.

Instead, we now seek to determine the "invariants" of the transformation group G. This is achieved by employing a theorem from group theory. The invariants are obtained from $QI = 0$ where I is the invariant and Q is the operator

$$Q = \frac{\partial \bar{n}}{\partial a}\bigg|_{a=1} \frac{\partial}{\partial n} + \frac{\partial \bar{x}}{\partial a}\bigg|_{a=1} \frac{\partial}{\partial x} + \frac{\partial \bar{t}}{\partial a}\bigg|_{a=1} \frac{\partial}{\partial t}$$

$$= -\alpha n \frac{\partial}{\partial n} - \beta x \frac{\partial}{\partial x} - \gamma t \frac{\partial}{\partial t} \tag{15.62}$$

The solutions of this first-order equation $QI = 0$ are obtained by solving the same "Lagrange subsidiary" equations that were written down when discussing characteristics, namely,

$$\frac{dn}{-\alpha n}=\frac{dx}{-\beta x}=\frac{dt}{-\gamma t}.$$
(15.63)

These "invariants" are the self-similar variables that we seek. One set of solutions of (15.63) is given by

$$\phi(\xi)=\frac{n(x,\,t)}{t^{\alpha/\gamma}} \quad \text{and} \quad \xi=\frac{x}{t^{\beta/\gamma}}$$
(15.64)

Recall from (15.61) that $\beta/\gamma=\frac{1}{2}$. One could combine the terms in (15.63) in a different combination to obtain a different set of solutions also.

Substitute (15.64) in (15.58) and obtain

$$D\frac{d^2\phi}{d\xi^2}+\frac{\xi}{2}\frac{d\phi}{d\xi}-\frac{\alpha}{\gamma}\phi=0$$

or

$$\frac{d^2\phi}{d(\xi/2\sqrt{D})^2}+2\,\frac{\xi}{2\sqrt{D}}\frac{d\phi}{d(\xi/2\sqrt{D})}-4\frac{\alpha}{\gamma}\phi=0.$$
(15.65)

Writing the equation in this latter form, we can recognize that its solution can be written in terms of complementary functions erfc $(\pm\xi/2\sqrt{D})$

$$\phi=A\,i^{2\alpha/\gamma}\,\mathrm{erfc}\left(\frac{\xi}{2\sqrt{D}}\right)+B\,i^{-2\alpha/\gamma}\,\mathrm{erfc}\left(\frac{-\xi}{2\sqrt{D}}\right)$$
(15.66)

where $i^{2\alpha/\gamma}$ is an ordering parameter $(i\neq\sqrt{-1})$ and

$$i^{-1}\,\mathrm{erfc}\left(\frac{\xi}{2\sqrt{D}}\right)\equiv\frac{2}{\sqrt{\pi}}\,\varepsilon^{-\xi^2/4D}$$

$$i^0\,\mathrm{erfc}\left(\frac{\xi}{2\sqrt{D}}\right)\equiv\mathrm{erfc}\,\frac{\xi}{2\sqrt{D}}$$

and

$$i^n\,\mathrm{erfc}\left(\frac{\xi}{2\sqrt{D}}\right)\equiv\int_{\xi/2\sqrt{D}}^{\infty}i^{n-1}\,\mathrm{erfc}\,t\,dt \quad (n=0,\,1,\,2,\,\dots).$$

At this stage, the parameter α/γ is still arbitrary. We shall specify it to satisfy boundary conditions or a conservation law. Note that (15.58) is a "third-order" equation with three boundary or initial conditions given while (15.65) is a second-order equation that requires only two boundary or initial conditions. Hence there must be a *consolidation* of these conditions. This can be seen below for the physically reasonable conditions.

$$\left.\begin{array}{r}n(x=\infty,\,t)=0\\ n(x,\,t=0^+)=0\end{array}\right\}\to\phi(\xi=\infty)=0$$
(15.67)

The third boundary condition could have one of two forms that would yield self-similar solutions. They are

$$n(x=0, t) = \text{constant} \tag{15.68}$$

or

$$\int_{-\infty}^{\infty} n(x, t)dx = \text{constant}$$

Combine (15.64) and (15.66) and obtain

$$n(x, t) = t^{\alpha/\gamma} \left\{ Ai^{2\alpha/\gamma} \, \text{erfc}\left(\frac{\xi}{2\sqrt{D}}\right) + Bi^{-2\alpha/\gamma} \, \text{erfc}\left(\frac{-\xi}{2\sqrt{D}}\right)\right\}$$

For any time t, the first condition given in (15.68), which states that there is an infinite source of n at $x=0$, implies $\alpha/\gamma = 0$. Therefore

$$n(x, t) = A \, \text{erfc}\left(\frac{\xi}{2\sqrt{D}}\right) = A \, \text{erfc}\left(\frac{x}{2\sqrt{tD}}\right).$$

where the constant $B=0$ in order to satisfy (15.67).

Let us assert that the second condition, which states that the quantity n must be conserved in space, must also be invariant under the transformation group G. This is called the *similarity postulate*. Therefore

$$\int_{-\infty}^{\infty} ndx = a^{\alpha + \beta} \int_{-\infty}^{\infty} \bar{n}d\bar{x}$$

and for this to be invariant

$$\frac{\alpha}{\gamma} = \frac{-\beta}{\gamma} = \frac{-1}{2}.$$

The self-similar solution is

$$n(x, t) = \frac{\phi(\xi)}{\sqrt{t}} = \frac{2A'}{\sqrt{\pi}\sqrt{t}} \varepsilon^{-x^2/4tD},$$

where the constant A' can be determined by substituting this in the integral in (15.68). You may recognize this solution as the "Green's function" for the diffusion equation. In Fig. 15.10 we sketch the response of these two solutions. A general comment can be made about this solution that seems to be in conflict with what we have already learned about waves, namely, a signal or density can appear at $x = \infty$ at a time $t = 0^+$. The amplitude may be very small but it is still there. Let us get ourselves out of this dilemma by looking at a nonlinear extension of (15.58).

In particular, let us examine the dimensionless nonlinear diffusion equation

$$\frac{\partial}{\partial x}\left[n^r \frac{\partial n}{\partial x}\right] - \frac{\partial n}{\partial t} = 0 \qquad (r > 0) \tag{15.69}$$

and let $D=1$. If we apply the group G defined by (15.60), we find the self-similar

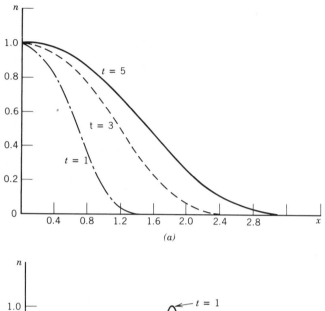

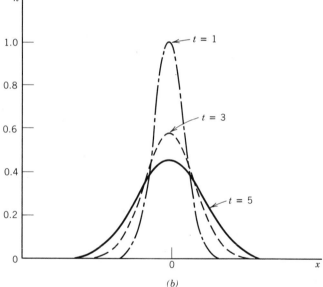

Fig. 15.10.   Diffusion of a quantity n due to (a) a step function source at $x=0$ and $t=0$ and (b) a delta function source at $x=0$ and $t=0$.

variables to be

$$\xi = \frac{x}{t^{\beta/\gamma}} \quad \text{and} \quad \phi(\xi) = \frac{n}{t^{(1/r)[2(\beta/\gamma)-1]}}.$$

The corresponding ordinary differential equation for $\phi(\xi)$ is

$$\frac{d}{d\xi}\left[\phi^r \frac{d\phi}{d\xi}\right] + \frac{\beta\xi}{\gamma}\frac{d\phi}{d\xi} - \frac{1}{r}\left[2\frac{\beta}{\gamma}-1\right]\phi = 0. \qquad (15.70)$$

If now

$$\frac{\beta}{\gamma} = -\frac{1}{r}\left[2\frac{\beta}{\gamma} - 1\right]$$

or

$$\frac{\beta}{\gamma} = \frac{1}{r+2},$$

(15.70) can be integrated once to

$$\phi^r \frac{d\phi}{d\xi} - \frac{1}{r+2}\xi\phi = \text{const.} \tag{15.71}$$

The boundary conditions that we shall employ are given in (15.67) with the additional requirement that there will exist a "sharp front" in the diffusion. We shall also impose the requirement that $\partial n/\partial x = 0$ at $x = 0$, which implies $d\phi/d\xi|_{\xi=0} = 0$ and the constant of integration in (15.71) is zero. The integral of (15.71) is

$$\phi(\xi) = \begin{cases} \left[1 - \dfrac{r}{r+2}\dfrac{\xi^2}{2}\right]^{1/r} & \xi \leqslant \xi_0 \\ 0 & \xi > \xi_0, \end{cases} \tag{15.72}$$

where

$$\xi_0 = \left[2\frac{r+2}{2}\right]^{1/2} \qquad r > 0.$$

The solution for $r = 1$ is

$$n(x, t) = \begin{cases} \dfrac{1}{t^{1/3}}\left[1 - \dfrac{x^2}{t^{2/3}}\dfrac{1}{\xi_0^2}\right] & \xi \leqslant \sqrt{3} \\ 0 & \xi > \sqrt{3}. \end{cases}$$

This solution is sketched in Fig. 15.11.

We see that fairly simple solutions can be found for nonlinear partial differential equation. These solutions may be "attractors" in that solutions with different boundary conditions may approach this self-similar solution. In the cases presented here, it was possible to integrate the ordinary differential equation that arose from the transformation. If this is not directly possible, one can resort to approximate or numerical methods. The problem, however, has been greatly simplified. One could reduce the ordinary differential equation further, and Sophus Lie was able to reduce it to an algebraic equation with further transformations.

To illustrate one approximate technique, we look at some ideas from boundary layer theory in fluid mechanics. Let us look at (15.65) with the constant $\alpha/\gamma = 0$

$$\frac{d^2\phi}{d\theta^2} + 2\theta\frac{d\phi}{d\theta} = 0, \tag{15.73}$$

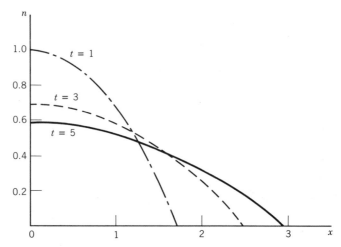

Fig. 15.11. Diffusion of a quantity n that satisfies a nonlinear diffusion equation.

where $\theta = \xi/2\sqrt{D}$. If ϕ and all its derivatives are zero as $\theta \to \infty$, one can integrate (15.73), at least formally, to write

$$\frac{d\phi}{d\theta}\bigg|_{\theta=\infty} - \frac{d\phi}{d\theta}\bigg|_{\theta=0} = -\int_0^\infty 2\theta \frac{d\phi}{d\theta} \, d\theta$$

or

$$\frac{d\phi}{d\theta}\bigg|_{\theta=0} = \int_0^\infty 2\theta \frac{d\phi}{d\theta} \, d\theta. \tag{15.74}$$

The philosophy of this technique is to make an "educated guess" on what the general solution should look like in order to satisfy the boundary conditions: $\phi = \phi_0$ at $\theta = 0$ and $\phi \to 0$ as $\theta \to \infty$. A function that would satisfy this would be

$$\phi = \begin{cases} \left[1 - \left(\dfrac{3}{2}\dfrac{\theta}{\theta_0} - \dfrac{\theta^3}{2\theta_0^3}\right)\right] & 0 \leqslant \theta \leqslant \theta_0 \\ 0 & \theta > \theta_0. \end{cases} \tag{15.75}$$

Other choices could be made. The value of θ_0 is found by substituting (15.75) into (15.74) and performing the integration.

We write finally

$$\phi = \phi_0 \left[1 - \left(\frac{3}{2\theta_0}\frac{x}{\sqrt{tD}} - \frac{x^3}{2\theta_0^3}\frac{1}{(tD)^{3/2}}\right)\right],$$

which is a reasonable approximation to the solution $\phi = \phi_0 \, \text{erfc}\,(x/\sqrt{tD})$ that we found before.

Approximate solutions have advantages in that they are fairly easy to obtain, follow from one's intuition, and can be used to check numerical procedures.

Problems

1. Sketch the solution given in Eq. (15.9) if $A=B=(1/3!)(gA^3/8l)=1$.

2. Consider a parallel electrical circuit consisting of an inductor L and a nonlinear capacitor ("VARICAP") $C(v)$.
   (a) Derive the equation describing the voltage oscillation.
   (b) Let $C(v) \sim C_0 + C_1 V$, where $C_1 < C_0$ and describe what the solution should look like.

3. Solve the nonlinear equation in Problem 2 using the multiple-time scale method.

4. Sketch the solution given in Eq. (15.13) for various values of $|A|^2$.

5. Follow the derivation that yielded the KdV equation. Carry it out to next order:

6. Derive the wave equation if the nonlinear capacitor had the relation

$$C(V) = C_0 V - C_N V^3$$

7. Sketch the solution of the NLS equation.

8. Using characteristics, find the solution of (15.47) if

$$u(x, 0) = \begin{cases} 1 - |X| & \text{for } |X| \leqslant 1 \\ 0 & \text{for } |X| > 1 \end{cases}$$

9. Using characteristics, find the solution of (15.48) if

$$u(x, 0) = \begin{cases} 1 - |X| & \text{for } |X| \leqslant 1 \\ 0 & \text{for } |X| > 1 \end{cases}$$

10. In Problem 9 find the time when a shock will form. (*Answer*: $t = 1$)

11. Find the self-similar solution of

$$\frac{\partial}{\partial y}\left[\frac{1}{\psi}\frac{\partial \psi}{\partial y}\right] - \frac{\partial \psi}{\partial t} = 0$$

for the condition that $\int_0^\infty \psi \, dy = \text{const}$.
(*Answer*: $\psi = [t(y^2/2t^2 + \text{const}]^{-1})$

12. Find the self-similar solution for the set of equations

$$\frac{\partial E}{\partial x} = \rho$$

$$\frac{\partial i}{\partial x} + \frac{\partial \rho}{\partial t} = 0$$

$$i = \rho E$$

$$\left(\text{Answer}: \quad E = \frac{x + x_0}{t + t_0}, \, \rho = \frac{1}{t + t_0}, \, i = \frac{x + x_0}{(t + t_0)^2}\right)$$

13. Starting from the equation

$$\frac{\partial \sigma}{\partial t} = A \frac{\partial \sigma}{\partial x} - B \frac{\partial}{\partial x}\left[\sigma \frac{\partial \sigma}{\partial x}\right] - C \frac{\partial^2 \sigma}{\partial x^2}$$

show that by change of independent and dependent variables that this can be transformed to

$$\frac{\partial \hat{\sigma}}{\partial \hat{t}} = \frac{\partial}{\partial \hat{x}}\left[\hat{\sigma} \frac{\partial \hat{\sigma}}{\partial \hat{x}}\right]$$

14. Find the self-similar solution of

$$\frac{\partial^2 E}{\partial x^2} = \frac{\partial}{\partial t}(\sigma_r E),$$

where $\sigma_r = \sigma_0 \exp[\beta\sqrt{E}]$ subject to

$$E(x=0, t) = 1, \qquad E(x=\infty, t) = E(x, t=0^+) = 0$$

You should use an approximate technique to integrate this.

15. For the equation

$$\frac{\partial f}{\partial t} = \frac{G}{p^2} \frac{\partial}{\partial p}\left[p^2 + \frac{\partial f}{\partial p}\right] + \frac{B}{p^2} \frac{\partial[p^3 f]}{\partial p}$$

show that the self-similar variables

$$\xi = \frac{p}{\varepsilon^{\alpha t}} \quad \text{and} \quad \phi(\xi) = \frac{f(p, t)}{\varepsilon^{\beta t}}$$

will allow it to be converted to an ordinary differential equation. If $\beta = -3\alpha$, solve this equation.

CHAPTER 16

Solitons and Shocks

16.1. Introduction

The reader is aware of the many famous rides on horses that have been taken in the course of history. Lady Godiva's ride and Paul Revere's ride are all well known. To this list we shall add the ride of John Scott Russell along the canal in Scotland as he described it in the following words:

> I was observing the motion of a boat which was rapidly drawn along a narrow channel by a pair of horses, when the boat suddenly stopped—not so the mass of water in the channel which it had part in motion; it accumulated round the prow of the vessel in a state of violent agitation, then suddenly leaving it behind, rolled forward with great velocity assuming the form of a large solitary elevation, a rounded, smooth and well-defined heap of water, which continued its course along the channel apparently without change of form or diminution of speed. I followed it on horseback, and overtook it still rolling on at a rate of some eight or nine miles an hour, preserving its original figure some thirty feet long and and a foot to a foot and a half in height. Its height gradually diminished, and after a chase of one or two miles I lost it in the windings of the channel. Such, in the month of August 1834, was my first chance interview with that singular and beautiful phenomenon . . . .

He called this wave the "great primary wave of translation" and also wrote: "The great primary waves of translation cross each other without change of any kind in the same manner as the small oscillations produced on the surface of a pool by a falling stone." These observations were reported in the middle third of the nineteenth century to the British Association for the Advancement of Science.

In addition to observing the soliton in the canal, Russell performed a series of experiments to study his observations. He used a shallow water tank with a piston at one end. In Fig. 16.1 we reproduce one of his figures, where $\xi(x, t)|_{max} = \delta h/h \sim 0(10\%)$.

At the end of the nineteenth century, Korteweg and de Vries were able to derive an equation to describe these waves. We encountered this KdV equation in Chapter 15 and we rewrite it in dimensionless form as

$$\frac{\partial u}{\partial t} + u \frac{\partial u}{\partial x} + \frac{\partial^3 u}{\partial x^3} = 0. \tag{16.1}$$

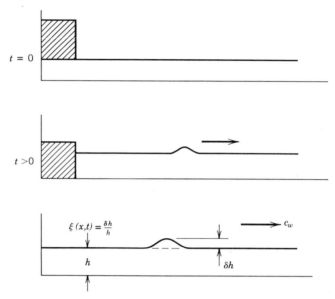

Fig. 16.1.   Sketch of a shallow water tank experiment and the evolution of the wave.

One of the first large-scale calculations to be performed on a digital computer was by Enrico Fermi, John Pasta, and Stan Ulam at Los Alamos in 1955, using the MANIAC computer. They investigated a large set of coupled non-linear difference—differential equations that would model a crystal structure. With a large amplitude sinusoidal excitation, they expected that harmonics would be generated because of the nonlinearity and that the energy would eventually thermalize. However, they found after some time that the energy all returned to the fundamental mode. This paradox, which will be discussed later, is called *FPU recurrence*.

In 1965 Norman Zabusky and Martin Kruskal numerically investigated the KdV equation and found that solitary waves solutions given by

$$u = \frac{c}{2} \operatorname{sech}^2 \left[\frac{\sqrt{c}}{2} (x - ct) \right] \tag{16.2}$$

could collide with each other and survive the collision. They coined the word *soliton* to describe this wave. There has been an exponential growth in interest in these waves among scientists, mathematicians, and engineers that continues to date. Textbooks on these waves are starting to appear and many features in nature are being described with soliton properties; for example, the giant Red Spot on Jupiter has been thought to be a soliton of a Rossby wave, waves on beaches in certain parts of the oceans have soliton properties, light waves guided by glass filaments have soliton properties, certain signals carried by nerve fibers are thought to be solitons, and so on.

As for solitary waves, some important properties of solitons can be deduced from (16.2). First, the product of the amplitude of the pulse c times the square of its width $(1/\sqrt{c})^2$ is a constant. Second, its velocity in the laboratory frame $= c_0(1+c)$ is amplitude dependent where c_0 is the linear velocity of the system. What distinguishes a soliton from a solitary wave is its ability to survive a collision with another soliton. A large-amplitude soliton can catch up with a smaller one and just pass through it.

16.2. FPU Recurrence

The easiest way to jump onto the "Soliton frame of reference" is to understand first the FPU recurrence phenomena. The fact that a difference scheme was used on the computer will introduce numerical "dispersion" in that there will be a minimum length that can be studied that is dictated by the difference length in the numerical scheme. Hence we should be aware that the problem will be *both* nonlinear and dispersive. As we shall focus on the KdV equation, we shall write the dispersion relation in dimensionless units as

$$\omega \sim k(1 - bk^2) \qquad \text{or} \qquad k \sim \omega(1 + a\omega^2), \qquad (16.3)$$

which is Eq. (15.19) of Ch. 15 and is shown in Fig. 16.2.

Let us assume a wave of the form

$$\phi = \phi_0 \sin [\omega_0 t - k_0 x] \qquad (16.4)$$

is excited at $x=0$ and propagates into the nonlinear dispersive media. Because of the nonlinearity, harmonics will be generated, that is, $2\omega_0$, $2k_0$; $3\omega_0$, $3k_0$, ... From Fig. 16.2 it is noted that a wave with $2\omega_0$ and $2k_0$ does *not* satisfy the dispersion relation (16.3) and will therefore not propagate. However, a signal $2\omega_0$, k^* does, and it can propagate. Therefore two signals exist at the same point

$$\phi_a = \phi_{a0} \sin [2\omega_0 t - 2k_0 x],$$

$$\phi_b = \phi_{b0} \sin [2\omega_0 t - k^* x]. \qquad (16.5)$$

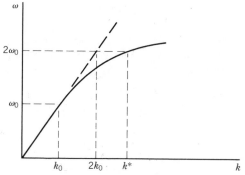

Fig. 16.2.   Linear dispersion curve in the long wavelength limit of Eq. (16.3). Dashed line is the curve $w=k$.

The sum of these two waves is (let $\phi_{a0} = \phi_{b0}$)

$$\phi_a + \phi_b = \phi_{a0} \{\sin [2\omega_0 t - 2k_0 x] + \sin [2\omega_0 t - k^* x]\}$$

$$= 2\phi_{a0} \sin [2\omega_0 t - \tfrac{1}{2}(2k_0 + k^*)x]$$

$$\times \cos [\tfrac{1}{2}(2k_0 - k^*)x].$$

This is an amplitude-modulated wave that is a minimum at $x = L$, where

$$\tfrac{1}{2}(2k_0 - k^*)L = -n\pi,$$

from which we write

$$\tfrac{1}{2}[2\omega_0 - 2\omega_0(1 + a(2\omega_0)^2)]L = -n\pi$$

or

$$L \sim + \frac{n\pi}{4a\omega_0^3}. \tag{16.6}$$

Additional modes could be included using this analysis, but the algebra becomes horrendous. This is the first example illustrating the delicate balance between nonlinearity and dispersion that is required in order to understand solitons.

16.3. Properties of Solitons

To understand the properties of a soliton from a mathematical point of view we could resort to advanced techniques, such as the "Inverse Scattering Transform" or "Spectral Transform." These are beyond the scope of this section and are detailed in several books dedicated to this subject. We shall just try and find a two-soliton solution to the KdV equation using a procedure given in the book of G. Lamb.

To do this we write down the one-dimensional time-independent Schrödinger equation as

$$\frac{d^2 y}{dx^2} + [\lambda - U(x)]y = 0 \qquad a \leqslant x \leqslant b \tag{16.7}$$

where λ is an eigenvalue, $U(x)$ is the potential well and the boundary conditions are at $x = a$ and b (which may be at $\pm \infty$). If $U(x)$ were zero and a and b were finite with the amplitude specified as $y = 0$ at $x = a$ and b, the solution of (16.7) would yield

$$y_n(x) = \sin [\sqrt{\lambda_n}(x - a)] \quad (n = 1, 2, 3, \ldots)$$

where

$$\lambda_n = \frac{n\pi}{(b - a)^2}.$$

Although it is easy to obtain this solution and the reader knew it already, it contains some important features that need to be further commented on.

The main new feature that may not have crossed your mind is that the eigenvalue $\sqrt{\lambda_n}$ does *not* change in time. You will think that this is not important since (16.7) did not contain any time variation. True, but we shall see that this is an important observation.

Potentials other than $U(x)=0$ can be used in (16.7) and still solutions can be obtained. For example, the potential $U(x)=-2\,\mathrm{sech}^2\,x$ with the boundary conditions $y(x=\pm\infty)=0$ leads to one eigenvalue $\lambda_1=-1$ with one solution $y=\mathrm{sech}\,x$. From quantum mechanics, we can interpret this as a particle confined in a potential well whose shape is proportional to $\mathrm{sech}^2\,x$ with the eigenvalue λ being proportional to the energy that the particle can possess.

Let us now ask the question, "What would happen to the solution of (16.7) if the potential well $U(x)$ depended on a parameter α (the parameter α could be time t), i.e., $U(x,\alpha)$?" If α varied, we might expect that the eigenvalue λ_n should also vary. The replacement of $U(x)$ by $U(x+\alpha)$ merely translates the potential and would have no effect on the eigenvalues. The potential would then satisfy the equation

$$\frac{\partial U}{\partial \alpha} - \frac{\partial U}{\partial x} = 0.$$

There are more complicated and more interesting equations that will leave the eigenvalues unchanged. As we shall see, one of these is the KdV equation

$$\frac{\partial U}{\partial t} + U\frac{\partial U}{\partial x} + \frac{\partial^3 U}{\partial x^3} = 0, \tag{16.8}$$

where we have made the obvious switch of the parameter α being replaced by time t.

To show this, we rewrite the time-independent Schrödinger equation as

$$\frac{\partial^2 y}{\partial x^2} - Uy = +\lambda y, \tag{16.9}$$

where U is the potential and $U=U(x,t)$. It is convenient to write this in operator notation as

$$Ly = +\lambda y \tag{16.10}$$

where $L\equiv(\partial^2/\partial x^2)-U(x,t)$, since we will want to do some differentiation. Therefore

$$\frac{d(Ly)}{dt} = L\frac{\partial y}{\partial t} + \frac{\partial L}{\partial t}\,y = +y\frac{\partial \lambda}{\partial t} + \lambda\frac{\partial y}{\partial t} \tag{16.11}$$

Also

$$\frac{\partial(Ly)}{\partial t} = \frac{\partial y}{\partial x^2 \partial t} - U\frac{\partial y}{\partial t} - y\frac{\partial U}{\partial t}$$

$$= \left[\frac{\partial^2}{\partial x^2} - U\right]\frac{\partial y}{\partial t} - y\frac{\partial U}{\partial t}$$

Therefore

$$\frac{\partial L}{\partial t} = -\frac{\partial U}{\partial t}. \tag{16.12}$$

We are now free to specify the temporal dependence of y. We choose it as

$$\frac{\partial y}{\partial t} = By, \tag{16.13}$$

where B is some linear differential operator that must be determined. Let us write the following equation

$$\left[-\frac{\partial U}{\partial t} + (LB - BL) \right] y = \frac{\partial \lambda}{\partial t} y \tag{16.14}$$

and expand.

$$-\frac{\partial U}{\partial t} y + LBy - BLy = \frac{\partial \lambda}{\partial t} y$$

$$-\frac{\partial U}{\partial t} y + L\frac{\partial y}{\partial t} - \frac{\partial(Ly)}{\partial t} = \frac{\partial \lambda}{\partial t} y$$

$$-\frac{\partial U}{\partial t} y + L\frac{\partial y}{\partial t} - L\frac{\partial y}{\partial t} - \frac{\partial L}{\partial t} y = \frac{\partial \lambda}{\partial t} y$$

$$0 = \frac{\partial \lambda}{\partial t} y$$

where we have used (16.12). Therefore if the operator B is chosen properly, then the eigenvalue will be independent of time! The choice of B is not a priori straightforward but requires some ingenuity.

$$\left[-\frac{\partial U}{\partial t} + (LB - BL) \right] y = 0. \tag{16.15}$$

The expression $(LB - BL)$ can be constructed such that it is devoid of differential operators and contains only U and its spatial derivatives. This will lead to a partial differential equation for $U(x, t)$ that, when satisfied, implies that $\partial \lambda / \partial t = 0$.

As a first example, let $B = a\partial/\partial x$ where a may be a function of U and its spatial derivatives. Hence

$$(LB - BL)y = \left(\frac{\partial^2}{\partial x^2} - U \right)\left(a\frac{\partial}{\partial x} \right)y - a\frac{\partial}{\partial x}\left(\frac{\partial^2}{\partial x^2} - U \right)y$$

$$= \frac{2\partial a}{\partial x}\frac{\partial^2 y}{\partial x^2} + \frac{\partial^2 a}{\partial x^2}\frac{\partial y}{\partial x} + ay\frac{\partial U}{\partial x}.$$

If the coefficients of $\partial^2 y/\partial x^2$ and $\partial y/\partial x$ are equal to zero, then $(BL - BL)y = ay(\partial U/\partial x)$. This implies that a is a constant. Therefore (16.15) becomes

$$\left(-\frac{\partial U}{\partial t} + a\frac{\partial U}{\partial x} \right)y = -\frac{\partial \lambda}{\partial t} y. \tag{16.16}$$

If we require that $\partial\lambda/\partial t$ be zero, then U must satisfy

$$-\frac{\partial U}{\partial t}+a\frac{\partial U}{\partial x}=0, \tag{16.17}$$

whose solution is $U(x+at)$. This solution was conjectured earlier.

Let us look at a slightly more complicated yet more interesting value for the operator B, namely,

$$B=a\frac{\partial^3}{\partial x^3}+f\frac{\partial}{\partial x}+g$$

where a is a constant. Using this, we compute as before

$$
\begin{aligned}
(LB-BL)y &=\left(\frac{\partial^2}{\partial x^2}-U\right)\left(a\frac{\partial^3}{\partial x^3}+f\frac{\partial}{\partial x}+g\right)y \\
&\quad -\left(a\frac{\partial^3}{\partial x^3}+f\frac{\partial}{\partial x}+g\right)\left(\frac{\partial^2}{\partial x^2}-U\right)y \\
&=\left(2\frac{\partial f}{\partial x}+3a\frac{\partial U}{\partial x}\right)\frac{\partial^2 y}{\partial x^2}+\left(\frac{\partial^2 f}{\partial x^2}+2\frac{\partial g}{\partial x}+3a\frac{\partial^2 U}{\partial x^2}\right)\frac{\partial y}{\partial x} \\
&\quad +\left(\frac{\partial^2 g}{\partial x^2}+a\frac{\partial^3 U}{\partial x^3}+f\frac{\partial U}{\partial x}\right)y.
\end{aligned} \tag{16.18}
$$

The vanishing of the coefficient of $\partial^2 y/\partial x^2$ leads to $f=-\frac{3}{2}aU+C_1$. The vanishing of the coefficient of $\partial y/\partial x$ where we have included the value for f leads to

$$\frac{\partial g}{\partial x}=-\frac{3a}{4}U+C_2.$$

Finally,

$$
\begin{aligned}
[LB-BL]y &=\left[-\frac{3a}{4}\frac{\partial U}{\partial x}+a\frac{\partial^3 U}{\partial x^3}+\left(-\frac{3}{2}aU+C_1\right)\frac{\partial U}{\partial x}\right]y \\
&=\left[\left(-\frac{3a}{4}+C_1\right)\frac{\partial U}{\partial x}+a\frac{\partial^3 U}{\partial x^3}-\frac{3}{2}aU\frac{\partial U}{\partial x}\right]y.
\end{aligned}
$$

From (16.14) we write

$$\left[-\frac{\partial U}{\partial t}+3\frac{\partial U}{\partial x}+6U\frac{\partial U}{\partial x}-4\frac{\partial^3 U}{\partial x^3}\right]y=-\frac{\partial\lambda}{\partial t}y,$$

where we have chosen the constants $a=-4$ and $C_1=0$. Let $\hat{x}=\bar{x}+3t$, $\hat{t}=t$ and the term within the brackets becomes

$$+\frac{\partial U}{\partial\hat{t}}-6U\frac{\partial U}{\partial\hat{x}}-4\frac{\partial^3 U}{\partial\hat{x}^3}=0, \tag{16.19}$$

since we desire $\partial\lambda/\partial t$ to be zero. Except for a numerical constant, (16.19) is the KdV equation (16.8).

In conclusion, if the potential well of the Schrödinger equation evolves according to the KdV equation, the eigenvalue will remain a constant. This procedure is called "finding the Lax operator," in honor of its originator.

We have seen that a potential of the form $2\,\text{sech}^2 x$ that satisfies the KdV equation will allow the eigenvalue of the Schrödinger to remain a constant in time. Let us relook at this and see if we can say anything about the interaction of two solitons. It turns out that we can obtain a formula that describes the interaction of two solitons, and it follows a technique devised by Bargmann.

Assume that there exist potentials for the Schrödinger equation

$$\frac{d^2 y}{\partial x^2} + (\lambda - U)y = 0 \tag{16.19}$$

such that the solution can be written in the form

$$y = \varepsilon^{i\sqrt{\lambda}x} F(\sqrt{\lambda}, x). \tag{16.20}$$

$F(\sqrt{\lambda}, x)$ is a polynomial in $\sqrt{\lambda}$. Let us choose the simplest case

$$y_1 = \varepsilon^{i\sqrt{\lambda}x}[2\sqrt{\lambda} + ia(x)] \tag{16.21}$$

and substitute this in the Schrödinger equation and separate terms in powers of $\sqrt{\lambda}$

$$\left\{ -\lambda[2\sqrt{\lambda} + ia(x)] + 2i\sqrt{\lambda}\left[\frac{ida(x)}{dx}\right] + \frac{id^2 a(x)}{dx^2} + \lambda[2\sqrt{\lambda} + ia(x)] \right.$$
$$\left. - U[2\sqrt{\lambda} + ia(x)] \right\} \varepsilon^{i\sqrt{\lambda}x} = 0.$$

These terms lead to the equations

$O((\sqrt{\lambda})^0)$:

$$\frac{id^2 a(x)}{dx^2} - ia(x)U = 0 \tag{16.22}$$

$O((\sqrt{\lambda})^1)$:

$$\frac{-2da(x)}{dx} - 2U = 0. \tag{16.23}$$

Eliminate U between (16.22) and (16.23)

$$\frac{d^2 a(x)}{dx^2} + a(x)\frac{da(x)}{dx} = 0 \tag{16.24}$$

whose first integral is

$$\frac{da(x)}{dx} + \frac{[a(x)]^2}{2} = 2\mu^2 \tag{16.25}$$

where $2\mu^2$ is the constant of integration. The substitution $a(x) = 2(d/dx)(\ln \omega(x))$

transforms (16.25) to

$$2 \frac{d}{dx}\left[\frac{d\omega(x)/dx}{\omega(x)}\right] + 2\left[\frac{d\omega(x)/dx}{\omega(x)}\right]^2 = 2\mu^2$$

$$\frac{\omega(x)(d^2\omega(x)/dx^2) - [(d\omega(x)/dx)]^2}{\omega(x)^2} + \frac{[(d\omega(x)/dx)]^2}{\omega(x)^2} = \mu^2$$

or finally

$$\frac{d^2\omega(x)}{dx^2} - \mu^2\omega(x) = 0, \qquad\qquad . \text{(16.26)}$$

whose solution is

$$\omega(x) = \alpha\varepsilon^{\mu x} + \beta\varepsilon^{-\mu x} \qquad\qquad \text{(16.27)}$$

Using the equation of $O((\sqrt{\lambda})^1)$, we write

$$U(x) = -\frac{da(x)}{dx} = -\frac{2d^2}{dx^2}[\ln(\omega(x))]$$

$$= -2\mu^2 \operatorname{sech}^2 \mu x \qquad\qquad \text{(16.28)}$$

if $\alpha = \beta$ and if $\alpha \neq \beta$,

$$U(x) = -2\mu^2 \operatorname{sech}^2(\mu x - \phi), \qquad\qquad \text{(16.29)}$$

where $\phi = \frac{1}{2}\ln \beta/\alpha$ and $\alpha = \alpha(t)$ and $\beta = \beta(t)$.

Substitution of this last expression into the KdV equation

$$\frac{\partial U}{\partial t} - 6U\frac{\partial U}{\partial x} + \frac{\partial^3 U}{\partial x^3} = 0 \qquad\qquad \text{(16.30)}$$

leads to

$$\frac{\partial \phi}{\partial t} = 4\mu^3.$$

Therefore

$$U = -2\mu^2 \operatorname{sech}^2[\mu x - 4\mu^3 t] \qquad\qquad \text{(16.31)}$$

or the single soliton solution.

Let us now extend this analysis to the case of two solitons. In this case, we choose

$$y_2 = \varepsilon^{i\sqrt{\lambda}x}[4\lambda + 2i\sqrt{\lambda}a(x) + b(x)] \qquad\qquad \text{(16.32)}$$

and substitute this in the Schrödinger equation. Define $a(x) = (d(\ln \omega(x))/dx)$ as before and solve the resulting fourth-order equation. After considerable algebra and with appropriate choice for all constants, we can finally arrive at

$$U(x, t) = -12\left[\frac{4\cosh(2x - 8t) + \cosh(4x - 64t) + 3}{[3\cosh(x - 28t) + \cosh(3x - 36t)]^2}\right] \qquad \text{(16.33)}$$

At $t=0$, the two solitons overlap and $U(x,0)=-6\,\text{sech}^2\,x$. A sketch of this collision is shown in Fig. 16.3.

The solitons have preserved their shape during the collision and John Scott Russell's observation has a mathematical foundation. Imagine what this would do to the insurance business or body repair shops if automobiles were solitons. This collision property has been used as a defining property for solitons, although it is now more common to give a more mathematical definition involving conservation laws.

The nonlinear Schrödinger (NLS) equation that was described in Chapter 15 also admits soliton solutions. In deriving the NLS equation, we introduced and made use of the ponderomotive force or ponderomotive potential. This introduces many new features to the soliton community. For example, in a plasma the ponderomotive force will create a density cavity in which a high-frequency Langmuir wave will be trapped. Such structures have been studied in the laboratory, have been created artificially in the ionosphere, and have been noted in the bow shock of Jupiter.

Solitons in one dimension are now reaching a state of maturity. In higher dimensions, both the mathematics and accompanying experimental verification are still in their formative stages. For example, if the effects of slight transverse inhomogeneity are introduced into the derivation of a slightly more general KdV equation, one can obtain the Kadomtsev–Petviashvili (KP) equation

$$\frac{\partial}{\partial x}\left[\frac{\partial U}{\partial t}+U\frac{\partial U}{\partial x}+\frac{1}{2}\frac{\partial^3 U}{\partial x^3}\right]+\frac{1}{2}\frac{\partial^2 U}{\partial y^2}=0. \qquad (16.34)$$

The KP equation is another nonlinear evolution equation that admits soliton solutions, one of which can be written as

$$U=6k^2\,\text{sech}^2\,[kx+ly-\omega t], \qquad (16.35)$$

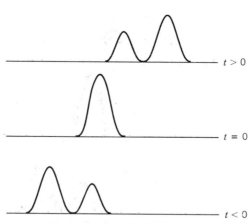

Fig. 16.3.   Collision of two solitons. The large amplitude soliton, which has a greater velocity, overtakes a small amplitude soliton and passes through it.

provided that

$$\omega = 2k^3 + \frac{l^2}{2k} = \omega(\bar{K}),$$

(16.36)

where

$$\bar{K} = k\bar{U}_x + l\bar{U}_y$$

is satisfied. Note that (16.36) is *not* the linear dispersion relation obtained by linearizing (16.34) with an assumed solution of the form

$$\psi = U_0 \varepsilon^{i(kx + ly - \omega t)}.$$

(16.37)

It has been noted that if (16.35) is one solution of (16.34), then a triad of solutions that satisfies

$$\bar{K}_1 + \bar{K}_2 = \bar{K}_3$$

$$\omega(\bar{K}_1) + \omega(\bar{K}_2) = \omega(\bar{K}_3)$$

(16.38)

is also a solution. A manifestation of this is that during the oblique collision of two solitons that satisfied (16.38), a new soliton would be created with an amplitude greater than the sum of the two colliding solitons. This collision has been noted in laboratory experiments using a water tank or a plasma chamber as the host environment and along various beaches throughout the world. There is a strong cross-fertilization between theory and experiment in this area of wave studies, and new effects are burgeoning forth almost monthly.

16.4. Shocks

Closely allied to the study of solitons is the study of shocks. The reader has probably read in the newspapers concerning the potential noise problems of a supersonic aircraft flying over inhabited land and may have then wondered if there existed a mathematical model to describe the phenomena. Since the early work of Mach, who demonstrated that if a projectile passed through a medium with a velocity greater than the velocity of sound, a cone of sound would be radiated at an angle from the projectile. This angle is related to the speed of sound and the projectile and is given by (see Ch. 8, Sec. 5)

$$\frac{C_s}{U_s} = \sin \theta.$$

(16.39)

Frequently in nature shocks can be easily created. The projectile of Mach may be a fast pusher, such as an eruption of a volcano or an explosion of a nuclear bomb in which energy is rapidly released at one point in space and time and rapidly expands—in fact, expands faster than the velocity of sound. The resulting transition between the undisturbed region ahead of the transition and the disturbed region trailing the transition is called a shock if a set of equations

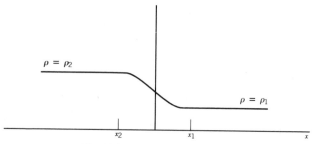

Fig. 16.4.   Structure of a shock.

can describe the physics ahead of the shock and the same set describes the region behind the shock; they must be connected in some sense through the shock. A simple first-order equation

$$\frac{\partial \rho}{\partial t} + \frac{d(\rho v)}{\partial x} = 0 \tag{16.40}$$

can be used to illustrate the derivation of such jump conditions. This fitting of the discontinuity is called the satisfying the Rankine–Hugenot relations.

In Fig. 16.4 a shock structure is illustrated. Let us integrate (16.40) across this transition

$$\int_{x_1}^{x_2} \frac{\partial \rho}{\partial t}\, dx + \int_{x_1}^{x_2} \frac{d(\rho v)}{\partial x} = 0$$

$$\frac{\partial}{\partial t} \int_{x_1}^{x_2} \rho\, dx + \rho v \Big|_{x_1}^{x_2} = 0. \tag{16.41}$$

If the transition does not change in time, we are left with the jump condition

$$(\rho v)\big|_{x_2} - (\rho v)\big|_{x_1} = 0. \tag{16.42}$$

More complicated and more general jump conditions can be obtained for sets of equations. In particular for gases, one would use the equations of continuity, motion, and energy and integrate them across the shock front.

Problems

1.   Show that the KdV equation (16.1) can be written as

$$\frac{\partial U}{\partial t} + aU \frac{\partial U}{\partial x} + b \frac{\partial^3 U}{\partial x^3} = 0$$

through a rescaling of all variables.

2.   Derive the FPU recurrence length for the NLS equation.

3.   Solve the linear Schrödinger equation (16.7) if $U(x) = ax^2$.

4. Let the operator $B=(a\partial^2/\partial x^2)+(b\partial/\partial x)+c$, where a, b, and c are constants in (16.15). Compute the resulting differential equation as we have done in deriving the KdV equation (16.19).

5. Carry out the detailed calculation that led to the two-soliton interaction equation (16.33).

6. From (16.38) show that the oblique collision of two equal amplitude solitons can produce a new soliton whose amplitude is four times the amplitude of the colliding solitons. You must use the fact that the soliton has $(\text{amplitude})(\text{width})^2 = \text{constant}$.

7. As in Problem 6, use (16.36) and (16.38) and find the critical angle θ_r for this resonance. Show that it equals

$$\theta_r = 2 \tan^{-1}\sqrt{2A_1}$$

8. Discuss the Mach cone formula given in Eq. (16.39) and sketch it.

9. Solve (16.40) if $v=\alpha\rho$. Sketch this solution.

10. Derive the jump conditions for the equation of motion and energy transport.

Appendix A

A.1. Fundamental Physical Constants

Speed of light	c	$\sim 3.00 \times 10^8$ m/sec
Permittivity	ε_0	8.85×10^{-12} F/m
		$\sim \frac{1}{36}\pi \times 10^{-9}$ F/m
Permeability*	μ_0	$4\pi \times 10^{-7}$ H/m
Elementary charge	e	1.6×10^{-19} C
Electron rest mass	m_e	9.11×10^{-31} kg
Proton rest mass	m	1.67×10^{-27} kg
Boltzmann's constant	k_B	1.38×10^{-23} J/K
Gas constant	R	8.31 J/mol·K
Absolute zero	$0°$K	$-273°$C
Planck's constant	h	6.63×10^{-34} J·sec
Avogadro's number	N_0	6.02×10^{23} mol^{-1}
Gravitational constant	G	6.67×10^{-11} N·m^2/kg^2
Gravitational acceleration	g	9.81 m/sec^2

A.2. Definition of Standards

Meter — 1,650,763.73 wavelengths of the transition $2p_{10} - 5d_5$ in ^{35}Kr atom

Kilogram — the mass of the international kilogram in Paris, France

Second — 9,192,631,770 vibrations of the hyperfine transition 4,3–3,0 of the fundamental state $^2S_{1/2}$ in ^{133}Cs atom

Coulomb — 1.0 A·sec

Ampere — if two equal straight, infinitely long currents 1 m apart exert a force of 2×10^{-7} N/m on each other, the current is defined to be

---

*The vacuum permeability is an assigned constant, while the permittivity is a measured constant. See the definition of 1 ampere current in Section A.2.

1 A. [Since the magnetic field due to an infinitely long current is $\mu_0 I/2\pi r$, the force per unit length between the currents is $\mu_0 I^2/2\pi r \,(\text{N/m})$.] Therefore $\mu_0 = 4\pi \times 10^{-7}$ H/m is an assigned constant introduced to define 1 A current.)

A.3. Derived Units of Physical Quantities

Quantity	Symbol	Dimensions	Derived Units
Mass	m	M	kg
Volume mass density	ρ_v	ML^{-3}	kg/m³
Surface mass density	ρ_s	ML^{-2}	kg/m²
Linear mass density	ρ_l	ML^{-1}	kg/m
Velocity	$\mathbf{v}$*	LT^{-1}	m/sec
Acceleration	$\mathbf{a}$	LT^{-2}	m/sec²
Momentum	$\mathbf{P}$	MLT^{-1}	kg·m/sec
Force	$\mathbf{F}$	MLT^{-2}	N = kg·m/sec²
Energy (work)	E	ML^2T^{-2}	J = kg·m²/sec²
Torque	τ	ML^2T^{-2}	N·m = kg·m²/sec²
Charge	q	Q	C
Current	i or I	QT^{-1}	A = C/sec
Current density	$\mathbf{J}$	$QT^{-1}L^{-2}$	A/m²
Electric field	$\mathbf{E}$	$MLT^{-2}Q^{-1}$	V/m
Electric displacement (electric flux density)	$\mathbf{D}$	QM^{-2}	C/m²
Electric flux	Q	Q	C
Electric potential	V	$ML^2T^{-2}Q^{-1}$	V = kg·m²/sec² C
Magnetic field (magnetic flux density)	$\mathbf{B}$	$MT^{-1}Q^{-1}$	T = Wb/m²
Magnetic flux	ϕ	$ML^2T^{-1}Q^{-1}$	Wb
Magnetization force (magnetic field intensity)	$\mathbf{H}$	$QL^{-1}T^{-1}$	A/m
Capacitance	C	$Q^2T^2M^{-1}L^{-2}$	F = C/V
Inductance	L	ML^2Q^{-2}	H = Wb/A
Resistance	R	$ML^2T^{-1}Q^{-2}$	Ω = V/A
Conductance	G	$M^{-1}L^{-2}TQ^2$	$\mho = \Omega^{-1}$ = A/V
Resistivity	η	$ML^3T^{-1}Q^{-2}$	Ω·m
Conductivity	σ	$M^{-1}L^{-3}TQ^2$	$\mho \cdot m^{-1} = \Omega^{-1}m^{-1}$

*$\mathbf{A}$ (boldface letter) indicates vector quantity.

Appendix B

B.1. Trigonometric functions

$\sin(-\theta) = -\sin\theta$ (odd function)

$\cos(-\theta) = \cos\theta$ (even function)

$\tan(-\theta) = -\tan\theta$ (odd function)

$\sin(\alpha \pm \beta) = \sin\alpha \cos\beta \pm \sin\alpha \cos\beta$

$\cos(\alpha \pm \beta) = \cos\alpha \cos\beta \pm \sin\alpha \sin\beta$

$\tan(\alpha \pm \beta)\, \dfrac{\tan\alpha \pm \tan\beta}{1 \pm \tan\alpha \tan\beta}$

$\cot(\alpha \pm \beta) = \dfrac{\cot\alpha \cot\beta \pm 1}{\cot\beta \pm \cot\alpha}$

$\sin\alpha + \sin\beta = 2\sin\dfrac{\alpha+\beta}{2}\cos\dfrac{\alpha-\beta}{2}$

$\sin\alpha - \sin\beta = 2\cos\dfrac{\alpha+\beta}{2}\sin\dfrac{\alpha-\beta}{2}$

$\cos\alpha + \cos\beta = 2\cos\dfrac{\alpha+\beta}{2}\cos\dfrac{\alpha-\beta}{2}$

$\cos\alpha - \cos\beta = -2\sin\dfrac{\alpha+\beta}{2}\sin\dfrac{\alpha-\beta}{2}$

$\sin\alpha \sin\beta = \tfrac{1}{2}[\cos(\alpha-\beta) - \cos(\alpha+\beta)]$

$\sin\alpha \cos\beta = \tfrac{1}{2}[\sin(\alpha+\beta) + \sin(\alpha-\beta)]$

$\cos\alpha \cos\beta = \tfrac{1}{2}[\sin(\alpha+\beta) - \sin(\alpha-\beta)]$

$\cos\alpha \cos\beta = \tfrac{1}{2}[\cos(\alpha+\beta) + \cos(\alpha-\beta)]$

$\sin 2\alpha = 2\sin\alpha \cos\alpha$

$\cos 2\alpha = \cos^2\alpha - \sin^2\alpha = 2\cos^2\alpha - 1$

$$= 1 - 2\sin^2\alpha$$

$\sin 3\alpha = -4 \sin^3 \alpha + 3 \sin \alpha$

$\cos 3\alpha = 4 \cos^3 \alpha - 3 \cos \alpha$

$\sin^2 \alpha = (1 - \cos 2\alpha)/2$ $\qquad \lim\limits_{x \to 0} \dfrac{\sin x}{x} = 1$

$\cos^2 \alpha = (1 + \cos 2\alpha)/2$ $\qquad \lim\limits_{x \to 0} \dfrac{\tan x}{x} = 1$

$e^{\pm i\alpha} = \cos \alpha \pm i \sin \alpha$

B.2. Calculus

An arbitrary constant of integration is not written.

$$\frac{d}{dx} \sin x = \cos x, \quad \int \sin x \, dx = -\cos x$$

$$\frac{d}{dx} \cos x = -\sin x, \quad \int \cos x \, dx = \sin x$$

$$\frac{d}{dx} \tan x = \sec^2 x$$

$$\int \tan x \, dx = \ln|\sec x|$$

$$\frac{d}{dx} \cot x = -\operatorname{cosec}^2 x$$

$$\int \cot x \, dx = \ln|\sin x|$$

$$\frac{d}{dx} x^n = nx^{n-1}, \quad \int x^n \, dx = \frac{1}{n+1} x^{n+1} (n \neq -1)$$

$$\frac{d}{dx} \ln x = \frac{1}{x}, \quad \int \frac{1}{x} dx = \ln|x|$$

$$\frac{d}{dx} (fg) = \frac{df}{dx} g + f \frac{dg}{dx}$$

$$\int f \frac{dg}{dx} dx = fg - \int \frac{df}{dx} g \, dx \quad \text{(integration by parts)}$$

$$\frac{d}{dx} e^{ax} = ae^{ax}, \quad \int e^{ax} \, dx = \frac{1}{a} e^{ax}$$

$$\int \frac{dx}{1+x^2} = \tan^{-1} x$$

$$\int \frac{dx}{\sqrt{x^2+a^2}} = \ln\left[\sqrt{x^2+a^2}+x\right]$$

$$\int \frac{dx}{\sqrt{a^2-x^2}} = \sin^{-1}(x/a)$$

$$\int \frac{x}{(x^2+a^2)^{3/2}}\,dx = -\frac{1}{\sqrt{x^2+a^2}}$$

B.3. Power Series

$$(1+x)^n = 1+nx+\frac{n(n-1)}{2!}x^2+\cdots$$

$$e^x = 1+x+\frac{x^2}{2!}+\frac{x^3}{3!}+\cdots$$

$$\ln(1+x) = x-\frac{x^2}{2}+\frac{x^3}{3}-\frac{x^4}{4}+\cdots$$

$$\sin x = x-\frac{x^3}{3!}+\frac{x^5}{5!}-\cdots$$

$$\cos x = 1-\frac{x^2}{2!}+\frac{x^4}{4!}-\cdots$$

$$\tan x = x+\frac{x^3}{3}+\frac{2}{15}x^5+\cdots$$

$$f(x) = f(a)+f'(a)(x-a)+\frac{f''(a)}{2!}(x-a)^2+\cdots \quad \text{(Taylor series)}$$

$$f(x+\Delta x) \simeq f(x)+\Delta x\frac{df}{dx} \quad (\Delta x \text{ small})$$

B.4. Laplace transform

$$F(s) = \int_0^\infty e^{-st}f(t)\,dt$$

$$f(t) = \frac{1}{2\pi i}\int_{Br} e^{st}F(s)\,ds$$

Br = Bromwich contour integral

$f(t)$	$F(s)$
$\dfrac{df}{dt}$	$sF(s)-f(0)$

$f(t)$	$F(s)$

$$\frac{d^2 f}{dt^2}$$ $s^2 F(s) - [sf(0) + f'(0)]$

$$\int_0^t f(t)\, dt$$ $\dfrac{1}{s} F(s)$

$f(t \pm a)$ $e^{\pm as} F(s)$

$e^{\pm \lambda t} f(t)$ $F(s \mp \lambda)$

$t^n f(t)$ $(-1)^n \dfrac{d^n}{ds^n} F(s)$

$t^{-n} f(t)$ $\displaystyle\int\int\underset{n}{\cdots}\int F(s)(ds)^n$

$\lim\limits_{t \to 0} f(t)$ $\lim\limits_{s \to \infty} sF(s)$

$\lim\limits_{t \to \infty} f(t)$ $\lim\limits_{s \to 0} sF(s)$

$\delta(t)$ (delta function) 1

1 $1/s$

t $1/s^2$

t^n ($n = $ integer) $\dfrac{n!}{s^{n+1}}$

$t^\alpha (\alpha > -1)$ $\dfrac{\Gamma(\alpha + 1)}{\alpha + 1}$ ($\Gamma = $ Gamma function)

e^{at} $\dfrac{1}{s - a}$

te^{at} $\dfrac{1}{(s - a)^2}$

$\ln t$ $-\dfrac{1}{s}(\ln s + 0.5772)$

$\sin \omega t$ $\dfrac{\omega}{s^2 + \omega^2}$

$\cos \omega t$ $\dfrac{s}{s^2 + \omega^2}$

$e^{\alpha t} \sin \omega t$ $\dfrac{\omega}{(s - \alpha)^2 + \omega^2}$

$f(t)$	$F(s)$

$e^{\alpha t} \cos \omega t$ $\qquad\qquad\qquad\qquad \dfrac{s-\alpha}{(s-\alpha)^2 + \omega^2}$

$1/\sqrt{t}$ $\qquad\qquad\qquad\qquad \sqrt{\pi/s}$

J_0 (at) (Bessel function) $\qquad\qquad \dfrac{1}{\sqrt{s^2 + a^2}}$

For a more complete table, see, for example, "Table of Laplace Transforms" by G. E. Roberts and H. Kaufman (W. B. Saunders Company, Philadelphia, 1966).

Bibliography

As waves are ubiquitous in nature, they are the subject of one or more chapters in many books. The following books will be useful to the reader who wishes to receive an alternative and/or more advanced treatment of the subject material covered in this text.

Ames, W. F., *Nonlinear Partial Differential Equations in Engineering*, Academic, New York. Volume I, 1965, Vol. II, 1972.

Baldock, G. R. and T. Bridgeman, *Mathematical Theory of Wave Motion*, Ellis Horwood Ltd., Chichester, 1981.

Halliday, D. and R. Resnick, *Fundamentals of Physics*, 2nd ed., Wiley, New York, 1981.

Karpman, V. I., *Nonlinear Waves in Dispersive Media*, Pergamon, Oxford, 1975.

Lamb, G. L., Jr., *Elements of Soliton Theory*, Wiley-Interscience, New York, 1980.

Lighthill, J., *Waves in Fluids*, Cambridge University Press, New York, 1978.

Lonngren, K. and A. Scott, *Solitons in Action*, Academic, New York, 1978.

Morse, P. M., *Vibration and Sound*, 2nd ed., McGraw-Hill, New York, 1948.

Ramo, S., J. R. Whinnery, and T. Van Duzer, *Fields and Waves in Communication Electronics*, Wiley, New York, 1965.

Whitham, G. B., *Linear and Nonlinear Waves*, Wiley-Interscience, New York, 1973.

Zeldovich, Ya. B., and Yu. P. Raizer, *Physics of Shock Waves and High Temperature Hydrodynamic Phenomena*, Academic, New York, 1966.

INDEX